Written for tumultuous times, this volume offers fresh and valuable psycho-analytic reflections on traumatized children's play and on how wars/climate change/pandemics inflect psychic development. Breaking with psychoanalysis's omertà around naming the genocide of Palestinian children and trans youth, Fiorella's bold and erudite collection tracks fantasies about blown-off limbs growing back, impending disablements by climate change/pandemics, and trans children's anxieties about being erased. It also necessarily reveals how adults fail children, children wounded not only by bombs, disease, and rising temperatures, but also by adults' lies, greed, narcissism, and inaction. Never has Winnicott's "there's no such thing as a baby" been such a searing indictment.

Avgi Saketopoulou, *Cypriot and Greek psychoanalyst on faculty at the NYU Postdoctoral Program in Psychotherapy and Psychoanalysis, New York, USA.*

This is a powerful and precious book. The contributors have gone undaunted into the darkest of places and seen some terrible sights. Someone had to do it and thanks to these brave psychoanalysts, now someone has. They have much to teach us.

Anne Alvarez, *Consultant, Child and Adolescent Psychotherapist, London, UK.*

It is an essential task for psychoanalysis to contemplate the precarity of human existence in a world under threat of extinction, a pandemic, and end-less wars. The social-political circumstances of our lives are as fundamental to who we are as the dynamics of our families of origins, doubly so when they are traumatic. This book explores the condition of the most vulnerable humans: children. It is written by people who have been meeting them in some of the most devastating trauma zones. It is a gift, and a demand that we pay close attention and do what we can.

Eyal Rozmarin, *psychoanalyst, New York, USA.*

Child and Adolescent Psychoanalysis in Times of Crisis

Child and Adolescent Psychoanalysis in Times of Crisis: War, Pandemic and Climate Change gathers a global cohort of psychoanalytic thinkers to consider the most pressing issues currently faced by young people worldwide.

Each chapter provides a theoretical exploration of our psychically and socially damaging collective reality and offers practical support to psychoanalysts working to create safe spaces in a world that feels fundamentally unlivable for many young people. Case studies span the Russia–Ukraine and Palestine–Israel conflicts, the COVID-19 pandemic, the opioid crisis, and forced displacement as a result of climate catastrophe. Contributors explore the active impact of these devastating events, including the pervasive hopelessness suffered by young people deprived of agency and a viable future, through first-hand accounts of working with children and adolescents in conflict zones. Drawing on the work of Wilfred Bion, D. W. Winnicott, Judith Butler, and others, these essays offer hope by showing the important role that psychoanalytic work can play in restoring the capacity for psychic development, resilience and meaning making.

Part of the Routledge Child and Adolescent Psychoanalysis series, this book is an essential read for all psychoanalysts, psychiatrists, caregivers, teachers, social workers, and academics.

Kristin Fiorella, PsyD, MFT, MFA is a psychoanalyst practicing in San Francisco where she sees children, adolescents, and adults. She is on the faculties of The Psychoanalytic Institute of Northern California and the San Francisco Center for Psychoanalysis. Her writings on the dialogue between Buddhism and psychoanalysis have been published in various journals, and she was awarded the American Psychoanalytic Association's Lee Jaffe paper prize in 2023.

The Routledge Child and Adolescent Psychoanalysis Book Series

Editor: Christine Anzieur-Premmereur
Co-Editors: Mary T. Brady
Christine Franckx
Fernando M. Gómez

The Routledge Child and Adolescent Psychoanalysis Book Series is devoted to manuscripts that illuminate the creative and challenging work of child and adolescent psychoanalysis and psychoanalytic psychotherapy. While we believe that the study of child psychoanalysis is relevant to all psychoanalysts - as the study of the unconscious is essential to all psychoanalysts – the particularities of child and adolescent work require that the setting adapt itself to the child and adolescent, not the other way around. We also see children and adolescents as quite sensitive to cultural and societal changes and catastrophes. Children and adolescents are like the canaries in the coal mine – particularly vulnerable to the presence of gases. For that reason, our Book Series is dedicated to excellent and creative clinical technique and theoretical work, writing that is sensitive to the setting, and writing that is perceptive of societal and cultural changes that affect children and adolescents.

The series editors are especially interested in selecting books which enhance the understanding and further expansion of infant, child, and adolescent psychoanalytic thought. Part of the mission of this international series is to nurture communication amongst psychoanalysts working in different models, in different languages, and in different regions of the world.

Series Editor Biographies

Christine Anzieu-Premmereur is a psychiatrist and psychoanalyst in New York City who works in private practice with adults, children, parents and their babies. A member of the Société Psychanalytique de Paris, she is on the faculty of the Columbia Psychoanalytic Center for Training and Research and is Assistant Clinical Professor in Psychiatry at Columbia University. She is the chair of the IPA Committee for Child and Adolescent Psychoanalysis (COCAP). With Vaia Tsolas, she is the co-founder of Pulsion Institute. She recently published *The Process of Representation in Early Childhood and Attacks on Linking in Parents of Young Disturbed Children. She co-edited with Vaia Tsolas A Psychoanalytic Exploration of the Body in Today's World: On the Body* (2017) and *A Psychoanalytic Exploration of the Contemporary Search for Pleasure: The Turning of the Screw* (2023).

Dr. Mary Brady is an adult and child psychoanalyst in private practice in San Francisco, USA. She is on the Faculties of the San Francisco Center for Psychoanalysis and the Psychoanalytic Institute of Northern California. She is Editor of *Braving the Erotic Field in the Treatment of Children and Adolescents* (2022). Her books, *Analytic Engagements with Adolescents* and *The Body in Adolescence* were published by Routledge in 2018 and 2016 respectively. She is North American Co-Chair for the Committee on Child and Adolescent Psychoanalysis (COCAP) of the IPA. She co-leads a Psychoanalysis and Film group.

Dr. Christine Franckx is an adult and child psychoanalyst and psychiatrist. She works in Antwerp in private practice for adult analysis and she has created an analytic psychotherapeutic center for early development (0-6 years). She is a Training Analyst of the Belgian Psychoanalytic Society, of which she has been the President (2016–2020). She is Editor of two books Eros op de scene (2021) and *Het kinderlijk trauma* (2023). She is trainer in Infant observation (Esther Bick). She is European Co-Chair for the IPA Committee on Child and Adolescent Psychoanalysis (COCAP).

Dr. Fernando M. Gómez is a child and adolescent psychoanalyst, psychiatrist and pediatrician. He works in Buenos Aires in private practice for children, adolescents and adults analysis. He was trained at the Asociación Psicoanalítica Argentina (APA), of which he has been Director of the Publications Committee (2016–2020) and of the Department of Children and Adolescents "Arminda Aberastury" (2020–2023). He is Latin American Co-Chair for the IPA Committee for Child and Adolescent Psychoanalysis (COCAP). He is a member of the Advisory Council of the General Directorate of Mental Health of the Government of the City of Buenos Aires. He edited a 4 volumes collection: *Pilares del Psicoanálisis Contemporáneo* (2017, 2018, 2019, 2020), *Psicoanálisis Contemporáneo Latinoamericano* (2017, coedited with FEPAL), and he is currently editing *Clinica e Investigación en el Psicoanalisis de bebés, niños y adolescentes. Nuevos horizontes, nuevos desafíos.*

Forthcoming Titles

Psychoanalysis with Adolescents and Children: Learning to Surf
Mary T. Brady

Child and Adolescent Psychoanalysis in Times of Crisis: War, Pandemic and Climate Change
Kristin Fiorella

Child and Adolescent Psychoanalysis in Times of Crisis

War, Pandemic, and Climate Change

Edited by
Kristin Fiorella

Routledge
Taylor & Francis Group

LONDON AND NEW YORK

Designed cover image: Getty Image © Westend61

First published 2026
by Routledge
4 Park Square, Milton Park, Abingdon, Oxon OX14 4RN

and by Routledge
605 Third Avenue, New York, NY 10158

Routledge is an imprint of the Taylor & Francis Group, an informa business

British Library Cataloguing-in-Publication Data
A catalogue record for this book is available from the British Library

ISBN: 978-1-032-85978-1 (hbk)
ISBN: 978-1-032-83582-2 (pbk)
ISBN: 978-1-003-52075-7 (ebk)

DOI: 10.4324/9781003520757

Typeset in Times New Roman
by Taylor & Francis Books

For all the children living in conflict zones

Contents

Tables

Contributors

Kateryna Abashkina, is a psychologist, psychoanalytic psychotherapist, candidate of the Ukrainian Psychoanalytic Society (study-group of the IPA), and member of IPSO. She has been in private practice in psychoanalytic psychotherapy with adults, adolescents, and children since 2009.

Alpatova Kateryna is a psychoanalytic psychotherapist, a candidate of the Ukrainian Psychoanalytic Society (study-group of the IPA), and a representative of Ukraine in IPSO. A full member of the Ukrainian Association of Psychoanalytic Psychotherapies (collective member of the EPPP) and head of the Kharkiv Society of Psychoanalysis and Psychotherapy, she works in private practice with adult patients. Her area of interest is Kleinian analysis, and in light of recent war, trauma and within supportive therapy.

Christine Anzieu-Premmereur is a psychiatrist and adult and child psychoanalyst in New York, and member of the Société Psychanalytique de Paris and Columbia University Psychoanalytic Center. She chairs the IPA Committee for Child and Adolescent Psychoanalysis (COCAP) and is co-founder of the Pulsion Institute. She has published on motherhood, addiction, infantile depression, playfulness in analysis, and psychosomatics. She co-edited with COCAP *The Infinite Infantile and the Psychoanalytic Task: Psychoanalysis with Children, Adolescents and their Families,* and with Vaia Tsolas *A Psychoanalytic Exploration of the Contemporary Search for Pleasure: The Turning of the Screw* in 2023.

Elizabeth Berger is a child psychiatrist living in New York City whose work focuses on mental health and well-being in Palestine. She is one of the founding members of the USA-Palestine Mental Health Network and has worked with colleagues in Palestine for many years developing and supervising clinical programs, engaging in healthcare policy planning, and supporting advocacy projects. With psychiatrist co-author Samah Jabr MD, she has published over a dozen textbook chapters and scholarly papers on mental health in Palestine. Dr Berger is Associate Clinical Professor at the George Washington University School of Medicine.

Mary T. Brady is an adult and child psychoanalyst in San Francisco. On the Faculties of the San Francisco Center for Psychoanalysis and the Psychoanalytic Institute of Northern California, she is recipient of the American Psychoanalytic Association's Roughton Award for her writing. Her book *Psychoanalysis with Adolescents and Children: Learning to Surf* has just been published by Routledge. She is editor of *Braving the Erotic Field in the Treatment of Children and Adolescents* (Routledge, 2022) and author of *Analytic Engagements with Adolescents* and *The Body in Adolescence* (Routledge, 2018 and 2016, respectively). She is North American co-chair for COCAP and is co-chair of the Committee of the Status of Women and Girls for APsA. For a decade she has co-led a psychoanalysis and film group.

Martha Bragin, PhD, LCSW is Professor at the Silberman School of Social Work at Hunter College, City University of New York, where she chairs Global Social Work and Practice with Immigrants and Refugees. Dr Bragin is an associate member of IPTAR, a member of the IPA Committee on the Humanitarian World and a member of the Alliance for Child Protection in Humanitarian Action the organization that sets global standards and conducts research on children and adolescents in emergencies. She joined the CUNY faculty after 30 years supporting United Nations agencies, governments, and non-governmental organizations to address the effects of violence and disaster on children, youth and families bringing global learning back to the US Dr Bragin is recipient of the International Psychoanalytic Association's Tyson Prize (2007) the Hayman Prize (2011 and 2021) for published work on traumatized children and adults. She maintains a private practice in New York City.

Mónica Cardenal is a training analyst of Buenos Aires Psychoanalytical Association and a former scientific coordinator. She is an associate professor in child and adolescent clinic psychology and psychiatry postgraduate training at the Italian Hospital University, Buenos Aires, and she is a coordinator of the Infant Observation Seminar, taught according to Mrs Bick's method, Tavistock model. She chairs the IPA Committee on Psychoanalytic Assistance in Crises and Emergencies, PACE; is a consultant of COCAP; and is academic advisor of the work program with street children and young refugees in Mexico and Ecuador, Fundación JUCONI. She is co-author and co-editor of the books: *Territorios postkleinianos. Una actualización de la tarea Psicoanalítica* (Teseo, 2020) and *The Infinite Infantile and the Psychoanalytic task. Psychoanalysis with Children, Adolescents and their Families* (IPA, COCAP and Routledge, 2022).

Benjamin Fife, PsyD is a psychoanalytically oriented clinical psychologist working in the San Francisco Bay Area. He has taught in the San Francisco Center for Psychoanalysis's Child and Adolescent Psychoanalytic Psychotherapy Program (CAPPTP) and has trained and supervised early career mental health clinicians in a wide variety of settings. Dr Fife is interested in the impact of the socio-political world on psychological theories, clinical practice, child development, and trends in mental health.

Kristin Fiorella, PsyD, MFT, MFA is a psychoanalyst practicing in San Francisco where she sees children, adolescents, and adults. She is on the faculties of The Psychoanalytic Institute of Northern California and the San Francisco Center for Psychoanalysis. Her writings on the dialogue between Buddhism and psychoanalysis have been published in various journals, and she was awarded the American Psychoanalytic Association's Lee Jaffe paper prize in 2023.

Christine Franckx is a child psychiatrist and training (child) analyst in the Belgian Psychoanalytic Society. She is EU co-chair of COCAP and a member of editorial board of the *Dutch Journal of Psychoanalysis*. She works in private practice in Antwerp. She is an Infant Observation trainer. She co-edited *Eros op de scene* (2021) on Joyce McDougall and *Kinderlijk Trauma* (2023) on Sandor Ferenczi.

Fernando M. Gomez is a medical doctor (Medical School, University of Buenos Aires, Argentina), paediatrician (Ricardo Gutierrez Children Hospital, Buenos Aires); and psychiatrist (Medical School, University of Buenos Aires). He is the author of many papers published in academic journals and presented at national and international congresses. He lives in Buenos Aires, where he has worked for 10 years at the Pediatric Mental Health Department of the Hospital Alemán. Today he has a private practice as an adult and child and adolescent psychoanalyst.

Antònia Grimalt is a training and supervising analyst of the Spanish Society (SEP-IPA) and of FEAP, former chair of the forum for child psychoanalysis (FEP), and a child and adolescent training analyst of EPI-CAP school. As a member of Monografies de Psicoanàlisi I psicoterapia she has been translating books by Bion into Catalan. She is the editor of the complete works of Pere Folch, the Spanish edition of Matte Blanco's *Thinking, Feeling and Being*, and the Catalan edition of *Formless Infinity* by Riccardo Lombardi. She was chair of the 2020 Bion International conference (Intuition) held in Barcelona and is a member of COCAP.

Caroline Hickman has a background in mental health social work and is an integrative psychosynthesis psychotherapist. She is a lecturer at the University of Bath researching children and young people's emotional responses to climate change in the UK, Brazil, the Maldives, Nigeria, Europe and the USA for 10 years examining eco-anxiety and distress, eco-empathy, trauma, moral injury, and the impact of climate anxiety on relationships. She is co-lead author on a 2021 quantitative global study into 10,000 children and young people's emotions and thoughts about climate change, published in *The Lancet Planetary Health*. A practicing psychotherapist and member of the Climate Psychology Alliance, she has been developing a range of therapeutic approaches to ecological distress including a psychological assessment model for eco-anxiety, and has delivered workshops in climate psychology, emotional resilience, and mental health internationally.

Samah Jabr is a consultant psychiatrist living in Jerusalem whose work focuses on mental health under occupation in Palestine. She has led mental health services in the West Bank, developed community-based care, and advocated internationally for a liberationist approach to psychiatry. She has collaborated on clinical programs, policy initiatives, and advocacy projects addressing the psychosocial impact of colonial violence. Dr Jabr is Associate Clinical Professor at the George Washington University School of Medicine.

Ryan LaMothe is a professor of pastoral care and counseling. Over the last 30 years he has published over 200 articles and book reviews, as well as 14 books. These publications address topics in psychoanalysis, psychology of religion, pastoral counseling, pastoral theology, and political pastoral theology. Among them are the following titles: *A Political Psychoanalysis for the Anthropocene Age, Pastoral Care in the Anthropocene Age, Care of Souls, Care of Polis, Pastoral Reflections on Global Citizenship, A Radical Political Theology for the Anthropocene Age*, and *Peril, Perseverance, and Perversion: Pastoral Reflections on Masculinity*. Professor LaMothe has lectured at universities in the US, Europe, and Asia. In 2017 he received the Springer Publishing Award, "Transforming the World One Article at a Time." In 2022 he gave the Jacob Neumann Lecture at Princeton Theological Seminary.

Alicia Beatriz Dorado de Lisondo is a child and adolescent psychoanalyst and professor in Sao Paolo. She is the winner of two FEPAL awards and the Dr. Jose Blerger Award. She created the SOS Project that won first place in the IPA World Prie in "Health in Community" in 2023.

Ana Belchior Melícias inaugurated training on the Bick Method in Portugal. She is a member of the Association Internationale pour le Dévelopment de l'Observation du Bébé selon Bick (AIDOBB) and is President of the Portuguese Association for Baby Observation – Bick Method (APOBB). In 2019, she was awarded the Rebeca Grinberg Prize (Accésit da II Edition) from the Madrid Psychoanalytic Association (APM). Institutionally, she has collaborated in numerous scientific activities and various PPS governing bodies, in particular, chairman of the Scientific Committee (2010–2011), vice-president of the PPS Board (2022–2023) and member of the Local Arrangements Committee (LAC) of the IPA for the 54th Congress in Lisbon (2024–2025). She divides her activity between private practice with babies, children, adolescents, and adults, and teaching, both academically and in psychoanalytic training. She regularly presents scientific papers and collaborates in various national and international scientific journals in Portuguese, English, German, Spanish, Catalan, and Polish. She has published several chapters and books as editor and/or author.

Kathleen Del Mar Miller, MFA, LCSW is a writer and a psychoanalyst practicing in New York City. Her writings have been published in various anthologies and journals, including Adam Phillips' *The Cure for Psychoanalysis, Psychoanalytic Dialogues*, and *Room: A Sketchbook for Analytic Action*. She was the 2020 recipient of the Symonds Prize from *Studies in Gender & Sexuality*, where she serves as co-editor for the Miscellany section. She currently teaches at the Institute for Contemporary Psychotherapy (ICP) and Manhattan Institute for Psychoanalysis (MIP).

Carlos Padrón is a licensed psychoanalyst from the Institute for Psychoanalytic Training and Research (IPTAR). He holds an MA in Philosophy with a Concentration in Psychoanalysis from the New School for Social Research, and an MPhil in Latin American Literature from New York University. He has written and presented on the relations between philosophy, the political, and psychoanalysis; on community psychoanalysis; and on clinical issues related to diversity. His essays have been published in *Division Review; Psychoanalytic Psychology; Out Art (Argentine Psychoanalytic Society); Trópicos (Psychoanalytic Society of Caracas); Journal of Infant, Child, and Adolescent Psychotherapy;* and the book *Psychoanalysis in the Barrios: Race, Class, and the Unconscious* (Routledge), among others places. Carlos is currently a faculty member at IPTAR and at the Blanton-Peale Psychoanalytic Training Program. He has also taught at the Silberman School of Social Work (CUNY). He has worked psychoanalytically in community settings and has a private practice in Brooklyn. He is the co-founder of the Center for Critical and Clinical Analysis www.cccacommunity.com).

Panu Pihkala (b. 1979) from the University of Helsinki is currently one of the world's most cited eco-anxiety researchers, and he serves as an advisor to many practical projects on eco-emotions. Pihkala hosts the podcast Climate Change and Happiness together with Dr Thomas Doherty (https://climatechangeandhappiness.com).

Adriana Prengler is the vice-president of the International Psychoanalytical Association (IPA). She is a clinical psychologist trained in Caracas, Venezuela with a private practice in the United States, in Seattle. She is also a training and supervising analyst at the Northwestern Psychoanalytic Society and Institute in Seattle and at the Caracas Psychoanalytic Society. She has written mainly about loss, trauma, emigration and clinical matters and maintains a private practice with adults, couples, adolescents and children in Seattle, Washington. She has presented papers in international psychoanalytic conferences in several cities in South and North America,

Europe, Australia, and China. She was founding chair of the IPA's Psychoanalysts Emigration and Relocation Committee, among many other positions in the International Psychoanalytical Association, the Psychoanalytic Federation of Latin America and her local societies.

Emanuela Quagliata is a training and supervising psychoanalyst with the Italian Psychoanalytical Society (SPI-IPA), and guest member of the British Psychoanalytical Society (BPS). She qualified as a child and adolescent psychotherapist at the Tavistock Clinic in London where she also earned a doctorate in recurrent miscarriage and infertility. She is a part-time lecturer in child and adolescent psychopathology at the University of L'Aquila's Institute of Child Neuropsychiatry, and she has been consultant at the Foetal Medicine Unit of the Gynaecological Dept. and Fatebenefratelli Hospital in Rome. She teaches at the Institute of Psychoanalysis (SPI) and at the Tavistock Training in child and adolescent psychoanalytical psychotherapy in Florence. She has been member of the IPA Committee on Child and Adolescent Psychoanalysis (COCAP) and Co-Chair for Europe of the IPA Committee Women and Psychoanalysis (COWAP). She has written extensively about child analysis, eating disorders, a volume edited with Lisa Baraitser on violence and the feminine, and a series of twelve books on psychoanalysis for parents, 101bambino (www.101bambino.com). She lives in Rome where she has her private practice.

Tetiana Stasiuk is a psychoanalytic psychotherapist from Kyiv currently living in Prague. She attended Taras Shevchenko National University and the Donetsk National Medical University. She is a member of the Ukrainian Association of Psychoanalytic Psychotherapists and a candidate of the Training and Supervision Program-3 at UAPP (collective member of the EPPP).

Anastasya Svinarchuk is a Ukrainian child psychiatrist and is currently the head of the Regional Children's Mental Health Center of the Vinnystia Regional Clinical Hospital. She is also a psychotherapist in private practice. She is a member of the Ukrainian Association of Psychoanalytic Psychotherapy and the coordinator of the International Child Development Program (ICDP).

Virginia Ungar MD is a training analyst at the Buenos Aires Psychoanalytic Association (APdeBA). She was the chair of COCAP and of the IPA Committee for Integrated Training. She is a member of the board of the *International Journal of Psychoanalysis*. She was awarded the Sigourney Award in 2023. She is the former president of the International Psychoanalytic Association (2017–2021), having been the first woman elected to that position.

Ruth Weinberg is a clinical psychologist working in private practice with children, families and adults. She is a training and supervising psychoanalyst at the Israeli Psychoanalytic Society where she served on the educational committee and as Head of the Child/Adolescent Track. She also teaches in the Kleinian and primitive mental states advanced programs at Tel Aviv University. She is a scientific editor of translations of psychoanalytic literature to Hebrew and published several articles in the *Journal of Child Psychotherapy.*

Acknowledgments

The majority of the chapters of this book were written by the authors for this project. There are seven exceptions. Martha Bragin's "Pour a Libation for Us: Restoring the Sense of a Moral Universe to Children Afflicted by Violence," was published by the *Journal of Infant, Child and Adolescent Psychotherapy* 18:201–211. Carlos Padron's "Other Lullabies: Attacks on Blackness, Confusion of Tongues, and the Loss of Play," published by the *Journal of Infant, Child and Adolescent Psychotherapy* 21: 97–107. Kathleen Del Mar Miller's "Caring for Cryptids: Welcoming the More-than-Human into Psychoanalytic Treatment," published in *Psychoanalytic Dialogues* 34: 164–172. Caroline Hickman's "Eco-anxiety in Children and Young People: A Rational Response, Irreconcilable Despair or Both?" published in Vakoch, D. and Mickey, S. (2022) *Eco- Anxiety and Planetary Hope: Experiencing the Twin Disasters of COVID- 19 and Climate Change*, Springer Nature, Switzerland. My paper "The Climate Crisis and the "Unnatural Body: Onto-epistemological Possibilities of and Threats to the Genders of Children and Adolescents,"was published in the *Journal of the American Psychoanalytic Association's* online first publication, and will appear in that publication in print at a later date. Panu Pihkala's "Climate Anxiety, Maturational Loss, and Adversarial Growth," was published in the *Psychoanalytic Study of the Child* 77: 369–388. Ryan LaMothe's "Dwelling and the Climate Crisis: A Developmental Perspective and its Implications" was an elaboration of ideas that originally appeared in "On Being at Home in the World: A Psychoanalytic- Political Perspective on Dwelling in the Anthropocene Era" that was published by *Psychoanalytic Review* 107: 123–151. I am grateful to these journals and to Springer Nature for permission to reproduce these works.

Zeina Azzam allowed me to use a stanza from her haunting poem "Write My Name" published by Vox Populi on October 30, 2023. I appreciate her generosity and her interest in this project.

Preface

Kristin Fiorella

Ours is a tortured and torturing world. More children and adolescents live in conflict zones today than at any other time since the Second World War. In this globalized and virtual age, the unassimilable horror of violent conflict is not hidden. We see the brutal realities of children and adolescents trapped in war. As witnesses, we still have the protection of non-thinking and dissociative states, but we cannot claim ignorance. Simultaneously, our climate crisis makes clear not only that we are more than capable of destroying the human and more-than-human world, but also that such knowledge does not move us to radically transform our way of living.

Sartre and Merleau-Ponty both observed that there are moments in history when the structure of the world comes into view and gets called into question. These moments come when immense destruction exposes the psychic[1] and sociopolitical structures that enact our world and simultaneously obscure many of its dimensions. Judith Butler writes that these times "indict the very world in which that destruction proved possible. It is as if such a possibility had lain outside the thinkable until a certain historical time, but once it became thinkable, it was directly apprehended as a possibility of the world itself" (Butler 2022).

This book suggests that ours is one such time. The pandemic and the climate crisis disclose that our world is at once radically shared, and due to racist histories and inequalities, radically divided. The intense cruelty faced by so many children and adolescents growing up in the midst of war is justified, too often, with claims that "we had no choice."

Globally, one in five children live in conflict zones. For those of us who bear witness, their suffering indicts our fragile humanity. UNICEF warns that we seem to be slipping into a world where brutality towards children in war is somehow acceptable.

Origin of the Book

As we were emerging from the intensity of the pandemic and its accompanying movements for social justice, the IPA leaders chose *Mind in the Line of Fire* as the theme for the 53rd Congress in Cartagena in 2023. The Committee on Child and Adolescent Psychoanalysis (COCAP) inaugurated a series of short essays, *Children's Minds in the Line of Fire*, to think about the various psychic, social and ecological challenges that face children and adolescents. We alternated each month between three regions – South America, North America, and Europe. Three phenomena emerged prominently: war, the pandemic, and climate change. Many of the short essays described children and adolescents reacting to these three realities with overwhelming rage, terror, and grief. Some of the children and adolescents were living through incomprehensible violence. Even when they physically survived, children absorbed utter devastation and potential psychic annihilation. The psychoanalytic authors who participated sought ways of helping these young people find psychic meaning and to revive agency and subjectivity. Several authors took up the analytic insight that, within the complex psychic task of apprehending traumatic loss or experience, children and adolescents need an Other who can grieve their experience and receive their annihilation.

The committee decided to create this book in an effort to explore some of the complexities, terrors, and possibilities for children and adolescents growing up in this apocalyptic time. Some of the short essays from our initial series were transformed into the chapters that follow. Each part also contains one or two recently published papers. Each part has its own introduction with brief descriptions of the chapters comprising that section.

From various theoretical vantage points, many of the chapters that follow explore the psychic work required to give form to the unbearable and to weave meaning in its midst. These three central phenomena – war, pandemic, and climate crisis – all illuminate the ways in which sociocultural material and discursive conditions enable and constrain ways of thinking, feeling and acting. Many norms and ways of organizing life, sedimented with violent histories and inequalities, are proving radically destructive to children and adolescents. In order to approach the complexity of these three domains and their impact on children and adolescents, several authors draw on philosophers and critical theorists, in addition to psychoanalytic thinkers. This book argues that children and adolescents need a world that truly grieves their suffering, refuses to acclimate to unacceptable violence towards them, and insists upon and fights for the value and dignity of all their lives. Some of the psychoanalysts contributing to this collection drew on other disciplines to examine what kind of work and thought would support this effort as well as what might obscure it and justify the continuing abjection of certain groups of children.

Some psychoanalytic readers may be unfamiliar with the postcolonial and critical theory utilized in this collection. In the following section, in order to orient the reader to potentially unfamiliar concepts, I will briefly explore Judith Butler's work on grievability and precarity (Butler 2016), as well as its implication for children and adolescents. I will then describe Achille Mbembe's concept of necropolitics (Mbembe 2019). Both Mbembe and Butler (2022) argue that shared precarity and radical interdependence, both of which are highlighted by the pandemic and the climate crisis, offer the fragile possibility for a more just and less destructive future.

Grievability and Apprehending Precarity

For Judith Butler, who we are as a global collective in times of war can be rendered by considering whose lives count as lives and whose deaths are grievable. An ungrievable life is one that was never counted as a life in the first place, at least not when such a life is considered a threat to other lives that are classified as grievable. When one population is defined as a direct threat to another whose lives are valued, the first population will not appear as "lives" but as a threat to life. These threats can then be killed, even with extreme violence, without provoking horror. Outrage over violent death can be reserved for the ones that I feel are closer to my life, the ones who are like me (Butler 2016).

Two aspects of Butler's description are important here. The first is how grievability is governed by norms that structure who qualifies as recognizably human and what violence has impunity. When these norms are regulated through nationalism and/or race, those who share my national affinity and racial identity are valued and human subjects. Another group, seen as a threat to the first, can then be killed. These norms need not have moral or ethical consistency. In the US, where I live, we have a long history of justifying violence with the logic of racism. This history is sedimented into many of our norms and continues to justify racist violence within our country and elsewhere. When there is a perceived threat to our national interest, our norms regularly highlight the violence of the other, while minimizing or ignoring our own. We do not formally acknowledge much less grieve the lives of those killed by our country in times of war. As a national collective, we do not apprehend those lives as *life*. This is an example, and many other countries have their own, of what Butler calls norms of recognition- norms that govern who is a valued human subject and what violence is permitted.

The second relevant aspect of Butler's description of grievability is how norms of recognition bypass the apprehension of life as life. For Butler, as for many psychoanalysts,[2] apprehending life as life means apprehending precariousness as an inextricable feature of living. If norms of recognition determine that one's value is a given, then one can certainly have an intellectual acknowledgment of precarity- death, impermanence, the vulnerability of oneself and everyone one loves- without *apprehending* these realities, at least at times. If others are subject to violence but my status as a valued subject is secure, I do not have to emotionally absorb that we are all vulnerable to destruction, just as we are all capable of destroying. We are ultimately existentially bound together in that, even though different collectives are exposed to radically different degrees of precarity (Butler 2022). The pandemic and the climate crisis reveal this existential collective precarity, as I will explore shortly.

For many, it is difficult, although certainly possible, *not* to apprehend the precarity of children. When we see children who are terrified, weeping over dead parents, or themselves lying dead, it is harder to deny a visceral, pre-discursive response to their agonizing vulnerability. The psychic work required to disavow this is more strained than with terrorized adults. One means by which we may cut our visceral tie to these children is by consciously or unconsciously rendering these children as monstrous: *These children are not like my children. They are violent children, and they will threaten me now or grow up to do so.* In this paranoid/schizoid world, one need not feel much pain at their destruction.

For someone who does not experience such a stark split, I speculate that witnessing severely traumatized children who could be killed at any moment stirs terror of non-being. On some level, it evokes glimmers of the agonizing anxieties associated with living in a world of extreme vulnerability with an Other who refuses to apprehend one's life as life. For some witnesses, it is perhaps more terrifying that one's children could be subject to such dehumanizing erasure than oneself. Many of us witnessing the extreme violence and precarity of children in war cannot fully imagine their terror and suffering. Nonetheless, I suggest that we cannot fully escape the existential terrors that witnessing evokes. Living through experiences of extreme violence can easily shatter representational capacities, and one can slide into a horrifying void without meaning. Existentially, we are all vulnerable to such a loss of our psychic functioning. Witnessing another's erasure evokes this recognition and stimulates our own anxieties of non-being (Winnicott 1974; Durban 2017; Botella & Botella 2004). Rather than wrestling with this, one might manage this existential recognition with a remote "knowing about" the suffering of these children that sutures over anxieties of non-being. Such an internal compromise also could calm despair, and ambivalence or even horror with one's collective, if it is participating or indifferent. It is then possible to preserve a sense of oneself as connected to one's collective while also being a moral agent saddened by the violence. One can feel that these children's fate is unfortunate, without the despair and outrage of apprehending that their destruction is radically unjust, (Butler 2016).

There are many rich psychoanalytic descriptions of the psychic work required to stay open to annihilation. Winnicott (1974), Bion (1970), Ofra Eshel (2019), and the Botellas (Botella & Botella 2004), among others, describe the analyst's potential openness to non-verbal perception of psychic devastation. Such receptivity involves some willingness to experience despair, outrage and disintegrating anxieties. Could we call on these capacities while witnessing the devastation of children and adolescents by violence? Psychoanalyst Sue Grand writes that despair is an antidote to pretense (Grand 2000). She suggests that despair, as opposed to feeling something is merely unfortunate, is a necessary resistance to the collective dissociation that can allow cruelty to continue. If we called upon our clinical capacities to open ourselves to the despair and anxieties of non-being that we inevitably face while witnessing children and adolescents living in extreme violence, perhaps we can be moved to resist brutality towards them in whatever ways we can.

As several of the following chapters explore, the pandemic and the climate crisis highlight our collective precarity as well as our porous interdependence with the human and more-than-human world. They disrupt the deeply ingrained Western ontology of discrete bodies, separated at the skin, and distinct from the more-than-human world. Yet, the climate crisis in particular has made apparent how staunchly we disavowed interdependence with one another and the more-than-human. Our lack of collective mobilization in the face of ecological breakdown highlights that many of the norms of modern life, sedimented with histories of violence, domination and acquisition, blind us to our embeddedness in the world and with each other.

Both the pandemic and the climate crisis illuminate deep structural inequalities as well. The pandemic, for example, demonstrated that, although we all lived close to death, racist inequalities made some of us much more likely to die than others. Likewise, as the third section of this book takes up, wherever and whoever you are on this planet, the climate crisis is an existential threat, and yet, people living in the poorest of nations, despite their having done the least to contribute to ecological breakdown, are by far the most vulnerable. Our lack of transformation in the face of a clear threat to all life urgently calls into question some of the onto-epistemological means by which we organize life. Philosopher Bruno Latour, who appears in a couple of the climate chapters, argues that the modern composition of the world has failed, and that multiple recompositions will be required if we are to continue to inhabit the planet.

This reckoning with our entangled existence (Barad 2007) opens a fragile possibility for another politics of the world (Mbembe 2019), rooted in a planetary recognition of radically shared finitude and vulnerability. It is clear we are long way from such a world. This book is an effort to think about what is required to move in that direction. If such a planetary, radically shared world became possible, how could it not find the current level of brutality towards children and adolescents in war ethically unacceptable?

Necropolitics: Life as Death's Medium

Why is this level of destructiveness possible in our world? While the answers to such a vast question will be multiple and complex, several of the authors in this collection argue that many of the modern sociopolitical practices and structures that are currently devastating many children and adolescents and threatening their futures have arisen from colonial histories. These authors suggest that these violent histories and subsequent structural inequalities have blunted our collective capacity to recognize the value and dignity of all life, as well as to care for the more-than-human world.

Postcolonial scholar Achille Mbembe refers to the use of social and political power to determine those whose lives will be protected and those who will be allowed to die as *necropolitics*. Those who are disposable, which includes countless children and adolescents, are subjected to objective signs of brutality. Within the logic of necropolitical power, life, for the dispensable, becomes death's medium (Mbembe 2019).

Mbembe traces ways in which the early colonial period gave rise to experimentation with violence almost beyond limits. He argues that such experimentation with boundless cruelty had deep psychic and sociopolitical ramifications and ultimately paved the way for the Holocaust and modern genocidal ideologies. He writes that our time "consists in counting whatever is not oneself as nothing ... death is something to which nobody feels any obligation to respond" (Mbembe 2019). Mbembe also echoes the psychoanalytic insight that traumatizing violence begets more violence. He points out that traumatized victims of such violence sometimes later inflict horrific violence on those in weaker positions. Some of the chapters in this book take up questions of what psychic work is required to interrupt this cycle.

Mbembe points out that our climate crisis is also entangled with colonial histories.[3] As most readers likely know, along with the slave trade of African peoples, colonial conflicts often led to significant diminishment if not wiping out of Indigenous populations. They also created ecological mutations in conquered regions. Prior to the colonial wars and genocides, most people considered themselves embedded in the world. Colonial conquerors popularized a notion of "nature" as an inert, separate entity that could be dominated. This concept supported a logic of extraction, acquisition and hierarchy that led to exploitation of certain bodies and of the more-than-human world. Another postcolonial writer discussed in the climate section of this book, Amitav Ghosh, writes:

> *Subdue* was a key word in these conquests, recurring again and again in reference not just to humans but also to the terrain. Out of these processes of subduing and muting was born the idea of "Nature" as an inert entity, a conception that would in time become a basic tenet of what might be called "official modernity."
>
> (Ghosh 2021)

Ghosh demonstrates repeatedly the way that the muting of "nature" accompanied the making of categories of people and enacted a sense of the more than human world as an entity that we could endlessly extract from.

As some of the following chapters explore, many children and adolescents are attuned to and involved in efforts to recognize a common world based on shared precarity and vulnerability. They are attempting to create new cultural imaginaries that would be more responsive to our current crises. A recognition of entangled existence with human and more-than-human others (Barad 2007) has the potential to inspire a reconfigured global politics that could return to life what had been discarded to forces of death (Mbembe 2019).

Notes

1 One might think of Freud theorizing the death drive in the aftermath of the First World War.
2 For example, Money-Kyrle (1968) and Holzhey-Kunz (2014).
3 Many other scholars and a few psychoanalysts have also illuminated and explored this link. See, for example, Elizabeth Povinelli (2021), Amitav Ghosh (2016, 2021), Donna Orange (2017), and Katie Gentile (2021, 2024), among many others.

References

Barad, Karen. 2007. *Meeting the Universe Halfway*. Durham, NC: Duke University Press.
Bion, W. R. 1970. *Attention and Interpretation*. New York: Routledge.
Botella, Cesar & Botella, Sara. 2004. *The Work of Psychic Figurability*. New York: Brunner-Routledge.
Butler, Judith. 2016. *Frames of War: When is Life Grievable?*London: Verso.
Butler, Judith. 2022. *What World is This?: A Pandemic Phenomenology*. New York: Columbia University Press.
Durban, Joshua. 2017. Home, Homelessness and Nowhere-ness in Early Infancy. *Journal of Child Psychotherapy* 43: 175–191.
Eshel, Ofra. 2019. *The Emergence of Analytic Oneness: Into the Heart of Psychoanalysis*. New York: Routledge.
Gentile, Katie. 2021. Kittens in the Clinical Space: Expanding Subjectivity through Dense Temporalities of Interspecies Transcorporeal Becoming. *Psychoanalytic Dialogues* 31: 160–165.
Gentile, Katie. 2024. Destroying the "Human" for the Survival of the World: A Proposal for a Temporally Dense Psychoanalytic Subject. *Psychoanalytic Dialogues* 34: 533–541.
Ghosh, Amitav. 2016. *The Great Derangement*. Chicago, IL: The University of Chicago Press.
Ghosh, Amitav. 2021. *The Nutmeg's Curse*. Chicago, IL: The University of Chicago Press.
Grand, Sue. 2000. *The Reproduction of Evil*. New York: Routledge.
Holzhey-Kunz, Alice. 2014. *Daseinsanalysis*. London: Free Association Books.
Mbembe, Achille. 2019. *Necropolitics*. Durham, NC: Duke University Press.
Money-Kyrle, R. 1968. Cognitive Development. In D. Meltzer & E. O'Shaughnessy (Eds.), *The collected papers of Roger Money-Kyrle*, 416–433. Strathtay: Clunie Press, 1978.
Orange, Donna. 2017. *Climate Crisis, Psychoanalysis, and Radical Ethics*. New York: Routledge.
Povinelli, Elizabeth. 2021. *Between Gaia and Ground*. Durham, NC: Duke University Press.
Winnicott, D. W. 1974. Fear of Breakdown. *International Review of Psychoanalysis* 1: 103–107.
Yeats, W. B. 1996. *The Collected Poems of W.B. Yeats*, revised second edition. New York: Scribner Paperback Poetry.

Part I

War

Kristin Fiorella

Introduction

> Write my name on my leg, Mama
> When the bomb hits our house
> When the walls crush our skulls and bones
> our legs will tell our story, how
> there was nowhere for us to run
> – Zeina Azzam (2023)

As of the end of 2024, nearly one in five children live in a conflict zone. "By almost every measure," said UNICEF director Catherine Russell, "2024 has been one of the worst years on record for children in conflict in UNICEF's history – both in terms of the number of children affected and the level of impact on their lives."

These children and adolescents – in Palestine, Sudan, Ukraine, Syria, Lebanon, Haiti, and elsewhere – endure horrific realities, beyond what any child should have to imagine in their darkest, most annihilating fantasies. War, for them, is terror, dismemberment, sexual violence, and seeing parents and siblings crushed or decapitated. In some war zones, survivors have little hope of a significantly more livable future. In Sudan and Gaza, most schools have been destroyed and children face starvation. The agony of parents, who must watch these children wither away as their stares become empty, is incomprehensible.

The death toll in Gaza after fifteen months of war is estimated at more than 48,000. As of November 2024, 44% of those deaths are thought to be children. This makes the war in Gaza by far the most deadly for children in recent history. In just the first four months of the Gaza war, more children were killed than in all conflicts worldwide during the previous four years. The numbers of child deaths are staggering in other conflicts too. Children and parents who are not killed are subject to catastrophic levels of annihilation anxiety. Psychoanalysts know well that beyond a certain point, threat of annihilation collapses the mind and obliterates the capacity to know reality. Beyond all the deaths, disfigurements, and losses, in other words, we face a

DOI: 10.4324/9781003520757-1

global humanitarian crisis of severely traumatized young people for whom the fabric of life and psychic reality has been shredded.

In Palestine, as the poet Zeina Azzam attests in the verse quoted above, parents took to writing children's names on their legs in permanent marker so that, if a child were to be disfigured in a bombing, the body could be identified. I feel haunted by what must have passed between parents and their children as they carried out this task and then lived with such a horrifying signifier drawn onto their children's skin. Trapped in relentless violence, parents could offer no protection, only a marking to help in the possible locating and burial of each child should they be crushed. In the face of such onslaught, a needs assessment carried out by a Gaza-based NGO in November 2024 found that 96% of Palestinian children felt that death was imminent and 49% wished to die rather than continue to live in such agonizing terror (Borger 2024).

In Sudan, in addition to being subjected to starvation and violence, children and adolescents are at risk of being abducted and forced to become child soldiers in one of the two armies in the Sudanese civil war. Children who suffer this fate are typically indoctrinated and forced to kill and terrorize, sometimes within their own communities. When children are coerced to commit what psychoanalyst Sue Grand (2000) has called a "bestial gesture of survival," their sense of agency and subjectivity are often confounded. In Sudan as elsewhere, such children rarely had viable paths to dignified existence prior to abduction, and therefore commonly identify with their aggressors. This identification further compounds the stultifying guilt and dehumanization suffered by these children if they are subsequently rescued. Martha Bragin, in Chapter 2, explores precisely this in her writing about the rehabilitation of child soldiers in Sierra Leone.

With the climate crisis promising a world with fewer inhabitable spaces, there is certainly the strong possibility that wars will become even more common and devastating for more children. Philosopher Elizabeth Strengers (2015) warns of a "coming barbarism," due to the climate crisis, including more racist nationalism, horrific conflict, and lack of responsiveness to the Other who will not survive.

UNICEF, again, at the close of 2024: "Children's rights are being brutally violated – and we're tired of hearing excuses about it. We refuse to accept a deadly new normal." So, too, psychoanalysts, we must now ask ourselves, how can we *not* join humanitarian efforts responding to such unbearable realities? As Grand (2000) warns, dissociation is contagious and can help to reproduce cruelty. Who do we become as a field if we do not make ourselves available to these traumatized children and adolescents, emotionally receiving their suffering and refusing their annihilation any way we can? These children and adolescents are already telling us a story of trauma and horror, as illuminated by Zeina Azzam's poem quoted above. We need to absorb this story and resist their destruction.

Yet, what, if anything, can psychoanalysts offer in the face of such bleak and horrible realities? And what blocks our responsiveness in the first place? One thing psychoanalysis has to offer is insight about the mechanisms of dehumanization. We know too well that trauma, annihilation anxiety, rage, narcissism, and envy can all encourage the splitting and projection that renders the Other disposable. Some chapters in this section describe these processes and also the psychic work necessary to interrupt such dehumanization and the cycles of violence to which it contributes.

As a traumatized profession that has also learned something about listening to trauma,[1] psychoanalysts can potentially work to listen to what Cathy Caruth (2013, 2016) calls " the enigma of trauma's incomprehensibility" in the lives of these children and adolescents. Caruth reminds us of the psychoanalytic insight that trauma is unassimilable by its very nature. Extreme trauma involves a nexus of knowing and unknowing precisely because of a death of the self that was not experienced. As Freud and Winnicott illuminated, because the death happened but was not experienced, it continues to reappear. As it does, both the rupture in being and its ungraspable nature are repeated.

Psychoanalysts know also that it is sometimes possible to experience what has not been experienced if there is a receptive Other who can bear the annihilation that continues to occur. Then, under the best of circumstances, a child or adolescent can awaken to the reality of annihilation and therefore reach once again towards life. Without such a relational process, the child may be condemned to dissociated devastation too overwhelming to be lived and thus always repeating itself in an endless static loop.

Our field has a wealth of clinical and theoretical experience, some of which is explored in the chapters in this section, describing the restoration of agency, subjectivity and meaning. For generations, we have thought about how trauma can be so extensive as to leave a survivor lost and unable to return to the world of the living. This clinical knowledge suggests an ethical imperative that psychoanalysts join UNICEF's urgent entreaty to resist violence against children in whatever we can. Beyond physical death, extreme trauma can, in a matter of minutes, collapse a psychic life indefinitely, even irretrievably. When violence descends on children in war zones, these children may go from a terrified state that still has words and objects into a state that has neither. In Chapter 3, Ruth Weinberg explores Joshua Durban's concept of no-whereness (Durban 2017) in her work with Israeli children traumatized by the events of October 7th, 2023. In no-whereness, there is no one left to listen internally or externally and there is nowhere to go. No home, no objects, no self.

While it is possible to restore a sense of one's narrative history and to revive a capacity for meaning, it is important to heed Avgi Saketopoulou's reminder that trauma never gets healed in any final sense (Saketopoulou 2023). Our work as analysts is to help frozen energies recirculate and revive the capacity for meaning in the context of trauma's enduring imprint. Saketopoulou further reminds us of trauma's unassimilable nature and of the anxieties this evokes in the analyst, who must learn not to self-soothe with fantasies of mastery and resolution.

The examination of dehumanization is vital and necessary; taking refuge in such examination, though, betrays those who suffer. If we listen to their trauma from the site of incomprehensibility, it can arouse deep and agonizing anxieties and despair (Eshel 2019; Caruth 2016). To listen from this place of incomprehensibility does not condone a lack of responding. Rather, it suggests that we listen to what has died or is dying without suturing it "with the obscenity of understanding" (Lanzman 1995). Then, we might also hear the psychic life that still pulses in severely traumatized children and adolescents.

Finding meaning and the child in the ruptured fabric of a child's world is familiar to the child psychoanalyst. A child's play can bring to life richly layered emotional experiences in ways that are devastating and breathtaking. We meet many children and adolescents in the following chapters who are reaching for meaning within their experiences of loss, abandonment, and violence. Some of these children turn to each other to play out trauma and pain that they cannot fully grasp.[2] With psychoanalysts, with each other, and with their collectives, children are sometimes able to mourn and even give form to the nightmares they have lived.

Overview of Chapters

Part I begins with Samah Jabr and Elizabeth Berger's chapter about the traumatic effects on children and adolescents of the war in Palestine (Chapter 1). They utilize a psychoanalytic framework of emotional development and psychic life, but incorporate also the domains of human rights and public health as well as considerations of history and social justice. Jabr and Berger further describe the psychic impacts of genocide – the agonizing violence, demolition of schools and institutions, and the blocking of food and other necessities. A series of clinical vignettes illustrate aspects of trauma and injury characteristic of Gaza. Jabr and Berger discuss the centrality of justice, human rights and the collective integration of healing. They make recommendations for the massive scale of grieving, rebuilding and recovering from severe trauma that will be needed for the children and adolescents of Gaza in the aftermath of the war.

This is followed by Martha Bragin's chapter centered on her work with African colleagues in Sierra Leone helping to reintegrate child and adolescent soldiers who were kidnapped and forced to murder and terrorize their communities (Chapter 2). Bragin gives a clinical example of a painful and organic unfolding within the community that involves reparations on many levels, including for previous colonial violence and trauma subtending the civil war in Sierra Leone. Bragin uses a Kleinian lens to explore the ways in which she needed to confront her own intrapsychic violence and participation in structural violence in order to be a useful object for these children and adolescents. In order for the child and adolescent soldiers and their communities to begin to heal, they had to find some way to incorporate this terribly violent period into a narrative that both acknowledged the violence and restored some sense of moral order. To examine this collective task, Bragin uses the perspective of traditional Sierra Leonean healers as well as Jessica Benjamin's concept of the moral third.

In Chapter 3, Ruth Weinberg explores the psychological implications of October 7th, 2023, and the ensuing war in Gaza for Israeli children and adolescents. The events of October 7th, briefly described in the chapter, have specific characteristics responsible for their intense psychic impact. These include the scale of the massacre, the sadism and abuse that accompanied the murder of women and children, the taking of hostages, and the sense of abandonment by the Israeli government and army, both of which failed to intervene in time. Following a brief review of psychoanalytic literature about children in war, the author gives examples of short-term psychoanalytic interventions. The chapter examines the specific anxieties stirred by the events of October 7th: confusional, paranoid-schizoid, depressive and anxieties of being. Weinberg describes brief interventions that hold in mind the ways in which the trauma meets the internal unconscious reality of both parent and child.

Chapter 4 consists of the personal accounts of four psychoanalytic psychotherapist mothers whose babies were born during the Ukrainian war. The mothers formed a group, led by an experienced psychoanalyst, Emanuela Quagliata, to digest their traumatic experience and restore symbolic thinking. These mothers describe their early days with their babies in bomb shelters, and then later waking in the night to gauge how far away the bombs are; when to stay and when to leave; and how to build some semblance of normalcy for their babies and young children. The mothers describe the impact of the war on their maternal capacity and on the children's development. They also describe the life-giving function of the group, as a context for the gradual expression of painful and repressed emotions.

In Chapter 5, Mónica Cardenal describes working in Latin America with refugee children and adolescents fleeing still other wars in which they have suffered extreme violence. In Cardenal's account, these young people sometimes travel alone and sometimes with their mothers, not all of whom have survived the journey. Cardenal uses the work of Esther Bick to describe states of mind without an internal containing object capable of sustaining and unifying the self. She also describes the work of PACE (Psychoanalytic Assistance in Crises and Emergencies) in short term interventions with children and adolescents impacted by the war in Ukraine. She offers several clinical vignettes to highlight the challenges and complexities of these short term interventions.

Chapter 6 examines the impact on children of opioid use in families to highlight social conditions often overlooked in the oversaturated language of the "war on drugs." Ben Fife points out the role of drugs in maintaining some of the splits that war requires, both during and after the initial violent conflict. Fife offers several clinical vignettes to describe the impact of opioid use in families impacted by wars and utilizes Link theory to explore the matrix of socio-symbolic signifiers that these families are embedded in. Drawing on the work of Andre Green, he describes the way in which children in these families encounter absences in those they depend upon and the implications of this on these children's ability to experience need.

Chapter 7 utilizes a 1952 René Clément film, *Forbidden Games,* to ask if it is possible to mourn what cannot be represented. Ana Belchior Melícias explores what can be done externally and internally about trauma, death, and violence. Melícias uses Ferenczi's work to describe the psychic paralysis that follows overwhelming trauma and the potential erasure that can follow unless psychic work finds a way to represent violence and loss. In the film, a six-year-old child witnesses the murder of her parents during the Nazi occupation. Orphaned, she wanders in a state of non-reality until she is taken in by a ten-year-old boy. The two seek to mourn and restore a capacity for meaning through the potential space they create in cemetery artwork.

In Chapter 8, the co-authors, Mary T. Brady, Adriana Prengler, Ana Belchior Melícias, and Virginia Ungar explore the internalization and transgenerational transmission of political and personal trauma through the 1976 film, *Cria Cuervos.* The authors utilize the work of Haydee Faimberg to examine the way in which repression affects the mind across generations. The film's central character is an eight year old girl fractured by traumatic loss. The chapter examines the impact of family violence as well as the political violence of Francisco Franco's dictatorship on the mind of the young protagonist. The authors discuss the interpenetration of fantasy and reality as the child's psyche struggles to make meaning.

Finally, in Chapter 9, Antònia Grimalt explores the nefarious impact on children and adolescents of malignant manipulation of truth. It begins with a vignette from a woman who was a 10 year old girl at the time of the Chernobyl explosion. This woman describes how incomprehensible it was, to her ten year old self, that the entire government could lie about something so deadly. Grimalt then moves to the war in Ukraine and Putin's assertion that he is the savior, not an invader. She utilizes Bion to describe the necessity of truth for healthy psychic development. The balance between truth and lies is always precarious, and a child is particularly dependent on the minds of others for the structuring of her own relationship to truth. Grimalt takes up the lies we might all tell ourselves – there's nothing we can do, governments will sort it out, countries have to defend themselves – as we witness the unbearable suffering of children in war zones.

Notes

1   Several psychoanalysts have suggested that psychoanalysis has been traumatized by the Holocaust. See, for example, Dori Laub (2017).

2 There is a poignant example of this in *A House Made of Splinters*, a beautiful documentary filmed in a temporary orphanage on the frontline of the eastern Ukrainian war. Two young girls, Sasha and Polina, both severely neglected by alcoholic mothers, play with the tangle of love and hate they feel for their mothers and each other, as well as their complex identifications with their mothers and the war that has devastated their community. There are electric scenes between them, full of erotic and aggressive energy. While they both run into the unassimilable and dissociate, they keep re-finding each other. When they are separated due to Polina's adoption, they mourn the loss of the other.

References

Azzam, Zeina. 2023. Write My Name. https://voxpopulisphere.com/2023/10/30/zeina -azzam-andy-young-david-ades-three-poems-about-gaza.

Borger, Julian. 2024. Death Feels Imminent for 96% of Children in Gaza, Study Finds. *The Guardian*, December 11.

Caruth, Cathy. 2013. *Literature in the Ashes of History*. Baltimore: John Hopkins University Press.

Caruth, Cathy. 2016. *Unclaimed Experience*. Baltimore: John Hopkins University Press.

Durban, Joshua. 2017. Home, Homelessness and Nowhere-ness in Early Infancy. *Journal of Child Psychotherapy* 43(2): 175–191.

Eshel, Ofra. 2019. *The Emergence of Analytic Oneness: Into the Heart of Psychoanalysis*. New York: Routledge.

Grand, Sue. 2000. *The Reproduction of Evil*. New York: Routledge.

Lanzman, Claude. 1995. The Obscenity of Understanding. In Cathy Caruth (ed.), *Trauma: Explorations in Memory*. Baltimore, MD: Johns Hopkins University Press.

Saketopoulou, Avgi. 2023. *Sexuality Beyond Consent: Risk, Race and Traumatophilia*. New York: New York University Press.

Strengers, Elizabeth. 2015. *In Catastrophic Times: Resisting the Coming Barbarism*. London: Open Humanities Press.

Chapter 1

War, Trauma, and the Survival of Hope in Palestine

What We Learn from its Children

Samah Jabr and Elizabeth Berger

Introduction

The authors are psychiatrists who have worked together for many years. One of us (S.J.) is a psychiatrist, previously heading the Mental Health Unit at the Palestinian Ministry of Health in Ramallah, Palestine, and the other (E.B.) is a child psychiatrist in New York City. Our previous undertakings include writing about Palestine for the scholarly press as well as the development and implementation of training programs, clinical services, and policy planning initiatives in mental health. We speak here on the traumatic effect of war on children in Palestine primarily as clinicians, utilizing a psychoanalytic framework of emotional development and psychic life. But to fully illuminate the struggles of children in Palestine in the context of war, we see an urgent need to integrate the psychodynamic perspective with other spheres of knowledge – the domains of public health and human rights, as well as broad considerations of history and social justice.

Why, one may ask, are these wider perspectives necessary? We would reply that the psychological development of children is profoundly dependent upon the functional integrity of the family and of the society surrounding that child. Children are wounded not only by bullets and bombs but by the downstream consequences of damage suffered by their parents and their communities. To understand the totality of the trauma experienced by children and adolescents in Palestine, therefore, we must be alert to the harm to growing children when their families are devastated. Likewise, recognizing the crucial role played by supportive community institutions such as schools and healthcare systems, we must take inventory of the deprivations children experience when these vital structures are destroyed. In addition to these concrete injuries, we understand the growth of any individual child as deeply rooted in the integrity of intangible aspects of identity embedded in historical, cultural, religious, ethnic, and racialized images, narratives, and belief systems. War victimizes children not only through acts of immediate political violence and material deprivation, but through damage to these deep psychic structures conveying human attachment and human meaning – structures which are inherent in shared communal values.

DOI: 10.4324/9781003520757-2

We begin therefore with an overview of the context of Palestine – a context both present and past – before taking up the question of the mental health of its children.

We acknowledge at the outset that this chapter itself suffers from limitations imposed by war. A print volume passing through the publication process cannot reflect newsworthy events in real time. We have arbitrarily "stopped the clock" on our statistics nine months after October 2023, although the volatile situation in Palestine is constantly evolving. We further note that neither of us have entered Gaza under current conditions and that therefore we must depend upon other primary sources, anecdotes, and examples in the media to reflect its immediate reality.

Gaza and the West Bank 2023–2024

Since early October 2023, the massive military assault unleased by Israel upon the people of Gaza has shocked the world with its images of unremitting horror. The decimation of the civilian population has been of such magnitude that in January 2024, after less than three months of bombing, the International Court of Justice in the Hague, Netherlands "found it plausible that Israeli's acts could amount to genocide …"[1]

Subsequent findings reported by the United Nations' Office of the High Commissioner for Human Rights concluded that Israel has committed the war crimes of starvation, murder, intentional attack upon civilians, torture, forcible transfer, sexual violence, arbitrary detention, and outrages upon personal dignity. The Commission also determined that Israel was guilty of extermination – a crime against humanity – as well as gender persecution against boys and men. The Commission concluded that the enormous number of civilian casualties and pervasive destruction of infrastructure were the inevitable outcome of the Israeli intent to inflict the maximum damage.[2]

Further, Israel's total siege was viewed by the Commission as collective punishment – blocking the necessities of life such as water, food, electrical power, fuel, and humanitarian aid. These weaponized restrictive measures have inflicted grave harm to children and have brought about the death of children from starvation. The Commission further observed that sexual violence was part of the Israeli official standard operating procedure – a judgement based on the frequency, severity, and prevalence of these violations. The Commission identified war crimes committed in the West Bank as well, including torture, sexual violence, and attacks on personal dignity. The fact that Israeli forces encouraged and instigated violence perpetrated by Israeli settlers against Palestinians in the West Bank was also noted.[3]

After the initial ten weeks of assault, UNICEF determined that at least 1,000 children in Gaza had already lost one or both legs.[4] And after a period of less than four months since the beginning of this war, UNICEF reported that at least 17,000 children in Gaza had no adult taking care of them, having been orphaned or separated from their parents and from all members of their extended family.[5]

Considerably later, in July 2024, the official death count has now reached 37,953 persons, with more than 85,452 injured and over 10,000 missing. There are estimates that Israel has killed over 15,000 children – on average, killing 60 children per day. In the West Bank, 561 Palestinians have been killed, including 138 children, and over 5,300 injured.[6] However, a letter in the major British medical journal, *Lancet*, concluded that the actual death toll after eight months of assault – counting those killed by the imposed deprivation of food, water, and medical care as well as the number of those immediately killed – far outstripped these official figures. The estimate here was 186,000 dead, approximately 8% of the population of Gaza.[7]

And further, data assembled from the World Health Organization, the UN's Office for the Coordination of Humanitarian Affairs, and the government of Palestine indicate that eight months after the initiation of the Israeli assault, more than one half of all the homes in Gaza have been damaged or destroyed; 88% of school buildings have been damaged or destroyed, only 16 out of 35 hospitals operate with any functional capacity, and 267 places of worship have been destroyed or damaged.[8]

Media around the globe continue to display photos and videos of the city of Gaza reduced to rubble, its captive surviving families struggling to survive on foot and in refugee tents, the cries of the injured and the grieving, and the desperate doctors performing amputations without stretchers, clean water, anesthesia, or antibiotics.[9]

Those who wonder how these atrocities could take place will find the answer in the remarks by the Israeli Knesset member Tally Gottlieb, who declared that without imposing hunger and thirst on Gaza, Israel will be unable to recruit the necessary intelligence through bribery.[10]

As the resulting famine in Gaza intensifies, the UN reports that 50,000 children require immediate medical treatment for malnutrition.[11] The UN reports that 90% of children in Gaza lack the proper food for growth.[12]

Parents in Gaza have taken to writing their children's names in marking pen on their bodies, so that the child's identity might be known when the child is injured, killed, or orphaned. A poem in which a child asks for this signature has been widely circulated online.[13]

Palestine 1948–2023

These appalling circumstances are often framed in current Western media as if they arose suddenly and without precedent. Unfortunately, for the people of Palestine, these circumstances are all too familiar. An adolescent of 14 living in Gaza, for example, has already survived four previous wars.[14]

We have described and documented elsewhere the details of the Israeli occupation and ethnic cleansing of Palestine since 1948: a relentless

downward spiral of military and political strangulation, legal disenfranchise-
ment, economic constriction, enforced poverty and unemployment, pervasive
human rights violations, imposed medical neglect, and crushed dreams. We
have noted the politically driven degradation of Palestinian family cohesion,
personal identity, and its historical record.[15]

Among the many concrete manifestations of this historical multi-genera-
tional trauma, including pervasive Israeli control over Palestinian land,
resources, water, air space, and human movement, we choose to focus here on
a single representative aspect – the fact that since the 1960s, approximately
one-third of all Palestinian men have been subjected to detention by the
Israeli military. Detention is often arbitrary, imposed to punish relatives of
persons believed to be suspicious, and carried out without specific charges
being brought, without access to legal representation, without due process,
and without a sentence of specified length; sentences may be arbitrarily
extended. Many persons have been detained for decades or, in some cases, for
their entire lifetimes.[16]

There is extensive reportage of torture in detention (and the threat of torture of family members) as a routine measure to obtain false confession, to
enlist new personnel within the enormous network of covert collaborators
with the Israeli police, to intimidate families and communities, and to deter
and to silence the detained.

Palestinian children are frequently taken from their beds in the middle of
the night, hooded and shackled, and interrogated by military police. In our
personal experience interviewing these children, the threat or practice of
sodomy is commonplace. In one report, for example, 40% of children
detained in Jerusalem described sexual abuse.[17] Children who have revealed
the names of their comrades under torture are often afflicted with shame; we
have observed permanent personality change, school drop-out, and social
withdrawal in young people who have been violated in these ways. Current
statistics on child detention and torture indicate that as of July 2024, 640
Palestinian children are detained in the West Bank.[18]

The devastating mental health consequences of detention on the child or
adult detained, on the family whose member or members remain in detention,
and on the community fearful of reprisals and collaborators, are phenomena
that have not received sufficient public and professional attention. But given
that approximately one third of the men in Palestine have been detained –
often multiple times over years – one can at the very least see that detention
represents a massive public health problem.

Vignettes

The following vignettes highlight themes we observe within the traumatic
experiences of Palestinian children and their families surviving amid pervasive
political violence. The first group of vignettes emerges from clinical

encounters taking place in East Jerusalem and the occupied West Bank, concerning patients treated or supervised by one of us (S.J.); minor details have been changed to protect confidentiality.

Case One

A boy of 13 from East Jerusalem was brought to a psychiatrist by his mother, who reported that the youth's teacher had suggested a stimulant medication for his attentional problems. The mother explained that her son had developed the habit of laying his head upon his desk at school and failing to complete his classwork, although he had been an excellent student until recently.

The boy himself described that his school recently had instituted a new "Israeli curriculum" which omitted recognition of the people of Palestine and failed to acknowledge their history. He was offended by this curriculum. The act of laying his head on his desk was his way to assert resistance to a political narrative which he found humiliating. What his teacher had interpreted as a psychiatric disorder was, to him, a defiant effort to reclaim of his dignity. He was on strike.

Case One reveals particularly vividly the way that a youthful behavior may be perceived as a sign of personal pathology or as a symptom of an illness – even by familiar adults who are sympathetic to the young person – when the behavior is experienced from the inside as an act of personal agency, a response to a political challenge, and a demand for collective justice. Both the mother and the school system had framed their concerns in this inaccurate manner. Psychiatric evaluation however ruled out the presence of anxiety and depression, as well as ADHD and Oppositional Defiant Disorder. This youth was not opposed to authority in general but to a particular imposed ideology. What may appear to be adolescent rebellion or a sign of inattention may be better understood as an individual response to a social problem.

Case Two

A boy of 15 living in the West Bank was brought to the psychiatrist in the midst of a hypomanic state. He repetitively stole large amounts of money from his father and spent it on his friends. He then lied about these activities.

In an unrelated development, his 16-year-old brother was arrested and detained by the Israelis. This crisis precipitated a dramatic shift in the patient's mood. He became depressed and remorseful, complaining "I'm the one who should be in prison – not my brother!" To offer himself as a substitute atoned for his dual guilt – as the bad, thieving son and as a survivor of political violence. He wished to undo the war-related harm to others through self-sacrifice. The fantasy of exchanging his fate with his brother solved a number of problems.

Case Two reminds us that not all psychic suffering is generated by collective oppression alone. Rather, the course of individual psychiatric illness may be interwoven with the symbolism, needs, and impulses generated by the traumatic context.

Case Three

A pregnant woman in the West Bank, in treatment for anxiety and depression since October 2023, remarked, "I feel guilty to bring a child into our unsafe world." Her sentiment was echoed by several women presenting for out-patient treatment of peripartum and postpartum depression.

Many themes are embedded in these poignant reflections. These women expressed the fear that the genocide has damaged their capacity to provide a suitably libidinalized uterine "holding environment" for their babies in the most fundamental way. In their mental representation of the pregnant womb and the child within it, they grieved that they were unable to provide basic safety for the vulnerable creature entrusted to them and thus had already failed as mothers in their primary responsibility. In addition to their sorrow and shame, perceiving themselves as inadequate protectors of their babies, the mothers also experienced ambivalence about their own happiness. We see here the dynamic of survival guilt, as if partaking in the joy of motherhood was illegitimate – a stolen pleasure – in the context of the danger and deep grief experienced by other people. The helplessness of the survivor is also implied, "if I cannot protect my own child, how can I protect anyone at all?"

The interplay between the current war and psychiatric office practice offers a subtle barometer of the emotional climate during this period.

Immediately following October 2023, there was a significant fall-off in the number of out-patient office visits in Jerusalem. The checkpoints were closed and many surrounding towns were locked by military fences or threatened to be locked in this manner at any moment. After a time however, many patients returned to their customary psychotherapy appointments. Some these patients had previously filled their sessions with longstanding complaints about their dissatisfactions with their mother-in-law or their employer at work. No longer. These patients now castigated themselves for having been absorbed with such, as they described them, "trivial" problems when so many people were suffering and dying. They felt guilty to have complained of matters that now seemed to them to be minor.

Somewhat later, the psychiatrist began to see a different phenomenon – another group of patients effected by the war-related events. These were individuals suffering with depression, psychosis, or other chronic illness who had been clinically stable despite serious and persistent mental health disorders. Now they demonstrated signs of relapse with sleepless nights or agitated reactions to news media.

Case Four

A video from Rafah, quite famous on the internet, displays a man holding the body of his decapitated baby.[19]

A little girl of four living in the West Bank saw this video on the news and subsequently suffered frequent nightmares in which her own house was burning. Her psychological development apart from this trauma was robust and normal. This vignette is a reminder that trauma reverberates through families and communities – its impact can be fused with and embedded within ordinary developmental issues experienced by children and adolescents.

Case Five

An adolescent boy in the West Bank was seen for insomnia and anxiety. He complained that his parents frequently argued. He offered a banal example: his father wanted to purchase a refrigerator for their apartment, but his mother did not want a refrigerator. It emerged that some months before, the Israelis had killed the patient's brother and were withholding the return of his frozen corpse indefinitely – a common and longstanding practice which extends and intensifies the grief of the surviving family members.[20]

The deceased boy's mother could not look at a refrigerator without being overwhelmed with grief and rage.

Case Six

A therapist in the West Bank evaluated a laborer from Gaza who had sought work in the West Bank and had been arrested there by the Israelis. He reported that he could not sleep; he was afraid to close his eyes. He described that he had been detained for a period of 28 days with his hands tied behind his back in plastic handcuffs and his head covered with a hood. The therapist was previously experienced in the treatment of persons who had been tortured in Israeli military detention and had provided documentation on these cases for use in court. Nevertheless, the therapist could not believe this man's account – which was so extreme that it appeared that the patient surely must have been exaggerating. Soon however media reports substantiated that Palestinian prisoners were being routinely subjected to continuous blindfolding and other forms of torture more intense than the atrocities previously encountered.[21]

Here the therapist was able to understand that countertransference disbelief and denial, defenses against secondary trauma and survival guilt, had interfered with a full grasp of the patient's reality.

We have observed similar instances of initial denial in response to other news reports that at first had seemed too appalling to believe. Disbelief had met the following story, for example: a 15-year-old Palestinian boy in Gaza was captured by the Israelis and forced to wear a belt containing explosives, a camera on his head, and a rope around his waist. The boy was then pushed into a tunnel so that the camera could reveal hidden fighters to the Israelis, who were watching the camera feed at a safe distance. The plan was to blow up the boy, the fighters, and the tunnel. However, no fighters were found, so the boy was pulled out of the tunnel alive and later released.[22]

Case Seven

Since October 2023, an 11-year-old girl from the West Bank had gradually decreased her intake of food and barely ate during the Ramadan holiday. She had not yet reached menarche. This girl had previously excelled in school and was very neat and clean in her habits. She expressed guilt over children starving in Gaza and was angry when her father purchased edible luxuries such as meat or fruit. Later, she began to hoard cans of food, bottled water, and dried beans "to be ready when they starve us in the West Bank." In addition to these obsessive preoccupations, she wrote her name on her leg with permanent marker, so that her body could be identified when she died.

Case Eight

A 13-year-old girl in the East Jerusalem had lost weight since October 2023, claiming that something in her throat prevented her from eating, a complaint resembling the classic syndrome of globus hystericus. A comprehensive medical investigation had revealed no physical disorder that might have generated this sensation. Her father had been detained since the beginning of the assault on Gaza and the family had received no news about him since that time. She reported to the therapist that "prisoners live in horrible places. Both men and women are attacked in sexual ways and have no food."

Case Nine

In the West Bank, an adolescent boy of 17 had lost nearly 18 pounds since the detention by the Israelis of his friend in November 2023. He denied that he was upset by this, claiming "I'm not sad about it – prison makes men! I just have no appetite." He suffered from disturbed sleep and expressed anger in response to the famine in Gaza. In 2022, he had lost a significant amount of weight in reaction to a hunger strike launched by a group of Palestinian prisoners.

Cases Seven, Eight, and Nine (and other cases like them) represent the emergence of self-imposed food restriction among children and adolescents who present for treatment in the context of very prevalent news reports of imposed starvation – both of the civilian population of Gaza and of Palestinian prisoners – which often including highly disturbing skeletal images.[23]

None of these youngsters appeared overly concerned with personal body-image issues. Their restricted eating superficially resembled the eating disorders as conventionally described, but without the characteristic focus on thinness as an end in itself. Commonplace preoccupations seen in typical eating disorders, such as a strict standard of physical attractiveness, mirror-gazing, or exercise as a route to burning calories, were absent. The starved body – rather than seen as a sign of self-mastery, self-discipline, and beauty – was viewed by these young people with fright and pity. The image of starvation induced guilt rather than envy.

None of these young patients reported a personal history of sexual victimization, abuse, or neglect. We should note however that the issue of sexual violation of prisoners has also been prominent in news reportage – as one of these young patients articulated – so that although these children had not exposed to sexual victimization as individuals, awareness of this threat was widespread within the environment.

Here the inciting trauma for restricted food intake did not reflect the personal and/or family psychosexual pathologies which are theorized to be acted out and expressed symbolically on the body. The inciting trauma here is mass starvation and sexual assault as weapons of war.

These cases raise the broader question of the recognition of new psychological syndromes which need to be understood in the context of political violence.

Case Ten

The following is material did not emerge from a treatment case, but from a video posted by the *Al Jazeera* news organization: we see a tearful young girl of about 10 surveying the debris that had been her home, after it was destroyed by the Israelis in a bombing that resulted in the death of six Palestinians and injury to others. The translated caption identifies the town of Kafr Dan, lying west of the city of Jenin in the northern part of the West Bank. The young girl takes a key from a door-lock lying amidst the rubble, explaining that it is a souvenir to help her remember her home.

This case illustrates the importance of honored memories as a dynamic in the psychological survival of displaced children and refugee Palestinians generally, often concretized in the image of a key or – as here – in an actual key which is sometimes displayed on a wall in a place of honor or worn as a piece of jewelry. In this way, adaptive memories are recruited and curated to neutralize the traumatic impact of injuries and loss.[24]

The authors wish to explain that as we initially prepared this manuscript for publication, there was an ordinary functional link to the *Al Jazeera* video referenced in the footnote above. The authors were then concerned to observe, at a point further along in the publication process, that the video had now been blocked from view in the United States; the computer screen shows instead a notice claiming, "after review, we restricted access to the content in the location where it goes against local law." The nonfunctional link and the entirety of the notification of restricted access appear in our reference list. The authors further remark that the silencing of speakers and writers, the assassination of journalists, and the blocking of information are routine facts of life in Palestine and that the blocked access to this video is a case in point.

The group of vignettes below explores the experiences of children in Gaza. Our personal and professional connections with our colleagues in Gaza – once robust – has been devastated by the conditions of war. As a result, we are currently unable to document clinical patient-related material, but present cases known to us through non-patient contacts or through online sources. Material featured on social media, unless otherwise noted, was originally posted through the *Al Jazeera* news organization.

Case Eleven

Here we see a boy of perhaps 6 seated on a chair being interviewed in Gaza by a video crew in a make-shift tent. He is interrupted suddenly by the loud roar of an aircraft flying low overhead. The little boy, moaning in fear, leaves his chair in panic to drift about restlessly while he scans the sky. We see here the origin of trauma in actual experiences of grave danger and the child's affect of helpless anguish.[25]

Case Twelve

A family we know was recently able to escape from Gaza through Egypt and relocate in a European country. The youngest, a boy of 8, had been very frightened during the escape and repeatedly vomited during the plane flight. Once the family relocated in safety, he continued to experience nightmares, insomnia, and anxiety. Walking down the street in his new neighborhood, he happened to observe a surveillance camera attached to an entrance to a fire station. He cried out with alarm that it was a "Jewish camera" spying on him like the cameras on Israeli military watchtowers.

His grandmother meanwhile fretted over the immense display of delicacies available at the local grocery store. The sight of chocolate made her wish she could provide chocolate for relatives remaining in Gaza. She reflected sadly that this was a luxury item, expensive and of low nutritional value. Her eye fell on a fat man who was also shopping in the store, a resident of this foreign city and a stranger. Seeing this man made her wish that her relatives in Gaza were as fat as he was, so that it would take longer for them to die of starvation.

Case Twelve illustrates the reactions of a Gazan family for whom the trauma of war continues to intrude psychologically despite displacement into a safer physical environment. Here, everything is a trigger and everyone is triggered. Ordinary sights become ready vehicles to stimulate reexperience of past threats. We recognize here too how it may be difficult for adults, who are themselves overwhelmed with traumatic losses and anxieties, to provide adequate emotional scaffolding to children who are also overwhelmed.

Case Thirteen

A video taken at a funeral in Gaza displays a man holding the body of his small son wrapped in a burial shroud, while a young boy of about 7 – the brother of the martyred child – stands beside him shielding the face of the dead little boy from the blinding sunlight. The translated caption indicates the brother's explanation, "He's afraid of the sun."

Case Thirteen introduces an important theme observed among children in Gaza – the adaptive impulse to protect, to help, and to comfort. The viewer cannot be certain whether the older brother actually holds the concrete belief that the dead body of his brother might still be troubled by the bright sunlight, or whether he is honoring the deceased through a tender and symbolic gesture, giving attention to the little brother's "needs" as if such needs could still be experienced and lovingly met.[26]

Case Fourteen

This video displays a weeping boy of about 5 with smears of blood on his face receiving emergency medical attention while lying on a blood-stained floor. The caption indicates that this child was injured in the Israeli bombing of the Al-Brig camp, located in the middle of Gaza. According to the translated audio recording, he is crying that he wants to survive his injuries so that he can help his mother.[27]

Here again we see children in extreme circumstances asserting the meaningfulness of their own agency as helpers and doers rather than as mere passive vessels in need of help. One may speculate that a measure of childish omnipotence and wishful role-reversal is at work in the mind of this injured little boy. At the same time, there is something sturdy and steadfast about his insistence that he continue to live because he wants to help people he loves.

Case Fifteen

This video displays the struggle of a small boy of about 4 carrying a very large plastic container of water. The translated caption indicates that he has already been displaced several times and is on his way to reach Khan Younis in the southern Gaza Strip.[28]

Case Sixteen

This Facebook video from Gaza shows another boy of about 10 dragging behind him what appears to be a child-sized wheelchair loaded with water bottles. One of the large water containers falls from the wheelchair and lands on his leg, causing him to stumble and fall to the road. The child steps away from the cart in tears and takes a seat sobbing nearby on a rubble-strewn ledge of a building.

This episode illustrates the moment of psychological collapse of the helper under the weight of the role. It suggests too that trauma sometimes expresses itself in small ways, rather than big incidents.[29]

Case Seventeen

This video from EverydayPalestine displays several children carrying water in Gaza, again featuring a small wheelchair used as a cart. One of these water-carrying children who falls to the ground seems to be no bigger than a toddler. Among the many accompanying posted comments, often signed by people in Gaza, is a description of the children as "fleeing from death to death."[30]

Case Eighteen

A video on TikTok features an adult interviewing a poised young boy of perhaps 10 or 11, asking what he will be when he grows up. The youngster calmly explains, "in Palestine we don't grow up – we don't get any older." He goes on to clarify that he might be shot, for example, simply walking in the street.[31]

Case Nineteen

This video shows a thin boy of perhaps 7 years being examined by medical personnel. He has an IV in one arm and what appears to be a bruise on his chest and abdomen. The translated caption states that the Israelis had arrested this child and then released him through the Netsarim military blockade in the Gaza Strip; he was later brought for treatment at the Al-Aqsa Martyrs Hospital south of the Gaza Strip. The child is described as not speaking at all although the video shows him carefully producing drawings with pen and paper.[32]

The reaction of mutism is familiar to clinicians who treat children following significant trauma.

Case Twenty

Another child receiving medical care is a little girl of about 4. In this video, she expresses the wish that her hand and her foot would grow back after losing them to exploding rocket fragments in Gaza.[33]

We find that the assumption that lost limbs would "grow back" is not uncommon among young child amputees in Gaza, reflecting perhaps not only a developmentally appropriate naivety about the possibility of limb regeneration but some level of denial and magical thinking as a defense against an intolerably painful reality.

Case Twenty-One

In this video, Abdul Hafiz Al-Najjar and two of his sons speak with great dignity about siblings and family members who were killed in the Khayyam massacre in Barfah. One of these children, a little boy of one-and-a-half, had been beheaded by the bombing and two other sons as well as a daughter and the mother of all of these children, in addition, were killed. A boy of about 9, one of the two surviving sons, witnessed his little brother's beheaded corpse. This young survivor speaks through tears about how sad he feels when he sees the clothing and toys that belonged to the dead child or looks at everyday things such as cats, trampolines, or balls that his little brother had loved.[34]

Case Twenty-Two

Here a little boy of about 8 in Gaza distributes free dates as a tribute to his father.[35] Adaptive activities which sustain children often reflect the traditional Palestinian values emphasizing family loyalty and charity.

Case Twenty-Three

A boy of about 9, here walking among other deportees from Gaza, cradles his pet kitten in his tee shirt.[36] This little boy wears two hearing aids, recalling the fact that blast waves from Israeli bombing and sonic-boom air raids in previous recent Israeli wars have induced an increased prevalence of sensorineural hearing impairment in Gaza.[37] Here the kitten both provides comfort and receives comfort in an emotionally adaptive exchange.

Case Twenty-Four

This video from @trtarabi displays a young girl of perhaps 12 in Gaza explaining that "science is important even in times of war," even though the caption indicates that her own school and home have been destroyed. The camera follows her walking resolutely amidst a landscape of unrecognizable rubble to find a school folder and to dust it off. She is then shown sharing school workbooks with a group of smiling children who wave at the camera.[38]

A strong commitment to education and other forms of self-development has long infused the spirit of the Palestinian resistance. A dedication to learning has always characterized Gaza, which is assessed as having one of the highest literacy rates in the world.[39] Even now, the children of Gaza cling to preparing themselves for a future. The constellation of hope, resilience, caring, agency, and collective purpose evident here is an aspect of *Sumud* – a word often translated as "steadfastness" that is a core feature of Palestinian identity.

Case Twenty-Five

A video from @translating_falasteen shows Qusai Jarad, an enthusiastic adolescent demonstrating how to craft a sewing needle from the metal keys affixed at the bottom of a container of canned meat. He looks confident and capable although still young enough that a mustache is just beginning to appear on his upper lip.[40]

This young man's entertaining explanation illustrates far more than a technique to create a useful needle and conveys energy, inventiveness, and warmth. Like many social media posts from Gaza, it is a morale-building act of collective resolve.

Case Twenty-Six

Renad Attalah, a girl of ten living in Gaza, has become an internet sensation through her cooking videos posted on Instagram, where she has developed half a million followers. In this translated news feature, Renad's adult sister Nourhan Attalah explains that the family had sought a way to cheer up the ten-year-old Renad during the current war by filming her while she was cooking. After four months, the local internet functioning was restored and these videos were uploaded. Now Renad announces cheerfully to the camera that they have had no drinking water for three days, but proceeds to demonstrate with infectious joy how to prepare "war lollipops" and other treats using simple ingredients such as leaves or donated cans and packages of pasta. She explains, "I didn't know I had this skill before the war started."[41]

Discussion

These vignettes illustrate both the immense agony inflicted on the children of Palestine as well as their immense strength. From a certain perspective, what these children have to say for themselves is surely more impressive than any interpretive commentary that others might say about them. From a professional perspective, however, we can ask what current mental health services are available to these children in Palestine and what considerations should guide us going forward.

The status of mental health services for children in Palestine is a mixed picture. There is a longstanding mandate to prioritize the needs of children and several key mental health initiatives have been put in place by the Ministry of Health to achieve this goal.[42]

Mental health services are free or nearly free in Palestine and innovative programs for children have been integrated into primary healthcare and into schools. All the same, healthcare in Palestine has suffered chronically from insufficient funds, insufficient and unevenly trained personnel, and insufficient infrastructure. And notably, the presence of foreign NGOs, although well-intentioned, has often contributed to the fragmentation of these services through its adherence to donor-driven agendas. Despite these problems, a basic system of mental health care for Palestinian children and youth in East Jerusalem and the West Bank is largely functional at this time.

Gaza too had developed an array of multifaceted mental health services including programming in schools. Clearly, however, the current genocide has succeeded in all but destroying the delivery of mental health services in Gaza. The immediate need is not for trained clinicians or medications but for water and food. The urgent necessity is the end of genocidal war.[43]

Yet even in this current catastrophic moment, mental health supports to children are still being rendered. The Gaza Community Mental Health Programme (GCMHP), despite the current devastation, continues to offer psychological first aid and group sessions as well as toll-free telephone services.[44] Harrowing realities as well as the professional responses to these realities are documented in GCMHP's comprehensive recent report.[45]

Our friend and colleague Dr. Yasser Abu Jamei, the General Director of the GCMHP, has recently explained, "When relatives are killed, we try somehow to calm the child and then ask questions: What are you going to do tomorrow? What are you going to do the day after tomorrow?"[46]

Notably too, in Gaza at this time there are a multitude of spontaneous activities through which adults engage children and adolescents in psychologically supportive groups. One artisan creates elaborate puppets from debris and cast-off items to the delight of children,[47] while another man provides training in boxing to boys and girls.[48] The uploading of social media material from Gaza and its opportunity for commentary from a global audience is itself an example of psychological support that arises naturalistically from the population in Palestine under grave threat. There is a creativity and wisdom in these inventive activities which we would do well to study.

Looking beyond the end of genocidal war to consider how to rebuild supports in the best interests of children, we must emphasize that the ultimate objective for the children of Gaza and the population of Palestine is the achievement of justice. Only through justice can there be the restoration of human dignity for those who have suffered from war and systems of oppression, with their manifold humiliations, injuries, and losses. Importantly, only through justice can the perpetrators come to grips with the full human reality including their culpability, and thereby recover their own dignity.

The understanding that mental health rests upon justice is increasingly recognized today through recognition of the principles of liberation psychology.[49] While a vision of justice on earth may appear very far from the landscape of our world today, we must remain hopeful of crafting a genuinely liberatory system of governance, healthcare, and mental health care on the level of public engagement and comprehensive policy planning. In that regard, Palestinian scholars have been world leaders, developing a Palestinian liberation psychology on the level of theory and everyday praxis, while colleagues internationally contribute to this work through their solidarity, advocacy, and scholarship.[50]

The case of Palestine thus can help us resolve the challenge of integrating liberation psychology into standard ways of approaching diagnosis, case formulation, treatment planning, and academic teaching of these clinical skills.

Integrating liberation psychology with traditional mental health requires a readjustment of the Western emphasis on individualism, which tends to locate diagnosis and therapy within each patient in isolation from a social context in such a way that personal adjustment (to an unexamined social context) is assumed to be a fundamental goal. This assumption then pathologizes the individual rather than opening the possibility that the context itself is pathological. It thus closes off important avenues for self-knowledge, agency, and interconnection with others.

The strength of Palestine, however, is its collectivity. The deep human need for collective meaning has always been a central value in Palestine and continues to express itself today with even more urgency. A collective model of psychological support, already implicit in the psychological life of Palestinian people, indeed cannot be suppressed – in our view, policy planning needs to go "with" this force and not attempt to atomize and quell it. Further, addressing the mental health of children scarred by war through harnessing collective modes of repair and growth is more cost-effective than a case-by-case approach to trauma, permitting the limited resource of professional specialist interventions to be devoted to the smaller number of children who truly require it.

Effective broad-based psychological repair must be a collective endeavor involving grieving and rebuilding, taking stock of traumatic loss as well as regenerative energies. We specifically recommend mental health services that:

- Encourage group participation among children in expressive activities such as play, art, drama, sports, and written expression.
- Work with groups of young people facing similar traumatic experiences – amputees, orphans, ex-prisoners – perhaps led by an adult who also has had these experiences.
- Provide structured opportunities for children to engage in authentic "helping" through age-appropriate action as public-spirited members of the community.
- Enhance community supports to children through initiatives that strengthen parenting skills, family stability, schools, religious life healthcare facilities, recreational spaces, historical archives, and cultural activities.
- Embody the ideals of public health and human rights, integrating international cooperation in solidarity with Palestine under the leadership of Palestinian mental health expertise.
- Participate in a global alliance to advance justice, well-being, safety, and the rights of children everywhere.

Conclusions

The genocide in Gaza, having led so far to the unnatural death of approximately 8% of its population, threatens to wreak its most enduring damage on the psychological life of its surviving children. The immediate responsibility of concerned international citizens must be to stop the genocide and the occupation that it serves. Restoring full justice to the people of Palestine in their long struggle to achieve it is a yet more difficult step, but it is integral to our commitment to Palestine and to all human beings in their quest for dignity and meaning.

Notes

1 United Nations Office of the High Commissioner of Human Rights, "Gaza: ICJ Ruling Offers Hope for Protection of Civilians Enduring Apocalyptic Conditions, Say UN Experts," press release, 31 January 2024, www.ohchr.org/en/press-releases/ 2024/01/gaza-icj-ruling-offers-hope-protection-civilians-enduring-apocalyptic#:~: text=The%20ICJ%20found%20it%20plausible,under%20siege%20in%20Gaza%2C %20and.

2 United Nations Office of the High Commissioner of Human Rights, "Israeli Authorities, Palestinian Armed Groups are Responsible for Crimes, Other Grave Violations of International Law, UN Inquiry Finds," press release, 12 June 2024, www.ohchr.org/en/press-releases/2024/06/israeli-authorities-palestinian-armed-groups -are-responsible-war-crimes.

3 United Nations Office of the High Commissioner of Human Rights, "Israeli Authorities, Palestinian Armed Groups are Responsible for Crimes, Other Grave Violations of International Law, UN Inquiry Finds," press release, 12 June 2024, www.ohchr.org/en/press-releases/2024/06/israeli-authorities-palestinian-arm ed-groups-are-responsible-war-crimes.

4 UN News, "'Ten Weeks of Hell' for Children in Gaza: UNICEF," press release, 19 December 2023, https://news.un.org/en/story/2023/12/1144927.

5 UNICEF, "Stories of Loss and Grief: At least 17,000 Children are Estimated to be Unaccompanied or Separated from their Parents in the Gaza Strip," press release, 2 February 2024, www.unicef.org/press-releases/stories-loss-and-grief-least-17000-chil dren-are-estimated-be-unaccompanied-or.

6 AJLabs, "Israel–Gaza War in Maps and Charts: Live Tracker," *Al Jazeera*, 9 October 2023/updated 3 October 2024, www.aljazeera.com/news/longform/2023/ 10/9/israel-hamas-war-in-maps-and-charts-live-tracker.

7 Rasha Khatib., Martin McKee, and Salim Yusuf, "Counting the Dead in Gaza: Difficult but Essential," Lancet 404, no. 10449 (20 July 2024): 237–238, doi:10.1016/S0140-6736(24)01169-3

8 AJLabs, "Israel–Gaza War in Maps and Charts: Live Tracker," *Al Jazeera*, 9 October 2023/updated 3 October 2024, www.aljazeera.com/news/longform/2023/ 10/9/israel-hamas-war-in-maps-and-charts-live-tracker.

9 Hiba Yazbek, Bilal Shbair, Cassandra Vinograd, and Abu Bakr Bashir, "'A Hells- cape': Dire Conditions in Gaza Leave a Multitude of Amputees," *New York Times,* 17 June 2024, www.nytimes.com/2024/06/17/world/europe/gaza-amputation- hospitals.html

10 Middle East Monitor, "Israeli Knesset Member Says Food, Water and Fuel Should be Cut Off from Gaza," Middle East Monitor, press release, 27 December

2023, www.middleeastmonitor.com/20231227-israeli-knesset-member-says-food-wa ter-and-fuel-should-be-cut-off-from-gaza.
11 Democracy Now! "50,000 Children in Gaza Need Immediate Medical Treatment for Malnutrition," press release, 17 June 2024, www.democracynow.org/2024/6/17/ headlines/50_000_children_in_gaza_need_immediate_medical_treatment_for_maln utrition.
12 Al Jazeera, "Nine Out of 10 Children in Gaza Lack Food for Growth: UNICEF," 6 June 2024, www.aljazeera.com/news/2024/6/6/nine-out-of-10-children-in-gaza-la ck-food-for-growth-unicef.
13 Zeina Azzam, "Write My Name," Vox Populi, 30 October 2023, https://voxpopulisp here.com/2023/10/30/zeina-azzam-andy-young-david-ades-three-poems-about-gaza.
14 "Four Wars Old," Visualizing Palestine, infographic image, June 2021, https://visua lizingpalestine.org/visual/four-wars-old.
15 Samah Jabr and Elizabeth Berger, "The Survival and Well-Being of the Palestinian People Under Occupation," in *The State of Social Progress of Islamic Societies: Social, Political, Economic, and Ideological Challenges*, ed. Habib Tiliouine and Richard J. Estes (Springer, 2016), 529–543, doi:10.1007/978-3-319-24774-8; Samah Jabr and Elizabeth Berger, "The Trauma of Humiliation in the Occupied Palestinian Territory," *Arab Journal of Psychiatry* 28, no. 2 (2017): 154–159, https://search.ema refa.net/en/detail/BIM-813397-the-trauma-of-humiliation-in-the-occupied-palestinia n-territ; Samah Jabr and Elizabeth Berger, "Mental Health Under Occupation: The Dilemmas of 'Normalcy' in Palestine," in *Global Mental Health Ethics*, ed. Allen R. Dyer, Brandon A. Kohrt, and Philip J. Candilis (Springer, 2021), 289–303, doi:10.1007/978-3-030-66296-7; Samah Jabr and Elizabeth Berger, "The Children of Palestine: Struggle and Survival Under Occupation," in *Handbook of Children's Security, Vulnerability, and Quality of Life: Global Perspectives*, ed. Habib Tiliouine, Denise Benatuil, and Maggie K. W. Lau (Springer, 2022), 279–296, doi:10.1007/978-3-031-01783-4; Samah Jabr and Elizabeth Berger, "Community Mental Health, Psychoanalysis, and Palestine," *International Journal of Applied Psychoanalytic Studies* 20, no. 2 (1 March 2023): 285–301, doi:10.1002/aps.1801.
16 Samah Jabr and Elizabeth Berger, "The Survival and Well-Being of the Palestinian People Under Occupation," in *The State of Social Progress of Islamic Societies: Social, Political, Economic, and Ideological Challenges*, ed. Habib Tiliouine and Richard J. Estes (Springer, 2016), 529–543, doi:10.1007/978-3-319-24774-8.
17 Tabatha Kinder, "Israel: 240 Palestinian Children 'Sexually Abused' in Jerusalem Detention Centres, Group Claims" *The International Business Times*, 22 November 2014, www.ibtimes.co.uk/israel-240-palestinian-children-sexually-abused-jerusa lem-detention-centres-group-claims-1476061.
18 New Arab Staff, "640 Palestinian Children Detained in West Bank Since October 7, Several Tortured," The New Arab 18 June 2024, www.newarab.com/news/ 640-palestinian-children-detained-west-bank-october-7.
19 Khaled Beydoun. "The 'Rafah Tent Massacre' and the Myth/Truth of Beheaded Babies." Substack.com. Pen>Sword 27 May 2024, https://khaledbeydoun.substack. com/p/the-rafah-tent-massacre-and-the-mythtruth.
20 Abeer Otman, "Fathers in and Against Pain; Father's Interruptions of Settler-Colonial Technologies of Loss" *Affilia* 38, no. 2 (25 September 2022): 244–262, doi:10.1177/08861099221126975.
21 CNN's International Investigations and Visuals teams, "Strapped Down, Blind-folded, Held in Diapers: Israeli Whistleblowers Detail Horror of Shadowy Detention Facility for Palestinians," CNN (updated 11 May 2024), www.cnn.com/2024/ 05/10/middleeast/israel-sde-teiman-detention-whistleblowers-intl-cmd/index.html.

22 Abeer Ayyoub, "Palestinian Says Israeli Soldiers Sent Him into Hamas Tunnel Strapped with Bombs," Middle East Eye 15 December 2023, www.middleeasteye.net/news/israel-palestine-war-soldiers-hamas-tunnel-bombs-sent.

23 Josh Breiner, "Israel Reduces Food for Palestinian Security Prisoners, Conceals Data, Sources Say," *Haaretz*, June 26, 2024, www.haaretz.com/israel-news/2024-06-26/ty-article/.premium/israel-reduces-food-for-palestinian-security-prisoners-conceals-data-sources-say/00000190-542e-de5e-abd0-ff7ee9580000.

24 Al Jazeera, [Video with restricted access], www.instagram.com/reel/C8HWnsAiirJ/?igsh=MXB2NzM1ajV6ajJ1Yg%3D%3D. The following statement now appears on a computer screen in the USA when the link, previously accessible, is clicked: "NOT AVAILABLE IN MY COUNTRY. We received a legal request to restrict this content. We reviewed it against our policies and conducted a legal and human rights assessment. After the review, we restricted access to the content in the location where it goes against local law. You can learn more about content restrictions in our Transparency Center."

25 Eye on Palestine (@fadi_badwan), "While Filming an Interview with a Child in Gaza...," Instagram video, 13 July 2024, www.instagram.com/reel/C9Xzp hBKWMB/?igsh=Y3BhMDBqYXpxMW5x.

26 Al Jazeera, "He's Afraid of the Sun...," Instagram video, 11 June 2024, www.insta gram.com/reel/C8EBtcpJHrH/?igsh=b2k3b2VnYXlsbHgx.

27 Al Jazeera, "I Just Want to Be Living for My Mother...," Instagram video, 10 June 2024, www.instagram.com/reel/C8CL6_0iTxQ/?igsh=NncwMWJkeWQ0c2ps.

28 Al Jazeera, "The Suffering of a Child Who Was Displaced Several Times...," Instagram video, 11 June 2024, www.instagram.com/reel/C8GYw6rqMvD/?igsh=OWU1bG93cGQ5YTR6.

29 "We Will Never Forget. We Will NEVER Stop Sharing Free Palestine!! by صفحة الجاليه العربية والإسلامية في امريكا," Facebook video, 1 June 2024, www.facebook.com/watch/?mibextid=w8EBqM&v=3956983684531084&rdid=zz6xcCzjJ8UMz5Fh.

30 Everydaypalestine, "We Have Grown Up without Realizing it ...," Instagram video, 12 June 2024 www.instagram.com/reel/C8INqD-Bv0b/?igsh=MTdhODVjbDljbn V2YQ%3D%3D.

31 Di Ramahi (@di_ramahi), "We (Children) in Palestine Don't Grow Up ...," TikTok video, 17 October 2023, www.tiktok.com/@di_ramahi/video/7291066399894621442.

32 Al Jazeera, "The Israeli Occupation Releases a Palestinian Child ...," Instagram video, 13 June 2024, www.instagram.com/reel/C8KGt0qIbcv/?igsh=MWJza GQ2cG5hNDdhbQ%3D%3D.

33 Al Jazeera, "I Want My Hand...," Instagram video, 24 June 2024, www.insta gram.com/reel/C8mW_A4pbCP/?igsh=cWdwZHdreDJmenh0.

34 Al Jazeera, "Ahmed my Son before Being Bombed...," Instagram video, 10 June 2024, www.instagram.com/reel/C8ChUK7OMqC/?igsh=MTJjdmVia2N0aGNkbA% 3D%3D.

35 Al Jazeera, "About the Soul of My Father...," Instagram video, 14 June 2024, www.instagram.com/reel/C8Nnr9bsgTR/?igsh=eW9xdm13d3Z5aHAw.

36 Al Jazeera, "A Child Carries His Cat...," Instagram video, 1 July 2024, www.insta gram.com/reel/C83lEN3hAve/?igsh=b3lqenpvOW5mdjJ2.

37 Raja Sharaf, "Hearing Loss in Gaza: Toward a Political Diagnosis," *Science for the People* 23, no. 1 (23 May 2020), https://magazine.scienceforthepeople.org/vol23-1/hearing-loss-in-gaza-toward-a-political-diagnosis/#:~:text=Large%2Dscale %20Israeli%20bombing%20and,to%20bend%20and%20stop%20working.

38 TRT (@trtarabi), "Science is important even in times of war...," Instagram video, 13 June 2024,
www.instagram.com/reel/C8K5vJXo1Be/?igsh=MThjeHZtMWdtcXQ3.

39 Diana Alghoul, "Palestine Has One of the Highest Literacy Rates Globally," *The New Arab*, 6 September 2018, www.newarab.com/news/palestine-has-one-highest-literacy-rates-globally.

40 Qusai Jarad (@qusi.gaza), "A Sewing Needle out of a Can of Meat?" Instagram video, 20 June 2024, www.instagram.com/reel/C8bneQmNzOy/?igsh=bzVzZm9vdWFncmtp.

41 "Gaza Girl Inspires with Cooking Videos despite Israel's Siege," Al Jazeera video news feed, 4 July 2024, www.aljazeera.com/program/newsfeed/2024/7/4/gaza-10-year-old-inspires-with-cooking-videos-despite-israels-siege.

42 Palestine Ministry of Health, *Child and Adolescent Mental Health National Strategy 2023 to 2028* (State of Palestine, Ministry of Health, n.d).

43 Samah Jabr and Elizabeth Berger, "Palestine Meeting Gaza's Mental Health Crisis" *The Lancet Psychiatry* 11, no. 1 (28 November 2023): 12, doi:10.1016/s2215-0366(23)00398-x.

44 Gaza Community Mental Health Programme (GCMHP), n.d., www.gcmhp.org; Sondos Alfayoumi, "How One Organization Is Providing Mental Health Care to Palestinians Living through Genocide," Mondoweiss (May 13, 2024), https://mondoweiss.net/2024/05/how-one-organization-is-providing-mental-health-care-to-palestinians-living-through-genocide/.

45 Gaza Community Mental Health Programme (GCMHP), "Nine Months of Israel's War on Gaza: the Mental Health Impacts & the GCMHP's Response – July 2024," 16 July 2024, https://gcmhp.org/publications/4/208.

46 Isaac Chotiner, "How Gaza's Largest Mental-Health Organization Works through War," *New Yorker*, 19 April 2024, www.newyorker.com/news/q-and-a/how-gazas-largest-mental-health-organization-works-through-war.

47 Sustainable Living (@sustainablelivingkw), "A moment of joy for the children of Gaza...," Instagram video, 17 May 2024, www.instagram.com/reel/C7EmMMyM3V7/?igsh=M3B0NHEyNnJxdWFj.

48 Translating Palestine (@translating_falasteen), "In Abu Ghazi camp, home to 65 tents...," Instagram video, 6 July 2024, www.instagram.com/reel/C9FQtbjtLYX/?igsh=OXd5N3JwZzkxczl4.

49 Frantz Fanon, *Black Skin, White Masks*, trans. Richard Philcox (Grove Press, [1952] 2008); Edward Said. *Orientalism* (Vintage, 1978); Paulo Freire, Pedagogy of the Oppressed, trans. Myra Bergman Ramos (Continuum, 1970).

50 Brian K. Barber, Carolyn Spellings, Clea McNeely, Paul D. Page, Rita Giacaman, Cairo Arafat, Mahmoud Daher, Eyad El Sarraj, and Mohammed Abu Mallouh, "Politics Drives Human Functioning, Dignity, and Quality of Life" *Social Science & Medicine* 122 (December 2014), 90–102, doi:10.1016/j.socscimed.2014.09.055; Rana Nashashibi, "Violence Against Women: The Analogy of Occupation and Rape – 'The Case of the Palestinian People,'" *Gender in Conflicts: Palestine, Israel, Germany*, ed. Ulrike Auga and Christina von Braun, (Lit, 2006) 183–190; Lara Sheehi and Stephen Sheehi. *Psychoanalysis under Occupation: Practicing Resistance in Palestine* (Routledge, 2022); Nadera Shalhoub-Kevorkian, *Incarcerated Childhood and the Politics of Unchilding* (Cambridge University Press, 2019); Lama Khouri, "Palestinian Clinicians Launch Mental Health Network after International Association Plans Summit in Israel," *Mondoweiss* 24 June 2019, https://mondoweiss.net/2019/06/palestinian-international-association/; Iman Farajallah, "Continuous Traumatic Stress in Palestine: The Psychological Effects of the Occupation and Chronic Warfare on Palestinian Children" *World Social Psychology* 4, no. 2 (May-August 2022): 112–120. doi:10.4103/wsp.wsp_26_22.

Chapter 2

Pour a Libation for Us

Restoring the Sense of a Moral Universe to Children Affected by Violence

Martha Bragin

In the late 1980s, when charged with creating a clinical model for a New York State Office of Children and Family Services (OCFS) court diversion program, I started to think carefully about the effects of violence on child and adolescent development. The youngsters who were involved in this program had been accused of offenses ranging from shoplifting and running away, to armed robbery, gang violence, and attempted murder. They were affected by three levels of violence: (1) structural violence that is baked into how people live their lives, (2) social violence that keeps the structures in place, and (3) interpersonal violence that happens between people – in relationships, families, and neighborhoods – where the political becomes deeply personal (Van Soest, 1997, p. 13). Working with these levels of violence is especially challenging, as South American psychoanalysts Janine Puget (2002) and Elizabeth Lira Kornfeld (1995) noted; when therapists are living through the same forms of structural and social violence as their patients, it is difficult for those therapists to recognize the intrapsychic effects of violence, and to assist their fellow survivors in the integration process. As a US citizen benefiting from white privilege, I too had been incubated in its violence as witness and beneficiary, and therefore I was affected by it in ways that created challenges to my ability to think about how to create systems that could help the children and families involved (Fine et al., 1997; hooks, 1998; Kovel, 1984).

Fortunately, while developing the court diversion program, I had the opportunity to work in solidarity with colleagues in Central America, Sub-Saharan Africa, Southeast Asia, and the Middle East who were addressing these very issues in their own countries. I was able to learn from thought leaders there and to think together with colleagues in the field. Armed with this support I was able to bring home my colleague's thinking in order to influence what became the diversion program's understanding of what was needed to help the children, youth, and families we met in New York.

This paper discusses the theoretical bases of the diversion program's clinical model, and how learning from colleagues working in the context of armed conflict in Latin America and Sub-Saharan Africa enriched and strengthened that theoretical basis. It further illustrates the value of psychodynamic theory

DOI: 10.4324/9781003520757-3

in helping to conceptualize a way to address the needs of children and youth from families affected by structural violence. Finally, the paper will introduce the specific application of the developmental concept of the moral third (Benjamin, 2016, 2017) to this population by creating a mental space in which the youngsters, their therapists and family members can contain and transform the nameless dread of extreme violence.

Theoretical Basis of the Juvenile Court Diversion Project

My Central American colleagues had the opportunity at that time to colla-borate with South American psychoanalysts exiled from their countries of origin because of the negative views the then-current Southern Cone of South America[1] regimes held toward psychoanalysis and psychotherapy. Several of this group paid particular attention to the effects of social violence on chil-dren and adolescents. Particularly important in developing this work were Julia Braun, Maren de Viñar, (2012), Marcelo Viñar (1989), and Juan Carlos Volnovich, (2002).

In the process of understanding the effects of violence, Latin American psychoanalysts began to critique the western concept of trauma healing as limiting in scope and effectiveness (Becker, 1995, Kornfeld, 1995). These authors suggested that using a disease model to address individuals whose suffering was in fact socially created and happened to them as part of a class or group of people, failed to bear witness to the cause of their pain, produ-cing further isolation (Becker, 1995; Kornfeld, 1995; Puget, 2002; Langer, 1989). The Latin American analysts insisted that it was important to do three things; to acknowledge the external causes of the suffering of the people that they were charged with helping; to recognize the fact that what happened to them did not happen to them as individuals alone, but to the entire society; and, at the same time, to try to make a deep sense of the intrapsychic effects of living in a world where the most hideous violence was constantly enacted before their eyes.

To address these issues, the therapists brought a politically informed use of Kleinian thought to their work with people affected by violence. They con-textualized suffering in social terms on the one hand, and addressed its dee-pest internal consequences on the other. My New York colleagues and I were able to benefit from their thinking, as I experienced it (Bragin, 2003, 2007). While much of the work was done with adults, James Garbarino and his colleagues working in Chicago and the middle east had begun writing about the ways that children in war zones and those affected by structural violence in the US might have something important in common (Garbarino, Kostelny, & Dubrow 1991). At the same time, applying a Kleinian approach to the day to day work was also being taught by child analyst Eugene Glynn, advisor to the Jewish Board of Family and Children's Services, who provided guidance on treating adolescents suffering from extreme levels of violence and loss.

Melanie Klein's "Criminal Tendencies in Normal Children" (1927) and "On Criminality" (1934) were part of our early learning and thinking (both included in Klein 1948). In these essays, Klein quoted children talking about cutting people up, making sausages from their insides, and eating them. She reminded readers of Jack the Ripper and early twentieth-century criminals stalking London, acting out in real life her young patients' play – not because the children had heard adults talking about such crimes but because the crimes represented enactments of what Klein believed all children fantasize. Klein's concepts, drawn from her analysis of primitive mental states, were helpful as we struggled with being deeply aware of and alive to the terrible realities of the world these young patients experienced, often daily.

One of Klein's particularly useful concepts was that of the *positions*, which I review here for convenient reference (Klein, 1975a, p. 93). The positions are different from the developmental stages because they are meant to represent universal mental states which develop in the early years to assist with mastering the struggles between aggression and libido, love and hate, goodness and destructiveness (Segal, 1974). People then return to these states, not only in childhood, but throughout life from time to time. In fact, they not only recur, but they also interact with one another as part of a dynamic inner process that manages the vicissitudes of outer experience as they affect and are affected by the inner life (Steiner, 1988).

The earlier position, called *paranoid schizoid*, engages Klein's understanding of very early mental states. An infant's situation is a terrifying one – having been thrust into a world in which one is wholly dependent without any way to make sense of it all; there is only loving care to mediate, if in fact one receives that loving care. Klein theorized that here begins a complex dialectic of aggression and libido, love and hate. Without the aggressive drive exercising its mighty force to propel the infant from the womb, the infant does not live. Thus, with life in the world comes a powerful, innate aggression alongside the many discomforts and terrors of life, from gnawing hunger to the gastric distress of an immature body and outside irritations of wet or cold. The infant cannot distinguish among any of these. The satisfaction of a long and loving feed creates an ideal state of well-being in that distressed little being, bringing an adoration of the "good breast" that brings the milk and loving comfort. Simultaneously, the distress of the aggressive drive that rages against pains the feed cannot resolve (e.g., air temperature, noise, wet or soiled diapers) are associated with the other, magical breast, the "bad breast" that cannot soothe. Klein believed that during murderous early rages, babies can imagine the power to destroy the bad breast and possess the good. These rages and fears are split off, as are desires to murder or destroy the object that cannot bring the peace of the womb, because they are too enormous to experience as coming from within and therefore are understood as external threats of violence. With these splits can come a sense of power and control as well as a capacity to destroy the frightening since the frightening is seen as outside the self.

At some point, as the child grows in capacity, she begins to realize that the good breast and the bad, unsatisfactory breast are, in fact, attached to the same caregiver. Time after time, feed after feed, the same, loving objects return, trying to get it right, wishing for a loving reunion. Klein believed that sadness and mourning well up in young children for the omnipotent world of pure good and guilt-free murder. Out of this melancholy knowledge of an imperfect world comes the desire to return the love of these loving objects, to repair the terrible violence that has been done in fantasy. Love wins out but at a cost, and the enraged desire for destructiveness can be understood as part of the world at large, and all humans in it – child, mother, and other alike. Devotion to values and causes, as well as creativity and love of learning, come from this position and the loving feelings it enables through mourning the loss of perfection (Klein, 1975b, p. 187). These ideas helped our team at the court diversion program to think about the youngsters we met and indeed, about ourselves as struggling with the same issues. We wondered: Can we accept the violence in the world and in ourselves and still go on being?

As we developed the court diversion program's assessment model, we began to understand the youngsters' behavior as a way to express in action what they could not say in words, which involved the ways that violence had affected them so deeply that they could not speak its name. They often had family histories replete with losses of caregivers through immigration or incarceration, of sexual and/or physical abuse from both caregivers and strangers, and the intergenerational transmission of historical trauma that invades the psyche. This proved especially true for families affected by geno-cide, colonialism and the slave trade, such as the African American, Native, and Latino youngsters in the court diversion program (Apprey, 1999, 2014, 2016; Bragin, 2010, 2012; Vaughans, 2014; Winnicott, 1984).

We found that our young clients – especially those who carried secret knowledge of unspeakable violence in which they were witness, victim, and sometimes perpetrator – were unable to use words to tell us what they knew, so their behaviors had to speak for them. These experiences of enacted pri-mitive violence were, in fact, tolerable only in the paranoid/schizoid posi-tion – that early place where violent realities are split off, protecting both children and the goodness of their caregivers (Bragin, 2003, 2004, 2010). For symbols to be formed – be they art, word, or activities – and contradictory experiences tolerated, the depressive position was required (Segal, 1974). The youngsters had to be certain that they carried enough of good and we and their families enough tolerance of bad to be able to be in the same mental space. Our team soon learned that the only way to enable them to speak was to engage them in some sort of reparative activity that allowed them to connect with their own goodness and that of the world while toler-ating the terrible things that they had lived and that therapists and care-givers needed to know about (Bragin & Bragin, 2010; Bragin, 2012; Klein, 1975b; Winnicott, 1984).

In time, we learned that if we could get the youngsters to the point of being able to speak about their experiences, they would express that the very fact of knowing such terrible things separated them from the rest of the world. No one wanted to know what they knew. Sometimes they hoped that maybe they only imagined their secrets. They seemed to feel permanently alienated from the world of love and goodness and right; those who had first grown up with love and connection did not themselves want to know about the world into which they had just been inducted by harsh experience. Many expressed that they felt they could only be understood by the very people who had harmed them or by others who had participated in or survived the worst of the violent acts. Working with these youngsters at that time, the therapists, too, felt alienated by what we learned from the public discourse about who these youngsters were ("super predators") and what sort of treatment approaches they required to be restored to well-being.

It occurred to us that one of the most terrible things about acts of extreme violence is the way that they take scenes that belong to the realm of dream and fantasy and make them literal. Thus, these events become enactments of people's worst nightmares or force them to be actors in the nightmares of others (Bragin, 2003, 2004). The same mental mechanisms that prevent survivors from integrating and, using Wilfred Bion's (1962) term, *learning from experience*, also prevents therapists who listen from integrating their own violence and its connection to that of the survivors (Bragin, 2010). Yet without that connection, therapists leave their patients alone, lacking the recognition and accompaniment that is essential for healing to begin (Bragin, 2007). So, it became part of our work to acknowledge our own violence, internally in fantasy as well as structurally and socially, so that we could begin to connect with the children and families with whom we worked (Bragin, 2007, 2010).

At the diversion program, we eventually were able to offer each family a space to speak and a therapist who could listen and acknowledge the painful histories that had led to the moment when the family appeared in court. In many cases, we could facilitate reconciliations within those families, with communities, and with religious leaders. We offered a place in which suffering adults could share their own pain as well as the sorrow that came with understanding that they had been unable to protect their children from living these terrible experiences, or from bearing the pain of their ancestors. We worked to become a staff who could acknowledge our own part in the violence that had created the conditions for all of this, and therefore have compassion for both the children and the families. This took us a long way toward a process where healing could begin.

Limitations of the Theory

However, there was one thing that we could not do and that the Kleinian lens did not offer – one thing so transformative that none of us could imagine it – acknowledgement and an offer of reparation from the highest levels of the society that had created, organized and set up the conditions for the violence that the families, children and adolescents had experienced.

Fortunately, theory and practice from Sub-Saharan Africa, combined with shared experience with clinicians from Sierra Leone, helped to add this element. In the process of learning, I found a strong resonance with Jessica Benjamin's (2011) work on the Moral Third, which also has global origins (Benjamin, 2016, 2017).

Recognition, Acknowledgement, and the Road to Healing: Contributions from Sub-Saharan Africa

Many of Freud's discoveries regarding guilt and aggression were derived from material he had gleaned from other sources and noted in *Totem and Taboo* (1913). Specifically, he developed an understanding of the effects of war on warriors, and he described processes of community reintegration through his research into post-conflict rituals in Sub-Saharan Africa and other places in the global south (Freud, 1974, pp. 35–37). At the same time, African writers responded to the effects of colonial practices, anti-colonial wars and proxy wars on the continent and began to write about the psychological effects of colonialism and armed conflict, including the psychological meaning of ritual and tradition for healing (Fanon, 1963; Kenyatta, 1938; p'Bitek, 1998).

Late-twentieth-century Sub-Saharan African psychologists and other specialists questioned the application of western trauma discourse to the situation of former child soldiers in non-western societies (Honwana, 1998a, 1999, 2002, 20011; Efraime & Errante, 2012). Alcinda Honwana (1998b, 2002, 2011), a leader in this theoretical movement, studied and documented the efficacy of African traditions in addressing in the community-based reintegration of child soldiers, contrasting this approach to the western trauma discourse. A number of colleagues from Mozambique and Angola discussed the theoretical implications of combining traditional methods with group and social interventions (Efraime & Errante, 2012; Honwana, 1998b; Peddle, Monteiro, Guluma, & Macaulay, 2013; Monteiro, 1996). Tunde Zack-Williams (2006) discussed the specific role of ritual in community reintegration programs in Sierra Leone. In the section that follows, I introduce an example of ritual reintegration as I experienced it. Then, I examine the implications of this work for psychoanalytic theory and its implications for child and adolescent development in the US today.

Sierra Leone

In 2001, I had the opportunity to work with Samuel T. Kamanda, a young Sierra Leonean social worker in charge of a program to reintegrate child soldiers in his country. This collaboration offered me new ideas to refresh and advance the court diversion program team's thinking about healing children affected by violence.

The children in the Sierra Leone program were placed in what the humanitarian community called *Interim Care Centers* – places where children ages 16 and under who had been rescued from armed groups were sent – while their surviving family members were traced for reunion and arrangements for ongoing psychosocial support and reintegration were made. The work at the centers was not easy; children who had been forced to terrify their communities were being cared for and treated by members of those same communities. These youngsters had not been abandoned by their parents; indeed, for most, their fathers were mutilated, tortured, and/or killed in front of them, and the children were abducted at gunpoint. Then, the children were tortured, threatened, given drugs, and forced to commit atrocities as a way to bind them to their capturers. Following Klein's work (Klein 1948), one could see that the atrocities mimicked unconscious fantasy as well as common children's games. This carefully choreographed induction into the armed groups resulted in a terror and self-loathing that prevented thought or play (Bragin, 2003, 2004, 2012). So, while the Interim Care Centers required an intensive program of education, play, and group discussion, the most damaged, the most violence-involved, and the youngest of the children were unable to participate. Rather than thanking their rescuers, many continued to insist on their loyalty to the commanders who had abducted and tortured them, with bravado masking their terror of re-abduction, of guilt and punishment, and the sadness of facing all that they had lost. They feared to think of the normal life laid waste in the destruction of war. Above all, they were terrorized by the belief, instilled daily by their captors, that should their families be found even their own mothers would never accept them.

One group of four boys, ages 8–12, all named Mohammed,[2] were among the most difficult and uncommunicative. Mr. Kamanda asked for my thoughts regarding what might be useful in caring for them. We worked as a staff to use the boys' threats and bravado to create the reciprocal back-and-forth actions of greetings. We smiled and greeted them daily as we entered the schoolroom and the boys stayed outside. They scowled and told us they liked our possessions. We continued to smile and agree that our possessions were nice, but we were going to school now. They gave their most menacing looks in return. As they began to look for us daily to join them in the give and take, the youngest began to smile in response to our greetings. The beginning stages of play had begun, making us hopeful.

Then, the trouble began. The four Mohammeds were accused of stealing soap to sell in the market, and they had responded by setting out on foot down the road, hoping to reach the capital hundreds of miles away, where they could plead their innocence. Mr. Kamanda and I left the site and collected them by car. We brought them safely back to the Center and told them that could plead their case there. The staff and students were brought together in a solemn circle, opened by the customary recitation of Christian and Muslim prayers. Then, the staff laid out their concern that the boys had taken their own bars of soap, sold them in the market for money to buy drugs or liquor, and then asked for more soap weekly so they could do it again. The boys clearly stated that they were innocent; they simply had used up their soap before the month ended.

To determine whether it was possible to use up a full bar of soap so quickly, they and I each were given a new bar of soap. We were asked to take one shower a day and show our bars to the community at the end of the week. At the end of the first week, I had enough soap left for the remainder of the month, while theirs were almost completely gone.

"But we are dirty boys," they said. "We have done the worst things in the world, and we scrub, and we scrub, but we cannot get clean. Pour a libation for us, Mr. Kamanda, so that we can be clean again."

I looked in horror at my remaining bar of soap; it was really fat. How quickly Klein's (1975b) paranoid schizoid position sets in! They knew that they were dirty boys, having participated in the most horrible violence. But I, soothed by my participation in this reintegration effort, or perhaps in order to tolerate being in it, I had split off awareness that I was a dirty woman with my own internally violent fantasies. I had split off the knowledge of my own complicity as a tax-paying US citizen in the outsourcing of violence and tolerating its enactment in proxy wars around the world or in our prisons at home. I had split off my memory of the slave trade and the ensuing disorder and colonialism that had set the stage for the current war in Sierra Leone, while contributing so much to the wealth of New York City (Coates, 2018). How completely I had split off the knowledge that it was I who needed a good scrub!

For me to be useful as an object – in other words, for me to be able to help the staff to contain the terror and dread that comes with facing violence in the process of healing – I had to come to grips with and acknowledge not only my own very personal intrapsychic violence but also my broader responsibility. If I could not own the violence in myself and the ways I, like most humans, casually participate in structural violence that affects others (e.g., by paying taxes that I know support wars and prisons even as they support roads and postal service), I would leave the staff and the children alone, unaccompanied. I would be leaving them to see themselves as different, dirty people who survived a dirty war with no way for deep connection or understanding by anyone except the dead, those who survived with them, and those who were bound up in the immediate and interpersonal violence.

Similarly, in order for the staff to be able to heal themselves and help the children, and for the children to be able to heal, they had to be able to find some way to integrate and incorporate this terribly violent period of their lives into a narrative that could take them forward into life that both acknowledges the violence and forgives them. That outcome required a means for reparation, a connection to their own goodness, enabling them to symbolize their own experiences, formulate the necessary words, tell their terrible, sad stories, and journey forward.

Fortunately, the healer who served as officiant understood this need well. In fact, reparation was built into the communal ritual of libation. The libation in Sierra Leone is a pouring out of ceremonial liquid as an offering to the ancestors, asking that they attend and in so doing restore balance to the universe that has been disrupted by great wrong. After engaging the staff and community members, he called to me as an American trying to cleanse my own violence and that of my country. He smeared blood on my head as he had on the children's heads and washed it away with a part of the libation poured on the earth.

The pouring of a libation is a complex process; it involves permission from the village elders, the owners of the land and any buildings on it, the civil and traditional authorities, and traditional and contemporary religious leaders (i. e., Animist, Muslim, and Christian). The Interim Care Center itself had to be washed clean and painted. A sacrificial animal, suitable for eating, had to be purchased, and the traditional healer who would officiate must be given a contribution. All of the civic and traditional leaders had to agree to be present along with the community members. The preparation takes weeks during which everyone involved is aware that a great rupture in the community's fabric and in the souls of those involved has been acknowledged and must be now be repaired. On the day of the libation, Muslim and Christian prayers must be said, drums beaten, the traditional gods of the land invoked, the animal slaughtered, the blood poured onto the ground, and washed with water. It is only then that the spirits of the ancestors would come to dance as the slaughtered animal, at last seasoned and cooked, is served in a stew to the attendant community members, in a symbolic act of reparation.

The Moral Third

The concept of thirdness refers to the process of co-creation between therapist and client that reaches beyond the binary of subject and object or doer and done to. Thirdness allows a shared space to be entered by all parties, in which all aspects of psychic life can be experienced and addressed. Thirdness specifically acknowledges the reality that within each relationship every possible opposite exists; yet, therein lies a capacity to move beyond opposites to the dialectical position that the difference between seeming opposites is represented not by compromise but by the creation of something both transcendent and new (Benjamin, 2016, 2017). Nonetheless, within the third can lie violent introjects that become to terrifying to manage.

The Moral Third is a developmental concept that expresses the way in which the sense of a moral world is at the heart of psychic development. It is the co-created space in which violations of moral law, of common humanity, and of all-too-imaginable terrible things enacted in the world can be witnessed together, acknowledged, addressed, and repaired. In this space in which all parties are both victim and perpetrator, violation is not othered or pushed out. Instead it is owned and acknowledged, creating space for the repair of ruptures. The mental space represented by the moral third makes the restoration of the self in the world possible:

> Specifically, the Moral Third is the position from which the violations of lawful behavior and dehumanization can be witnessed or repaired. It is a fragile position, hard for both individual and collectives to maintain. It is from the position of the Moral Third that we acknowledge violations, suffering, indignities, and the debasing of some humans to elevate others. What makes that position of acknowledgment possible? What prevents it? We must admit that we observe in ourselves continually the breakdown and restoration of the capacity to hold the connection with suffering, including our own.
>
> (Benjamin, 2016, p. 7)

The Moral Third then is a principle, a construction, and a vehicle; it involves making repair, acknowledging harm, recognition and witnessing – all the ways that people try to put right what was wrong. Most important, it provides a container, in Bion's (1962) sense; a mental space in which unthinkable and unimaginable horrors can be transformed into that which can be thought, can be imagined, and therefore can be survived.

The incident of the soap had allowed the children to call upon the memory of a loved and lawful childhood, which had been irreparably broken. They remembered the libation, the means of repair, for incidents of rupture, and they had called on that memory in the hope for restoring a just universe in which they could have a place. Equally important, Mr. Kamanda and the staff members, as therapists, recognized and responded to this request, acknowledging the boys' articulated need and, at the same time, encircling them within the restored community as a whole. The libation provided an opportunity for all of the community authorities to come together, from the chief of police to the traditional healer, and acknowledge collective responsibility for the terrible events of the war, and specifically their inability to protect these boys from being recruited, despite their best efforts, in a country once known as the Athens of West Africa.

Discussion

One special significance of Benjamin's (2016, 2017) concept of the Moral Third for therapeutic work with youngsters affected by violence today is that it organizes the interstices of a space in which there is not only human connectedness but also room for acknowledgement and witnessing, and with it the capacity for mutuality and repair – all the ways humans must continuously work to put right what is wrong. At the same time, it allows for the reality of integration first invoked by the depressive position: that within everyone is a saint and sinner and that there is no space beyond the realm of repair. Perhaps it was for this reason that families brought their children to the court diversion program – to try to make right the dangerous situations that their youngsters were experiencing and for which contemporary communal ritual leaves little room. Certainly, it was for that reason the court diversion program was created. The courts, by asking for the program to be created, had in effect charged clinicians to create a space where moral life could be enacted, not as punishment, but as an agreement that society, communities, and families have with their children, namely, that ruptures are reparable within a just society, if that society will take responsibility for its role in creating the harm that led the young ones astray in the first place.

The repair of the self – that is, recovery from the disintegration that occurs when knowing and living through the worst of horrors and realizing that humans do this – cannot occur in an isolated context of healing but requires the participation of all those responsible (Apprey, 1999, 2014, 2016). Such participation allows the integration of those who have been cut off from the magnetic chain because they cannot bear to know what they have found out about others and about themselves.

Our children in the streets today, whether united in formal movements or less formal ones, are asking for a libation; that is, they are asking for the Moral Third to be invoked, an acknowledgement, beyond the family, by all levels of society, that we have failed them. It is the job of child and adolescent therapists and of society itself to acknowledge the historic and present wrongs and to set things right. Through their memories of individual loved and lawful childhoods, however difficult, these youngsters call for a world in which that which was broken can be repaired.

Is it possible to learn from the children, families and healers of Sierra Leone that a transformative response is achievable? Is it possible that we too can acknowledge our creation of the seeds of violence, our failure to protect and provide a vehicle for repair? Is it possible that we can begin to work toward the acknowledgement that our young people are asking for, and with it join them in the task to repair the self, the community and the nation?

Postscript on Sierra Leone

- Ninety-eight percent of the 6,914 demobilized child soldiers (including the four Mohammeds) were successfully reunited with their surviving family members, who accepted them, as they all were helped by community healers to recover together, slowly, over time.
- Samuel T. Kamanda, the Director of the Reintegration Program in Sierra Leone, died of malaria in 2004 while on mission in Liberia as Psychosocial Technical Advisor to the program to demobilize child soldiers there.
- The Special Court for Sierra Leone (SCSL) was set up in 2002 to address the serious crimes against civilians committed during the civil war. It was

the first such tribunal to sit in the country where the crimes took place and to have an outreach program that engaged survivors in testimony. In 2013, it became the first such court to complete its mandate and transition to a residual mechanism.

- Out of 13 indictments, 10 persons were brought to trial, three having died prior to trial or sentencing. One of those indicted fled the country and may be at large. The remaining 9 were convicted and sentenced to terms of imprisonment ranging from 15 to 52 years. The Residual Court monitors completion of these sentences.
- Since Sierra Leoneans already knew that retributive justice is a necessary but insufficient placeholder for the restoration of the lawful third. a Truth and Reconciliation Commission was convened simultaneously. During its 3-year lifespan, it took the testimony of 9,000 victims and perpetrators in ceremonies designed to make them whole and prepared to live together again. In addition, traditional reconciliation ceremonies such as "Fambol Tok" and other justice practices were initiated.

Order has returned to Sierra Leone. Testimony was given, perpetrators convicted, and child soldiers reintegrated into families and communities. The universities reopened and the school and health systems resumed. The scourge of child deaths in the diamond mines was ended by Sierra Leonean activists. Communities mobilized to defeat the Ebola virus. Today's adults, once the children affected by the conflict, still seek to heal the wounds of war in order to develop an ever more just and peaceful society (Decker, 2018).

Notes

1 Argentina, Brazil, Chile, Paraguay, and Uruguay are sometimes referred to as the Southern Cone countries of Latin America. All five experienced political dictatorships during the 1980s.
2 In Sierra Leone's Muslim community, it is customary to name one or more sons Mohammed, to honor the prophet, and then give each child additional unique names to identify them specifically. The children would then be called by their unique names.

References

Apprey, M. (1999). Reinventing the self in the face of received transgenerational hatred in the African American community. *Journal of Applied Psychoanalytic Studies*, 1(2), 131–143.
Apprey, M. (2014). A pluperfect errand. *Free Associations*, 66, 16–29.
Apprey, M. (2016). Representing, theorizing and reconfiguring the concept of transgenerational haunting in order to facilitate healing. In S. Grand & J. Salberg (Eds.), *Trans-generational trauma and the other* (pp. 32–53). New York: Routledge.

Becker, D. (1995). The deficiency of the concept of posttraumatic stress disorder when dealing with victims of human rights violations. In R. Kleber, C. Figley, & B. Gersons (Eds.), *Beyond trauma: Societal and cultural dynamics* (pp. 99–114). New York: Plenum Press.

Benjamin, J. (2011). Facing reality together discussion: With culture in mind: The social third. *Studies in Gender and Sexuality*, 12(1), 27–36.

Benjamin, J. (2016). Psychoanalysis, non-violence as respect for all suffering: Thoughts inspired by Eyad El Sarraj. *Culture & Society*, 21(1), 5–20.

Benjamin, J. (2017). *Beyond doer and done to: Recognition theory, intersubjectivity, and the third*. New York: Routledge.

Bion, W. R. (1962). *Learning from experience*. Northvale, NJ: Jason Aronson.

Bracken, P. J., & Petty, C. (Eds.). (1998). *Rethinking the trauma of war*. London: Free Assn Books.

Bragin, M. (2003). The effect of extreme violence on the capacity for symbol formation: Case studies from Afghanistan and New York. In J. Cancelmo, J. Hoffenberg, & H. Myers (Eds.), *Terror and the psychoanalytic space: International perspectives from Ground Zero* (pp. 59–67). New York: Pace University.

Bragin, M. (2004). The uses of aggression: Healing the wounds of war and violence in a community context. In B. Sklarew, S. Twemlow, & S. Wilkinson (Eds.), *Analysts in the trenches: Streets, schools and war zones* (pp. 169–194). Hillsdale, NJ: Analytic Press.

Bragin, M. (2007). Knowing terrible things: Engaging survivors of extreme violence in treatment. *Clinical Social Work Journal*, 35(4), 229–236.

Bragin, M. (2010). Can anyone here know who I am? Creating meaningful narratives among returning combat veterans, their families, and the communities in which we all live. *Clinical Social Work Journal*, 38 (3), 316–326.

Bragin, M. (2012). So that our dreams will not escape us: Learning to think together in time of war. *Psychoanalytic Inquiry: A Topical Journal for Mental Health Professionals*, 32(2), 115–135.

Bragin, M., & Bragin, G. (2010). Making meaning together: Helping survivors of violence and loss to learn at school. *Journal of Infant Child & Adolescent Psychotherapy*, 9(2), 47–67.

Coates, T. N. (2018). *We were eight years in power: An American tragedy*. New York: One World/Ballantine.

Decker, C. (2018). *Beyond justice: the need for transgenerational healing in Sierra Leone*. (Unpublished master's thesis). Silberman School of Social Work at Hunter College, New York.

de Viñar, M. U. (2012). Political violence: Transgenerational inscription and trauma. *International Journal of Applied Psychoanalytic Studies*, 9(2), 95–108.

DiAngelo, R. (2018). *White fragility: Why it's so hard for white people to talk about racism*. Boston, MA: Beacon Press.

Efraime, Jr., B., & Errante, A. (2012). Rebuilding Hope on Josina Machel Island: Towards a Culturally Mediated Model of Psychotherapeutic Intervention. *International Journal of Applied Psychoanalytic Studies*, 9(3), 187–211.

Fanon, F. (1963). *The wretched of the Earth*, trans. R. Philcox. New York: Grove Press.

Fine, M., Weis, L., Powell, L. C., & Wong, L. M. (Eds.). (1997). *Off white: Readings on race, power, and society*. Florence, KY: Taylor & Francis/Routledge.

Freud, S. (1974 [1913]). Totem and taboo. In J. Strachey (Ed. and Trans.), *The standard edition of the complete psychological works of Sigmund Freud*, vol. 13 (pp. 1–162). London: Hogarth Press.

Garbarino, J., Kostelny, K., & Dubrow, N. (1991). *No place to be a child: Growing up in a war zone.* Lexington, MA: Lexington Books.

Hartmann, W. E., Wendt, D. C., Burrage, R. L., Pomerville, A., & Gone, J. P. (2019). American Indian historical trauma: Anticolonial prescriptions for healing, resilience, and survivance. *American Psychologist*, 74(1), 6.

Honwana, A. (1998a) Discussion guide 4: Non-Western concepts of mental health. In UNHCR, *Refugee children: Guidelines on protection and care.* Geneva: UNHCR.

Honwana, A. (1998b). Okusiakala ondalo yokalye: Let us light a new fire. www.child-soldiers.org/psycho-social/english.

Honwana, A. (1999). Challenging western concepts in trauma and healing. *Track Two: Culture and conflict*, 8(1), n.p.

Honwana, A. (2002). Negotiating post war identities: Child soldiers in Mozambique and Angola [abridged]. In G. Bond & N. Gibson (Eds.), *Contested terrains and constructed categories: Contemporary Africa in focus.* Boulder, CO: Westview Press.

Honwana, A. (2011). *Child soldiers in Africa.* Philadelphia, PA: University of Pennsylvania Press.

hooks, b. (1998). Racism and feminism. In C. Eze (Ed.) *African philosophy: An anthology.* Malden, MA: Blackwell.

Kenyatta, J. (1938). *Facing Mount Kenya.* London: Secker and Warburg.

Klein, M. (1975a [1952]). Some theoretical conclusions regarding the emotional life of the infant. In M. Klein, *The writings of Melanie Klein, vol. III: Envy and gratitude and other works, 1946–1963* (pp. 61–93). New York: Free Press.

Klein, M. (1975b [1957]). Envy and gratitude. In M. Klein, *The writings of Melanie Klein, vol. III: Envy and gratitude and other works, 1946–1963* (pp. 176–236). New York: Free Press.

Klein, M. (1948). *Criminal Tendencies in Normal Children in Contribution to Psychoanalysis.* London: Hogarth Press.

Kornfeld, E. L. (1995). Development of treatment approaches for victims of human rights abuses in Chile. In R. Kleber (ed.), *Beyond trauma* (pp. 115–131). New York: Plenum Press.

Kovel, J. (1984). *White racism: A psychohistory.* New York: Columbia University Press.

Langer, M. (1989). *From Vienna to Managua: The journey of a psychoanalyst.* London: Free Association Books.

Monteiro, C. (1996). Cultural issues in the treatment of trauma and loss: Honoring differences. Paper prepared for the Christian Children's Fund, Richmond, VA, July 31.

Osofsky, J. D. (Ed.). (1998). *Children in a violent society.* New York: Guilford Press.

p'Bitek, O. (1998). The sociality of self. In C. Eze (Ed.) *African philosophy: An anthology.* Malden, MA: Blackwell.

Peddle, N., Monteiro, C., Guluma, V. & Macaulay, T. (2013). Trauma, loss and resilience in Africa: A psychosocial community-based approach to culturally sensitive healing. In K. Nadar, N. Dubrow & B. H. Stamm (Eds.), *Honoring differences* (pp. 121–147). New York: Brunner/Mazel Publishers.

Puget, J. (2002). The state of threat and psychoanalysis: From the uncanny that structures to the uncanny that alienates. *Free Associations*, 9, 611–648.

Segal, H. (1974). *Introduction to the work of Melanie Klein*, 2nd edition. New York: Basic Books.

Steiner, J. (1988). The interplay between pathological organizations and the paranoid schizoid and depressive positions. In E. Spillius (Ed.), *Melanie Klein today: Developments in theory and practice* (Vol. 1 pp. 324–342). New York: Routledge.

Stevenson, B. (2015). *Just mercy: A story of justice and redemption*. New York: Spiegel & Grau.

Van Soest, D. (1997). *The global crisis of violence: Common problems, universal causes, shared solutions*. Washington, DC: NASW Press.

Vaughans, K. C. (2014). Disavowed fragments of the intergeneration of trauma from slavery among African Americans. In K. Vaughans & W. Spielberg (Eds.), *The psychology of Black boys and adolescents*, vol. 2 (pp. 563–577). Santa Barbara, CA: ABC-CLIO.

Viñar, M. (1989). Pedro or the demolition: A psychoanalytic look at torture. *British Journal of Psychotherapy*, 5, 353–362.

Volnovich, J. C. (2002). El default con nuestros hijos: la desesperación por no cobrar. El dolor de no poder pagar. *Consultado 6.*

Winnicott, D. W. (1984). Aggression and its roots: Aggression. In C. Winnicott, R. Shepherd, & M. Davis (Eds), *Deprivation and delinquency* (pp. 88–91). London: Tavistock/Routledge Publications.

Zack-Williams, T. B. (2006). Child soldiers in Sierra Leone and the problem of demobilization, rehabilitation, and reintegration into society: Some lessons for social workers in war–torn societies. *Social Work Education*, 25(2), 119–128.

Chapter 3

Psychoanalytic Thoughts on Evacuated Parents and Children during the First Weeks of the October 7th War

Ruth Weinberg

October 7th, 2023, a sabbath, was a holy day in Israel. Early that morning, as families were just rising from sleep, an unprecedented, large-scale, multi-front attack on the country, began. Thousands of armed terrorists infiltrated into Israeli territory through the ground, the air, and the sea, under the cover of a barrage of rockets, launched from Gaza into southern Israel. The attack, which began at dawn, lasted the whole day, and involved mass atrocities: acts of murder, sadistic rapes, and torture of Israeli civilians, including infants, children, teenagers, adults, and the elderly. These brutalities, mainly committed in quiet rural communities and at a rave music festival, were filmed by the terrorists. Videos containing graphic details of the events were uploaded to the internet in real time and were broadcast on social media in order to spread fear among the entire population. Many Israeli teenagers were exposed to these videos. Within a day, approximately 1,200 people were killed, and around 350 were injured – a number of casualties that, relative to the population size, would be equivalent to 45,000 people in the United States. Additionally, 250 people, including many infants, children, teenagers, women, and the elderly, some severely injured, were kidnapped and taken to Gaza. As a result of these events, the October 7th War began, which, as of the writing of these lines, continues to claim many lives and cause significant suffering to children and teenagers in both Israel and Gaza.

This chapter presents a brief review of classical psychoanalytic writings on children in war, two brief psychoanalytic interventions with evacuated Israeli girls and their mothers, and concludes with my own thoughts on the effect of war on the internal world and mental functioning of children. This chapter is based on my clinical work during the first two months of the war in Israel, through brief interventions in hotels where evacuees were accommodated. I believe that much more will be written about the damage inflicted on the minds of children, teenagers, and evacuees, both on the Palestinian and Israeli sides, as well as on the nature of interventions suitable for this situation in the near and far future. While this paper was written, 250,000 Israelis, who lived in areas that were attacked, were displaced. This would be equivalent to a displacement of 7,500,000 people in the US.

DOI: 10.4324/9781003520757-4

The First Weeks of the Evacuation of the South

In the early days of the war, amid rocket threats from Gaza and the fear of further infiltrations by terrorists from the area, approximately 130,000 children and their families from the south were evacuated to hotels in central Israel. The panic, confusion, and shock, in addition to the helplessness, experienced by the displaced families were immense. Psychologists, from across the country, arrived at these hotels, offering their help where needed. At this early phase, the main goal of the interventions was to provide emotional support for children and their parents and to enable a mental space for processing the intense emotions and anxieties that the first days of the war evoked. Interventions aimed to restore parents' ability to consciously position themselves as responsible and containing adults for their children and prevent deterioration in the mental state of both these children and adults.

All children who came from the southern communities experienced traumatic events. Some endured prolonged terror. They hid silently for hours on end, while terrorists were heard shouting in Arabic in their own homes, as well as storming their kibbutz, shooting and burning down neighboring houses. Some were exposed to the horrifying aftermath of the massacre and witnessed the mutilated bodies of family members and neighbors, some of whom had been tortured. Many lost their loved ones in the attack. Aside from children directly involved in the events of October 7, many other children, throughout the southern region, experienced life-threatening situations and suffered severe anxiety during the days that followed. Children were confined in the "safe rooms" for many hours, unable to attend to their basic needs, and witnessed their terrified parents trying to block doors that could not be securely locked. Many teenagers were in youth apartments in the kibbutzim without the presence of their parents, waiting for the Israeli army to rescue them, but to no avail. All were connected to WhatsApp groups and began receiving messages, videos, and pictures of the mayhem happening outside their homes, experiencing intense fear that they might be next. Only after two days did the scale of the disaster became apparent, and it took a full week for the Israeli military to completely regain control over the southern region of the country.

Following the disaster the hotels that housed the evacuees were flooded with donations of art supplies, clothing, and food. Celebrities also came to entertain the children and families. For these shocked, bereaved, and grieving communities, including families of hostages whose fate was unknown, this abundance exacerbated the evacuees' experience of derealization.

Children in the hotels showed many symptoms: sleep disturbances, eating difficulties, emotional regulation issues, motor restlessness, hyperactivity, and separation anxiety. Some children slept for many hours upon arriving at the hotels. Many were scared to leave the hotel for fear of rockets and terrorists, despite the distance from the borders and the relative safety in Tel Aviv. Some struggled to adapt to the urban landscape, which is so different from the rural open one in the south of the country. Some children expressed their fears verbally, while others showed social difficulties, tendencies toward outbursts, and somatization resulting from the stress. Several teenagers tended to withdraw, staying in their rooms and refusing to participate in activities or talk to professionals.

Principles of first aid treatment in extreme and traumatic situations primarily include relieving stress, offering support and emotional connection with the person in distress, reducing guilt and shame, strengthening the sense of agency and competence, and offering psychoeducation about natural responses in such situations. During those first days, many psychoanalysts found themselves learning basic guidelines for first aid help from experts on trauma online, while also continuing to think psychoanalytically during the encounters with children and adults. During those first months, I came to characterize short-term psychoanalytic intervention in situations of mass trauma, as an intervention which takes into account the state of internal objects, pays attention to the existence of unconscious anxieties triggered by the events, and observes the mental organization following the crisis while also offering first aid support. The analyst's special task is to examine how the harsh reality was perceived *unconsciously*. Addressing unconscious anxieties and the resulting defense mechanisms erected contributes to the effectiveness of the initial support and facilitates a movement toward more adaptive mental organizations and symbolic processing of mental content. Thus, psychoanalytic intervention, in the immediacy of the traumatic event, even if it is short termed, could potentially enable a more complete long-term adaptation.

Psychoanalytic Literature on Children and Adolescents in War

In the introduction to a small booklet on war neurosis that followed the Fifth Psychoanalytic Congress in Budapest, Freud wrote that war brings with it neurotic illnesses which he termed war neuroses. These often disappear with the removal of external conditions: "It is, however, a significant fact that, when war conditions ceased to operate, the greater number of the neurotic disturbances brought about by the war simultaneously vanished" (Freud, 1919, p. 207). He emphasized that war neuroses are traumatic neuroses resulting from a sudden and frightening event and not from conflicts within the ego. Freud understood that war neuroses stem from extreme external conditions rather than a lack of love or frustration of sexual impulses, but he also believed it was impossible to separate external dangers from those arising within the ego. Internal struggles within the ego concerning sexuality and libidinal conflicts, he concluded, might contribute to and intensify war neurosis. For instance, the pre-war ego can conflict with the new, more aggressive ego that developed during the war, threatening the existence of the old ego (Freud, 1919). In his writings, Freud mainly referred to soldiers, many of whom were adolescents when they enlisted, whose protective mental envelope was damaged, resulting in trauma. Freud's theory suggests an interaction between internal factors and external reality that determines the response of the child and adolescent to this reality. Addressing this interaction is the most significant contribution of psychoanalytic thought to understanding traumatic situations. *While many first aid protocols offer general principles to strengthen coping mechanism, the contribution of psychoanalytic thought lies in considering the interplay of the conscious and the unconscious internal world.*

It was the reality of the First World War in Europe, that shaped Freud's concept of the death drive. His groundbreaking work, "Beyond the Pleasure Principle" from 1920, opens with the veterans of The Great War, whose traumatic dreams of the battlefield repeatedly take them back to the horrifying scenes of death and destruction they have witnessed. Freud's understanding of repetition compulsion, as it relates to the pleasure principle, and the death drive, deeply inspired the work of Melanie Klein.

Klein, who made the death drive central to her work, observed that pathological conditions in children are influenced by sadistic aggressiveness, destructive aggression, and envy towards the good. Like Freud, Klein's theory is also based on the understanding that unconscious materials influence the perception of reality while also being influenced by it. Since unconscious phantasies, the mental equivalent of drives, underlie all mental processes and activities, they are also tied to the child's perception of external reality. These phantasies are projected onto the outside, coloring the perceived reality. When Klein moved to Scotland in 1941, a very difficult period during the war, she treated Richard, a 10-year-old boy who was evacuated, for several months. Although Richard shared with Klein information he had read and heard about the "real" war, she interpreted his material in the context of anxieties emanating from his internal world. It was the analysis of his internal world that Klein believed would strengthen Richard's ability to cope with the harsh external reality of the war. She wrote:

> Mrs. K interpreted that his fear of the dangers she would be exposed to in London was greatly amplified by other fears. She reminded him that when he mentioned that his mother was going home to stay with his father and Paul, he seemed very angry and worried; he might be afraid that he would attack his mother or harm her because of his anger and jealousy.
>
> (Klein, 1961, session no. 25)

Klein understood that due to the interplay between external reality and unconscious phantasies, life events involving loss, frustration and fear strengthen the death drive and exacerbate anxieties. When children are exposed to a very cruel reality, their inner world becomes even more frightening. Persecutory external objects combine with threatening internal objects and increase their potential threat. Conversely, "if the child can rely on good objects in his inner world, he will be able to survive a harsh reality" (Klein, 1944).

One of the most upsetting and painful consequences of war that parents and children face is the loss of home due to the necessary evacuation of civilian populations from combat zones. Winnicott (1984a [1939]), who was responsible for evacuated children during the Second World War, referred in a letter he co-wrote with John Bowlby to the "emotional blackout" resulting from the loss of home. He describes this "blackout" as a rupture. More than reflecting sadness of loss, it is similar to the devastation of primary depression that occurs when a child who loses the nipple feels as though he has lost a part of his mouth (Winnicott, 1984d [1963]). The loss of home and many loved ones, if experienced as a loss of part of the self and body, is thus significantly different from the pain of separation from a loved object.

Durban (2017), similarly to Winnicott, distinguishes between the mental registration of being homeless and being in a state of "nowhere-ness." The experience of being homeless is associated with the internalization of the mother as an internal space and the presence of unconscious phantasies that arise around the loss of the object. It is related to the ability to long for the object, the internal-mother-as-home, and to experience the object as destroyed but also lovingly remembered. "Being in nowhere-ness," on the other hand, is a state related to primary anxieties about the loss of an integrative self. A three years old girl, evacuated to a hotel after October7, whose tired and grieving parents left her alone in the playroom every day despite her constant crying, illustrated this very well. In the care of many other adults, she was inconsolable – sobbing, drooling, and unable to stand on her own feet even though she could easily walk independently. Her reaction marked the loss of a home and an internal parental anchor, experienced as a real skeleton. This was a devastating loss of an integrative bodily and self experience.

The home, therefore, is not merely a concrete entity; its existence is connected to the deepest layers of an integrative self. It possesses continuity, an internal space as well as a boundary, and differentiation from the external world. It is also a portal for communication and the passing on of psychic material. As Guston Bachelard notes: "The home has an objective dimension, and it is geographically and materially anchored in reality, but its essence goes far beyond objective boundaries and is linked to the core of our happiness" (Bachelard, 2020 [1994]).

Melanie Klein also addressed the impact of evacuation and displacement on the psyche and on the unconscious phantasies of children. She observed that young children might experience the loss of home as a punishment for their actions, leading to feelings of guilt and anxiety about parental retribution. A girl might experience her removal as the result of competition with her mother. The loss of home, which also involves separation from close individuals, may trigger anxiety about the loss of the object and fear of dependency on others (Klein, 1944). A report on evacuated children in Cambridge in 1942 showed that, despite the danger of bombings, over 75 percent of them returned home due to a deterioration of their emotional world and their severe longing for their parents, etc. (Isaacs, 1941).

During the events of October 7th, homes were shot at, broken into, looted, and set on fire. Parents and children were removed from their homes both because of the danger posed to the area and because of the destruction of the buildings, which became no more than monuments to the terrible catastrophe. The children who were evacuated lost their homes, friends, relatives, daily routine, and the entire landscape of their childhood in one day. Many families were left without a father due to the mandatory recruitment of men for reserve military duty. Mothers left with their children were themselves grieving, overwhelmed, and anxious about the fate of their loved ones, homes, and livelihoods. Many parents lost their ability during those days to parent their children as well as their internal infantile objects. Thus, many children lost also the parental anchor as an internal home – anchoring the child in time and space.

These changes in the reality and lives of many people were also reflected in the unconscious. In the unconscious phantasy, there is a connection between the home, the parental object, and the child's body and psyche. The home and the parent serve as a skeleton or an envelope for the self, as well as a home for the child's psyche. Therefore, the role of the parent in difficult times is critical for preserving the child's mental functioning. An artistic example of this can be seen in the 1997 Italian film *Life is Beautiful* directed by Roberto Benigni. In the film, Guido, a Jewish father, tries to conceal the horrors of the Holocaust from his young son Joshua by conveying the experience of the concentration camp as a competitive war game. He explains to his child that they must hide, and at the end, they will win a great prize. When Joshua meets his mother after the camp's liberation, he is in good spirits. Unaware of the horrors that took place, and happy to see her, he thinks that he has won the game. He believes that the ride on the American tank with the commander who liberated the camp is the prize he was awarded. The story told in this film is an extreme and poetic example of the fact that external reality perceived by children at a young age is greatly mediated by the parent. In reality children do feel the ominous quality of extreme circumstances, and they do see and feel the anxiety of the parents. Truth cannot be twisted or evaded completely, but the parent has a crucial role in mitigating the anxieties that arise in the face of harsh realities. Without the mediation of the parents, children are exposed to violent materials their young minds cannot digest. The presence of a thinking parent could possibly ease the burden of the difficult external reality.

Anna Freud and Dorothy Burlingham established three shelters for children during the Second World War (Freud & Burlingham, 1943). They observed that children who experienced the war alongside calm parents, who reassuringly mediated the events, were less psychologically damaged, even when exposed to bombings, as compared to children who were evacuated and not under threat of bombings but lacked a calming mediating figure (see also Ornstein, 2012). Winnicott (1984b [1940]) found that children under the age of five are only indirectly affected by the war, depending on how they perceive their parents. Young children, due to cognitive limitations, often do not under-stand what is happening in the war outside but are primarily influenced by crises within the home. Furman (1974) found that children in the Holocaust who wit-nessed horrific situations experienced their parents as a source of support and help regardless of their limited ability to influence the terrible reality. Gampel (1988), interviewed child survivors from the Holocaust and found that those who were in the company of a parent who was experienced as protecting them did not lose hope. Research findings, thus, indicate that parents have a decisive role in their child coping with catastrophic events. Hence, it is of utmost importance, when helping children in distress, to support their parents.

It is clear that when parents are nervous, devastated, impatient, anxious or angry, they are less able to take care of their child. Parents know that too. However, the impact of the parent's mental state is also expressed through less conscious osmotic transmission of psychic materials between parent and child, with each coping according to his or her psychological structure, available defenses, and age-appropriate functional level. This osmotic trans-mission is because, in extreme events, young children regress to a very early mode of attachment to the parent. This connection is more akin to an infan-tile attachment where the pair functions as a single psychological unit. Betty Joseph writes about this mode of psychological functioning:

> There is a relation between the mother and child – a reciprocal rela-tionship – so intimate that, whatever the complications it may bring into our psychology of individuals, it may be prudent sometimes to consider the psychology of dependent couples as a topic worthy of special study. In any case in this particular human relationship, when it comes to the treatment of the problems of the one, it seems that it is frequently most profitable to consider as inseparable the problems of the couple. ... Put briefly the question is "When does a person begin to be 'self-contained'?" or viewed more administratively "when is 'the patient' attending a clinic one person and not (however small the unit) a 'social aggregate'?"
>
> (Joseph, 1948)

Following are two cases from my work with two mothers and their daughters, exemplifying this dynamic. Both involve functional and rela-tively healthy mothers, who were affected by the harsh reality of October 7th and its aftermath. The interventions presented here demonstrate an attempt to understand the unconscious interpretation of the external rea-lity by parent and child and to initiate healthier emotional communication between them so as to enable the adaptation to the new situation and to prevent the development and consolidation of pathology. As is shown in these two cases, working with the "social unit" of parent-child allowed for a quick shift in the children involved.

Case 1: Karen and Zoe – Loss of Home as Loss of the Good Object

I met with Karen and her five-year-old daughter Zoe for several sessions in my office. Karen, Zoe and her two-and-a-half-year-old son, were evacuated after the massacre on October 7th. Her husband was enlisted to reserve duty that same Saturday. Although her fellow kibbutz members were evacuated to a hotel in The Dead Sea, Karen chose to return to Tel Aviv, where her family lives. Karen, a high functioning woman, moved after the first week from the hotel to an apartment she rented using the financial aid she received from the state. Seemingly not in need of help herself, she heard that I was assisting with evacuated children and asked to see me because she was worried about her daughter. I felt in her request that Karen was sad, even though she did not express it verbally. She mainly told me about the new daily routine of the children and herself.

Karen believed that Zoe did not experience any of the horrific events of the massacre that Saturday, since the parents made sure they concealed the events and spoke in English, a language the children do not understand. In our conversation, Karen did not address any sense of loss of home, of family life or of displacement. She talked about daily difficulties with Zoe. She described Zoe as a very energetic and happy child but also as an extremely demanding one. The new teacher in the temporary pre-school Zoe was attending complained that she hits others. She tended to have crying fits that intensified when she did not get what she wanted, even over "unreasonable" things. Karen gave as an example Zoe's responding with tears and screams when she requested that she leave outside a dirty branch she found in the park and carried home. Given their state and needs, I suggested seeing them together in my office.

The Session with Karen and Zoe

Zoe was happy to come with her mother and was immediately preoccupied with Barbie dolls she found in the room. Her manner of speaking imitated joy and cheerfulness but felt fake. She was not satisfied with the outfits of the dolls. She wanted a beautiful dress and nothing I suggested pleased her. I wondered if, on a deeper level, Zoe was searching for the lost internal aesthetic primary object (Meltzer & Williams, 1988). She insisted on making other beautiful dresses for the Barbies. I understood this as her way of saying that her inner world was not well. There was a problem with her mental "wrapping," and she was struggling to accept the current situation and wanted things to be better.

There was no mentioning of the war, and as time passed, the absence of play or playfulness became more evident. Zoe insisted on cheerful activity ("Oh, what beautiful colors, oh, what beautiful fabrics") while arbitrarily selecting and arranging the dolls on chairs ("These are dolls in the kindergarten")

without delineating "kindergarten" or "home" spaces and without any emotional context or an organizing narrative. Zoe seemed emotionally disconnected
from the play, obsessively arranging dolls and chairs. After receiving the Barbies in crepe dresses made by her mother at her request, she announced that
Barbie married Ken and that it was morning and they had just woken up. She
then added in a cheerful tone, "and he suggested they eat lunch." The mother
echoed what she had just said: that they were waking up in the morning, and
Zoe immediately corrected herself: "Oh, of course, breakfast." Was it too
painful for her to think about the October 7th morning?

Throughout the session, although Zoe had a cold and struggled to breathe,
she did not blow her congested nose even once. It seemed that both her body
and her psyche were burdened with materials that had no way of being dis-
charged or evacuated. Only when the session ended did she cry, finding it
hard to get up and leave the room.

After Zoe and Karen left, I felt the weight of their predicament. I realized
that, like Zoe, I had tried to maintain a good and cheerful atmosphere for her
and her mother. Knowing how hard it was for both, I understood that I was
collaborating with some sort of fake and internal command to function without
addressing the loss and the sadness. I felt I was witnessing a manic defense
leading to emotional disconnection, confusion, and unprocessed pain in both
child and mother. For Zoe, the manic defense failed because the loss of home
was experienced unconsciously not only as painful but also as a loss of a part
of her physical self, leaving her disoriented and anxious in "no-whereness."
This was also expressed in her play of trying to find chairs and sit all the dolls
while being unable to situate them in a clearly defined location. The separations
from loved ones, the loss of home and routine as an organizing structure in her
life, and the loss of the mother as a containing object led to regression to an
obsessive, active, and two-dimensional mental functioning (Meltzer, 1975) with
violent outbursts. In this state, Zoe did not experience herself as anchored in
time (lunch in the morning) and in bounded spaces with inside and outside (no
marked and delimited spaces in her play.) She experienced a sort of primary
anxiety expressed as restlessness that was met with a two dimensional adhesion
as a defense – to surfaces (room), to constant activity, to objects (mother, the
branch unconsciously symbolizing herself as a branch cut off from the tree).
Separations thus felt as interruptions in the continuity of experience. The
mother, in her way, also clung to activity and functioning to avoid the many
painful feelings the catastrophe had evoked. Both mother and child lost so
much in one day, but the loss and longing were not felt as pain.

Session with Karen

When I met with Karen she complained about her worsening difficulties with
Zoe. I asked her about her feelings since the evacuation to Tel Aviv. Gradually her communication conveyed more feelings of loss and loneliness. It was

evident that she felt there was no place for her sadness. She didn't feel she could "burden" her brothers, whose sons were enlisted in the ongoing war in Gaza. For similar reasons she felt unable to share her feelings with her husband who, too, was enlisted, and as part of his duties in the battlefield, barely slept. "He's not here even when he's here," she added, describing how he forgot that their close friends had fled the country. These statements reflected both her reality of being left alone with the children, but also her feelings that she was a burden to her brothers and forgotten by her husband, who had become detached.

I wondered about her connection with the rest of the community. I learned that there was a kibbutz gathering scheduled for the next Saturday, and she was debating whether to attend with the children because she was not sure if she was still considering the kibbutz her home. She had moved to the kibbutz because of her husband, who lived there. It seemed that the anxiety Karen experienced, of losing her home and being abandoned by her good objects, led to a rejection of the "home," followed by a sense of homelessness, uprootedness, and isolation. In such a state, it was also impossible for Zoe to long for her lost home.

The loss of home and the abrupt separation from her husband triggered in Karen a need for support but also fears of object abandonment due to dependency needs she perceived as excessive. She was supposed to manage on her own. On the other hand, she felt angry for being abandoned and had fights with her husband the few times they met on his return from duty. I asked about her sense of transience and lack of belonging, and Karen shared that her mother had died during her childhood. Following this early loss, her father took off and left her in the care of her older siblings for a whole year. When he then remarried, he took her to France with him, leaving her older siblings behind. Karen had been close to her father, who always comforted her, and he had died two years before the October War.

Karen's loss of home after the October 7th attack, that followed the forced separation from her husband and the friends who left the country, brought up emotional memories of earlier losses and abandonments (Klein, 1940). These latter, in turn, threatened the very existence of an internalized good object. In the conversation, Karen realized that she was trying to be like her father, to "take care of things." Unconsciously she identified with an abandoned and forgotten child, anxious that her needs had killed her mother and chased away her father. In the session, we could see how her needs were linked in her mind to loss and rejection, and she was able to make space for the longing, sadness, anger and isolation she experienced ever since the traumatic attack. I suggested another session with her and Zoe the following week.

Second session with Karen and Zoe

Karen sat on the chair and Zoe looked at the toys. Karen said she thought Zoe had a better week and then recounted the difficulty Zoe had when her aunt came to pick her up from pre-school. Zoe said, "I just need Mom on rainy days." I suggested that maybe when it rains, we feel sadder and when we're sad, we want Mommy. Zoe listened. It seemed there was more room for sadness in the meeting since my last encounter with Karen. Zoe brought the Barbies and the dollhouse, and we sat to play. She said that the couple who got married in the previous session had a baby girl named Nini who is four months old. I heard this communication as indicating an emerging connection to an internal infantile object. Her narrative, which appeared now within a demarcated space, the dollhouse, told of a mother who left the house to pick up Nini earlier. Nini, she explained, had a terrible day at her new school and wanted her mother to pick her up because she did not feel well, got wild, and then hit the other children.

I said: "Something must have bothered her."

ZOE: "She started hitting friends unintentionally. She screamed so much that her throat hurt."

R: "She felt bad inside, and she had to let it out."

Z: "They told her the rules. Then she screamed and kicked people."

R: "Did she feel strong that way?"

Z: "No. Just … ah … just she remembered she did something bad in the previous school. [Zoe gets up and plays Nini. She places her hand on her hair ribbon.] I remembered in the previous school, I pushed Adam, and he broke his head and didn't come to school for seven days."

R: "Oh, Nini must have been very scared."

Z: "Then she started screaming. She thought Adam was dead. Then when she played outside before the end of the day, she felt her stomach hurt."

R: "Nini, you didn't kill Adam, you just hurt him. Because children can't kill other children at school. They are protected. Do you understand that, Zoe?"

Z: "She goes home and her stomach still hurts."

R: "What's still bothering you, Nini? Is it Adam inside bothering your stomach?"

Z: "No. My stomach just hurts very, very, very much."

R: "What's bothering you very, very, very much?"

Z: "I was scared because Adam was my best friend."

R: "I understand, now you miss Adam, you want to know he's okay."

Zoe makes crying sounds for Nini.

R: "Why are you crying, Nini?"

Z: "Her stomach hurts even more."

R: "She is sad and misses her friend. What would make Nini feel better?"

Z: [Speaking as the Barbie mother] "She's boiling. Wow. I need to go to work, you (points at me), the sister, take care of her."

The mother leaves and then returns and says: "I don't let Nini watch TV when she's sick because once when she was in my belly, she watched a lot of TV and it made her stomach hurt." The mother leaves, and Nini cries and cries.

R: "Poor thing, she probably thinks everything happened because of her, that her mother leaves because of her. It's not her fault that her stomach hurts, it's not because of her!"

Z: [In a baby voice] "Mm … Mm … When are we going back to our house? Is there a war? [This is the first time she mentions the war.] When are we going back to our house?"

R: "Ah, now I understand what's bothering Nini! [I speak directly to Zoe.] That's why you're sad, Zoe. You miss your home, the kibbutz, your friends, and your father in the army, and it's not your fault that everything happened. It's because of the war."

I met with the mother once more. In the meeting, she reported that Zoe was doing better at school, and the teacher had not reported any issues in the week leading up to our meeting. It seemed to me that both the mother's recovery and Zoe's ability to connect to her pain and to relieve her unconscious guilt facilitated her adjustment to the new school. Karen also recounted a conversation with her husband who was now more supportive and shared that they decided, if possible, to return to the kibbutz in a few months.

Zoe's and Karen's ability to rely on others improved with the growing capacity to mourn and long. Both initially showed a regression to two-dimensional functioning, manifested in the collapse of the emotional internal space and in flat attachments to objects or actions. Meltzer (1975), observed in various cases of nonautistic patients a defense of two-dimensionality and emotional superficiality due to anxieties about object destruction and persecutory guilt arising in the inner world. For Zoe, this form of two-dimensionality and adhesive identification also served as self-containment and protected her from disintegration anxieties. Throughout the sessions, it was possible to see both mother and child trying to cope with painful longing, guilt, and anxiety about object destruction as well as abandonment. Addressing anxieties and acknowledging sadness and loss contributed to Zoe's ability to return to play, process emotional pain, and cope better in the new school. I will now address different types of anxieties observed in evacuated children.

Different Types of Anxieties Related to the War

During the war, there has been a rise in various types of unconscious anxieties among evacuated children. Klein famously identified two main types of anxieties in children that were exacerbated during wartime: paranoid-schizoid and depressive anxiety. Paranoid-schizoid anxiety involves fear of external attack related to the death drive projected outward. Depressive anxiety concerns the fate of the good object who's at risk due to the child's aggression. Anxiety about the fate of the object and the desire to protect it is fundamental to the ability to love and is associated with creativity tied to a deep internal desire to repair internal objects (Klein, 1940; Klein, 1946).

In addition to these anxieties, evacuated children exhibited confusion and anxieties related to the collapse of the basic split between safe and good versus dangerous and bad (Rosenfeld, 1950) triggered by a similar breakdown in reality. On October 7th, the basic distinction between the inside of the home as a safe place and the threatening outside was shattered. The massacre occurred in the "safe rooms," a name that is given to a sort of shelter within the house, first built inside homes in Israel after the Gulf War. These rooms were unlockable to allow quick entry during an attack, turning them into death traps, when the terrorists have invaded homes on October 7. The collapse of the boundary between inside and outside, good and bad, protected and exposed, can occur at different levels of mental development.

At the depressive level, a clear boundary between inside and outside allows for a reduction in projections and the preservation of the good object from attack. There is a decrease in paranoid-schizoid and depressive anxieties due to the ability to distinguish between internal aggression and outside reality (Klein, 1952). At times of acute stress, this ability can be impaired, leading to confusion between aggressor and victim evoking persecutory guilt.

In the paranoid schizoid position, the collapse of a clear boundary increases paranoid anxiety and creates confusional anxiety in the subject (Rosenfeld, 1950). Confusing the bad with the good can also serve as a defense which leads to identification with the bad object to avoid paranoid or depressive anxiety (Meltzer, 1975; Meltzer & Williams, 1988). In the playroom, a variation of such a manic defense was seen in children who chose to play as Hamas terrorists, identifying with the aggressor (Ferenczi, 1949; Freud, 1920). In contrast, other children identified with an internal good and protective object, insisting on playing as Israeli soldiers.

Another type of anxiety observed in these situations is anxiety of being (Durban, 2019). The rapid evacuation of children, the loss of significant figures, and the sense of danger evoke not only emotional darkness, as Winnicott described, but also early primitive anxieties of disintegration of the bodily self, loss of a sense of self-continuity in space and time, and orientation difficulties (see also Tustin, 1990; Winnicott, 1984c [1962]). These anxieties can also be experienced more consciously as nameless dread (Bion, 1962). One striking observation during my time with the children in the playroom was their constant restlessness and lack of bodily relaxation. Many children wandered around searching for a corner to settle in, to feel comfortable and relaxed. It was evident that they could not calm themselves and concentrate. This restless movement served both as a channel for evacuating anxieties and as a second skin formation, holding them and preventing complete disintegration (Bick, 1968). One of the older children demonstrated this disorientation in time and space when he could not say whether he was in Israel or abroad, despite knowing he had landed in Israel after the holiday vacation.

Durban (2021), following the COVID-19 pandemic, identified yet another type of anxiety in children, which he termed osmotic anxiety – fear of a threat whose location is unclear and therefore cannot be escaped. Danger is "nowhere and everywhere." Israeli children experienced a variation of osmotic anxieties triggered by external events. First, the state of Israel, perceived by many as a "safe home" for the Jews after the Holocaust, became an invaded and destroyed home in one day, as it was brutally attacked by a barrage of missiles as well as by thousands of terrorists calling clearly for the slaughter of Jews ("Yidbach al Yehud"). Missiles reached all areas in the country, with no way to predict where the next threat would come from. Second, the peaceful coexistence of Israeli Jews and Israeli Arabs was shattered as suspicions and mistrust grew between neighbors.

The inability to escape the horrors of October 7th was also due to the overwhelming, unceasing news coverage of the massacre both in media, social media and in public spaces. During that day and those that followed, it was impossible to protect the children from being exposed to harsh, age-inappropriate information and the experience was that danger was everywhere and nowhere was safe. The streets were flooded with posters of the kidnapped children. Uncanny props used at exhibits calling for the return of the hostages became sources of even more horror for children confronted with them. An example of such an exhibit consisted of giant teddy bears with bullet wounds, chained to benches in central streets in Tel Aviv with pictures of the children-hostages on them. In the image below, these terrifying objects were exhibited in the center of the city on Dizeng-off Square. No longer the soft comforting teddy bear children use as transitional objects, these wounded, bleeding, blindfolded, objects were a source of anxiety in and of themselves.

Osmotic anxiety is related at its core to the child's experience of being particularly permeable to the parent's state of mind. This porousness, which is normally present during early infancy or in states of symbiosis between parent and child, also appears in traumatic situations when the child relies on the connection to the real parental object to process an overwhelming reality. Following what came to be known as "The Black Sabbath", the evacuated children stayed close to their parents all the time, absorbing parents' anxieties and pain. At times, not only were these children's anxieties not contained, but, using an omega function instead of an alpha function, parents turned their children into a reversed container for parents' pain. I will now provide an example of this permeability of the child's mind to materials emanating from the parent's mental state.

Case 2: Judith and Mia – The Collapse of Distinction Between Inside and Outside, Good and Bad, Aggressor and Victim

One morning, the manager at the hotel told me about a mother and daughter who seemed low-spirited and suggested I approach them. I arrived at the arts and craft station and watched Mia, a gentle six-year-old girl, sitting and embroidering close to her mother. She worked quietly and withdrawn and passive, occasionally whispering something to her mother. When asked by the instructor about the colors she preferred, she did not seem to care what color the embroidery would be. Judith, her mother, sat next to her, quiet, tired, and listless. I wondered if both were also experiencing a sensory numbness characteristic of people who had gone through extremely dangerous situations. I introduced myself. Judith then told me that Mia had been clinging to her since the events of that Saturday. She was afraid that Mia had been traumatized. In response to my questions, Judith monotonously gave me some details about the kibbutz they came from, and the family members they were with. There were no casualties in her family or in the kibbutz, which was one of those which were not infiltrated by terrorists. The father continued to work in the south but returned to the hotel every evening. She recounted that on that Saturday, knowing what was happening in neighboring kibbutzim, they were in terror for hours. When evacuated they had to speed with the car so to avoid the risk of being attacked by Hamas terrorists that were believed to still hide in the area. When I suggested we have a quiet conversation, she preferred that I see Mia, as she did not want to talk. We scheduled to meet the following day at my office.

When we met the next day, while Mia was sitting on the carpet next to her mother, looking at toys, the mother quietly recounted the events of that Saturday. She felt intense anxiety when they were cut off from all information for several hours due to a power outage. She added that in the evening, she felt she had to "get out of there" the moment it was possible and was angry when her husband was slow when military forces came to evacuate them. Listening to the way she spoke, I felt that she wanted to "get out" of herself, to get rid of a feeling she could not contain for even one more moment.

Meanwhile, Mia sat by the dollhouse. She said with a flat affect and low voice, "This is a family's house," and started looking for figurines. The mother mentioned that Mia loved to play with Barbie dolls, but Mia chose warriors and superheroes. She said, "I'm bringing guards to protect the house. They are the good ones." She placed the figurines close to the house, searched through the box, and said in the same tone: "I'm bringing bad ones." The mother looked at me for reassurance feeling that her daughter was in an emotional distress. Then Mia moved some of the figurines that were previously the guards, out of the house, but very close to it, almost inside it: "They are the bad ones," thus enlarging and strengthening the group of the "bad ones." She then brought new figurines and said in the same monotonous tone: "These are the good ones." On the surface, there seemed to be a distinction between good and bad, but gradually I became confused. Who are the good ones? Who are the bad ones? And where are they located? There was no sense of two clear camps. Who was protecting the house and who was attacking it? I said: "It's hard to understand what is the good territory and what is the bad. Where is the boundary?"

It seemed to me that Mia was simultaneously expressing the terrible reality she witnessed, where external boundaries were broken by "bad people," and the state of the home of her internal objects. Her play represented her psyche in which ambiguity prevailed leading to a blurred distinction between good and bad, between protector and attacker, and between victim and aggressor. My intervention, seeking to restore and clarify a split, led Mia to mark a boundary with her hand: "Here," she said, "here is the border," and pointed very close to the house. Now, with the restored border marked, the confusion took on a new form. Mia struggled to decide who among the figures would be good and who would be bad: "He is good. No, he will be bad. He is also bad. No, he is good." It seemed that the presence of bad figures near the house, triggered anxiety about a confrontation and a need to deny the bad, once it was declared so. Mia's internal confusion, again, also sadly mirrored a national confusion: was the army protecting the kibbutzim or attacking them? Were the Palestinians, perceived by the kibbutz members as allies in coexistence, good, or are they terrible murderers? It also mirrored, as I later came to see, the mother's confusion between aggressor and victim. Mia's unconscious paranoid anxiety was exacerbated by the frightening reality confusion. The reality confusion was simultaneously a defense against anxiety. I said: "We need order here. Who is good and who is bad? Where is it dangerous and where is it safe? Who needs to be attacked, and who needs to be protected?" Following my words, Mia tried to make another distinction: "You be the bad ones."

Until the end of the session she kept taking good figurines for herself, and assigning me a villain figure for each of the good ones she took for herself. Mia fluctuated between a distinct split, where she projected aggression and murderousness outward and keeping the good protected inside of her, and states where she collapsed into confusion between the good and the bad, the inside and the outside. The confusion was both a result of the traumatic reality and a defense against it. When the split was restored, and a clear distinction between good and bad was established, Mia feared she might treat me unfairly and took care to behave "nicely." It was evident that now she feared her aggression would damage the good object. Her need to be fair and offer me a figurine for every one she took for herself suggested that Mia was frightened by any initiative or desire to triumph over the object or "kick out the bad ones."

Towards the end of our session the mother was worried and told me in English (so that Mia would not understand): "She never played with such figurines before." I felt that the mother was afraid of the change she observed in her daughter, and scared of the aggression she noticed. I wondered if the child was experiencing that anxiety and sadness that her mother felt as the result of her daughter newly expressed aggression. Was Mia also anxious about her aggression towards her inner object, given the mother's state? Or was she collapsing under the weight of the mother's guilt that permeated her without a boundary?

In Mia's case, in addition to the collapse of a clear split between good and bad, there was also a regression to a symbiotic ambiguous mode of being which reinforced the blurring of clear distinctions. Internal ambiguity and regression to a symbiotic connection with her mother were both a response to the severe external events and a way of avoiding the anxiety and aggression that arose within her in reaction to her mother's deteriorated state and the events she witnessed. Bleger (2013 [1967]) believed that in the core primitive ego of early life, distinctions are blurred, and there are no clear boundaries between the baby and the mother. There is a symbiotic ego and a psychic quality of ambiguity. Through developmental maturation processes, the psyche gradually responds to boundaries in reality and introjects distinctions between objects. In distressing situations, there is a regression to this earlier functioning of the psyche. Following the traumatic events of October 7th, Mia regressed to symbiotic ambiguity and passively clung to her mother. She did not express her needs, desires, or aggression.

I said: "Now, both of you are in Tel Aviv, safe and protected. There are no terrorists here." To facilitate a distinction between mother and child and to place what had no clear location, I added: "Mia sees you, Mom, worried and quiet, and she doesn't understand what's happening, and it probably worries her." I added: "And you and I, Mom, will think together about how to help Mia and keep her safe." I suggested to the mother that I would come the next day to meet her at the hotel, and she agreed, partly because she wanted to hear my opinion regarding Mia's psychological state.

The next day, Judith looked sad and said she found it difficult to acknowledge Mia's difficulties as they came up in the session. She felt intense guilt and feared she had caused irreversible damage to her child. How could she have prioritized economic considerations over her daughter's life? What a terrible mother she was, exposing her to such risks! She shared that she was originally from central Israel, and a year ago, she and her husband decided to move to the border area for economic reasons and because the father wanted to return to the place where he grew up. She then criticized her new workplace in the south as being "unprofessional" and added that she did not want to return there. I also learned that she had not been going outside the hotel due to fear of missiles and terror-ists, except for the necessary trip in a cab to see me the previous day.

Judith's guilt and self-criticism showed her confusion between the aggressor and the victim. It appeared she had lost the depressive distinction between inside and outside, between an internal object and an external reality that was beyond her control. This made her feel she could have prevented from her daughter the hours of fear and the evacuation under fire. Judith's internal lack of separation between inside and outside was also evident in her difficulty going out of the hotel although it was safe in Tel Aviv and there was enough time to go into shelter if needed.

The aggression outside and the aggression awakened in her internal world, due to the suffering she and her children experienced, increased her anxiety. She was overwhelmed with feelings that her daughter was now devastated beyond repair, resulting in persecutory guilt that threatened all her vitality. There was an insufficient distinction between the damaged and ruined internal object within her and her real daughter. Mia was not presenting any serious symptoms, but instead was frightened by the events of the Black Sabbath and by the transformation she sensed in her mother. She was not damaged but needed reassurance and time to heal. The internal persecutory guilt threatening Judith was projected outward as anger and blame towards her husband for hesitating during the evacuation and towards her workplace in the form of judgment and a desire to "cut" all ties with the south.

The work with Judith involved strengthening the distinctions between internal and external reality, both in regards to herself and as it was related to Mia. At the end of our conversation, Judith also agreed to come with me for a walk outside the hotel to face the anxiety of going out to the street.

A week later, I saw both of them in the playroom. The mother was at one end of the playroom with her younger son. Mia and her friend from the kibbutz came to play where I was sitting, at the other end of the hall. When her younger brother arrived, a bit later, she quickly reacted with impatience, and she did not want him to disturb them. I thought that her ability to part from her mother was a good sign. Her open expression of aggression towards her brother might indicate an ability to feel less guilt about her aggression and less fear of destruction. After her mother was able to reestablish the necessary separation from Mia, the child could return to a more separate existence.

The following week, Judith told me that they were about to join other kibbutz members, most of whom were staying at a hotel in the northern part of the country. The weakening of the mother's persecutory guilt allowed her to remain in a state of uncertainty about the future, to reconnect with the kibbutz's community, and thus also to alleviate her children's pain over the many loses they have experienced following that traumatic day, including the loss of their familiar social environment.

Summary

This chapter describes brief psychoanalytic interventions with higher functioning families, when war and evacuation challenged the internal world of both parent and child. Given the immediate need of men to enlist in the army, many mothers were left to care for their children on their own. This created an extra burden on these women's inner world, and consequently placed an additional weight on their children's psyches. The goal of the interventions was to support the mothers and strengthen their function as mediators of the harsh reality to the children by being a containing object/ home for their pain.

Brief interventions in times of emergency emphasize humane and emotional connection to the person in distress. Offering a calming presence and reinforcing their sense of control, alongside psychoeducation about what are the natural reactions to such horrific situations, diminish their anxiety and reduce their guilt. The interventions described in this chapter were supportive and based on the positive transference created within such a compassionate and psychoeducational relationship. They allowed the processing of grief and the restoration of the splits and distinctions necessary for normal mental functioning.

As I hope I have shown in the chapter, what makes the psychoanalytic intervention unique in these catastrophic situations is the fundamental understanding that the external trauma always meets the internal unconscious reality and that these are different in each child and parent. The state of the internal world can either contribute to or interfere with the ability to reestablish mental functioning after the trauma and to receive support.

In extreme external situations, it seems that there is sometimes a regression to an earlier state of mental functioning where the child's mental space is particularly dependent on the mental space of the parent. In this state, the children's psychological skin envelope, related to the bodily ego and their ability to hold themselves and their emotions, depends on the ongoing mental holding of the mother, which is also internalized (Bick, 1968). In dissociation following a traumatic event, memories get disconnected from the person's awareness. However, the children showed a generalized regression of functioning due to the difficulty processing what was happening and were not necessarily dissociated from it. In those situations, the child becomes more dependent on the parent for processing reality.

In the cases presented above, two variables emerged as axes requiring the analyst's attention during the intervention:

(A)   Unconscious anxieties and phantasies, that is to say paranoid-schizoid anxieties, depressive anxieties, and anxieties of being.

(B)   The level of regression in mental functioning. This could manifest itself in two ways:

    1   A movement from a place of separateness to symbiosis and ambiguity with no clear distinctions between good and bad, inside and outside; a collapse of splits.

    2   A movement from a playful tri-dimensional space and ability to express emotions and pain to a two-dimensionality as a way of self-holding.

The interventions discussed above suggest working both with the child and the parent in order to help children.

References

Bachelard, G. (2020 [1994]). *The poetics of space.* Beacon Press.

Bick, E. (1968). The experience of skin in early object relations. *International Journal of Psychoanalysis*, 49, 484–486.

Bion, W. R. (1962). *Learning from experience.* Rowman & Littlefield.

Bleger, J. (2013 [1967]). *Symbiosis and ambiguity.* Routledge.

Durban, J. (2017). Home, homelessness and nowhere-ness in early infancy. *Journal of Child Psychotherapy*, 43–42, 175–191.

Durban, J. (2019). Making a person: Clinical considerations regarding the interpretation of anxieties in the analyses of children on the autisto-psychotic spectrum. *International Journal of Psychoanalysis*, 100(5), 921–939.

Durban, J. (2021). Osmotic/diffuse anxieties, isolation and containment in times of plague. Unpublished paper.

Ferenczi, S. (1949). Confusion of the tongues between the adults and the child – (The language of tenderness and passion). *International Journal of Psychoanalysis*, 30, 225–230.

Freud, A., & Burlingham, D. (1943). *War and child.* New York: International Universities Press.

Freud, S. (1919). Introduction to psycho-analysis and the war neuroses. In J. Strachey (Ed., Trans.), *The standard edition of the complete psychological works of Sigmund Freud*, vol. 17 (pp. 205–216). Hogarth Press.

Freud, S. (1920). Beyond the pleasure principle. In J. Strachey (Ed., Trans.), *The standard edition of the complete psychological works of Sigmund Freud*, vol. 14 (pp. 69–102). Hogarth Press.

Furman, E. (1974). *A child's parent dies.* Yale University Press.

Gampel, Y. (1988). Facing war murder, torture and death in latency. *Psychoanalytic Review*, 75 (4): 499–509.

Isaacs, S. (Ed.). (1941). *The Cambridge evacuation survey: A wartime study in social welfare and education.* Methuen.

Joseph, B. (1948). A technical problem in the treatment of the infant patient. *International Journal of Psychoanalysis*, 29, 1–2.

Klein, M. (1940). Mourning and its relation to manic-depressive states. In M. Klein, *The writings of Melanie Klein, vol. 1: love, guilt and reparation*(pp. 344–369). Hogarth Press.

Klein, M. (1944). The effect of war on children. Translation of radio talk given on October 21. Melanie Klein archives, PP/KLE/C56.

Klein, M. (1952). Notes on the problem of evacuation. Melanie Klein archives, PP/KLE/C 94.

Klein, M. (1946). Notes on some schizoid mechanisms. *International Journal of Psychoanalysis*, 27(3–4), 99–110.

Klein, M. (1961). *Narrative of a child analysis: The conduct of the psycho-analysis of children as seen in the treatment of a ten year old boy.* The International Psychoanalytical Library.

Meltzer D. (1975). Adhesive identification. *Contemporary Psychoanalysis*, 11, 289–310.

Meltzer, D. & Williams, M. H. (1988). *The apprehension of beauty: The role of aesthetic conflict in development, violence and art.* Karnac.

Ornstein, A. (2012). Children in war zones, in genocides and living with the threat of terror. *International Journal of Psychoanalytic Studies*, 9(3), 273–279.

Rosenfeld, H. (1950). Note on the psychopathology of confusional states in chronic schizophrenias. *The International Journal of Psycho-Analysis*, 31, 132–137.

Tustin, F. (1990). *The protective shell in children and adults*. Karnac.

Winnicott, D. W. (1984a [1939]). A letter to the British Medical Journal (16 December 1939). In D. W. Winnicott, *Deprivation and delinquency*. Routledge.

Winnicott, D. W. (1984b [1940]). Children in the war. In D. W. Winnicott, *Deprivation and delinquency*. Routledge.

Winnicott, D. W. (1984c [1962]). Ego integration in child development. In D. W. Winnicott, *The maturational processes and the facilitating environment: Studies in the theory of emotional development*. Routledge.

Winnicott, D. W. (1984d [1963]). The mentally ill in your caseload (1963). In D. W. Winnicott, *The maturational processes and the facilitating environment: Studies in the theory of emotional development*. Routledge.

Chapter 4

Mind in the Line of Fire

Mothers during the War

Kateryna Abashkina, Kateryna Alpatova, Tetiana Stasiuk, Anastasya Svinarchuk and Emanuela Quagliana

Introduction

Four women. Four mothers. Four psychoanalytic psychotherapists united by the immense challenge of navigating motherhood in the midst of war.

I was asked to accompany them at this time in their lives and had the privilege of meeting with them once a week for almost three years. The original group was larger, but four of them contributed to the writing of this chapter.

Here are four unique yet deeply connected stories of bringing a long-awaited baby into the world and meeting its needs while dealing with a terrifyingly dangerous outside reality. These are four extraordinarily honest, heartfelt accounts of the emotionally overwhelming experience of motherhood in wartime.

For one of them, the war didn't start with Russia's full-scale invasion of Ukraine in February 2024 – it started much earlier. For Tetiana Stasiuk, the war began in the summer of 2014, when the armed forces of the Russian Federation invaded eastern Ukraine. This forced her to seek safety within her own country, moving with her two children. In February 2024, the conflict uprooted her again. This time she fled abroad with her children, into even greater uncertainty, holding her third son – a six-month-old baby – in her arms.

The war hit Kateryna Abashkina, who lived in southern Ukraine, when she was eight months pregnant. At the time, travelling thousands of kilometers to safety seemed as risky as staying in her town. Giving birth and spending six months with her baby in a frontline city, where enemy rockets destroyed civilian infrastructure and residential buildings, ultimately led to a difficult decision: to emigrate in search of physical safety and a chance to bond emotionally with her child.

For Kateryna Alpatova, the journey to motherhood had been a hard-fought battle. After years of struggle, her long-awaited pregnancy finally arrived. But the horrors of war robbed her of the ability to fully embrace this time of joy, or to emotionally prepare for her baby's arrival. At 29 weeks, she went into premature labor. Her son was born under constant medical supervision. Although the couple's resilience and faith in life helped their baby grow stronger, it came at an immense emotional cost against the backdrop of war.

DOI: 10.4324/9781003520757-5

Anastasia Svynarchuk also faced a long, emotionally and physically demanding journey to pregnancy. Two days after she felt her son's first kick in her belly, her world was shattered by explosions and the outbreak of war. An overwhelming desire to protect the life growing inside her drove Anastasia to leave the country for several months. Later, with a fierce determination to give birth in her hometown, she returned home to give birth to her child in the embrace of her beloved husband – the person with whom she had long dreamed of starting a family.

I often wondered if and how I could help. I could have used psychoanalytic theories to describe this navigation of a route, where you see on the screen the path to follow to reach the destination. But in this navigation there were only waypoints to be dealt with moment by moment, points where I always ended up feeling: here I am; miles away from the all-consuming tragedy that is war. I don't know what I can do or say to help these young colleagues, these young mothers.

I listened. I listened to the love that they pour into their words as they speak to their wonderful children, whose safety and tranquility lies within their arms. I tuned in to the lullabies sung to the beat of explosions and think of the poet Alda Merini's lines: "There's a place in the world where the heart beats, where you remain breathless for all the emotions you feel ... That place is your arms where the heart doesn't grow old, while the mind never stops dreaming." And may I add, where missiles don't whistle through the sky and buildings don't crumble to the sinister force of destruction.

Come night, I listened to the fearful thoughts that have lucrative roles in the movies that are their nightmares: To the guilt of those who stayed, for having robbed their children of a tranquil childhood; To the guilt of those who left, for having abandoned the others to suffer; and after all that, I listened to my own: I was here, in Italy, distant from the threat of losing all what I love and all that defined me.

I listened to the angst given by the all-too-evident signs of what is irreversible (perhaps as much reversible as not), and again to the trauma as much as to the bravery of fighting to be re-born every day: "Being born is not enough. We are born only to be born again, every day," as Pablo Neruda writes.

Once I was done listening, I thought, of what I would have done, what and how I would have felt. Thanks to the thread that connected us I felt the value of this discourse, and pure gratitude for all that I was given, trying each time to be of some use.

I was here, but I was also there.

Emanuela Quagliata

Formation of the Group and Difficulties in Writing

We all felt the need to find a circle of young mothers who could share their feelings and experiences of caring for young children in the midst of war. When a colleague suggested forming a group, it was a breath of fresh air for

all of us. Having a group of mothers who are also psychoanalytic psychotherapists was an incredible stroke of luck. Our group has been through a lot together.
It became closer and more intimate as difficulties arose and we tried to resolve
them. But at the beginning there was a sense of fresh air. We were not alone – we
felt contained by knowing there were other confused mothers with children in
their arms and a million questions. It helped us to gradually come to terms with
the overwhelming reality that had invaded our lives.

ANASTASIA SVYNARCHUK: The idea of creating a group, suggested by a
   colleague, was so unexpected and spontaneous for me. I did not realize
   that the group would eventually become a space where I could contain
   my experiences of motherhood and the war, which had become a strange
   daily reality for me. In this unity I found a place where my symbolic
   capacity was revived and where my child and I could develop. The
   difficulties of adapting to my new role as a mother were complicated
   by the fact that I had to develop my ability to adapt under conditions
   of chronic stress caused by the threats and dangers of the war.

TETIANA STASIUK: At the beginning of the war we left the country with our
   children. In conditions of total uncertainty about the future and in a new,
   unfamiliar environment, I acutely felt the lack of "my own" – people with
   similar experiences with whom I could share my feelings. Therefore, the
   group became an essential space for me to share my feelings and to give
   and receive mutual support during these incredibly challenging times.
   Communicating with other mothers gave me the understanding I so desperately needed. We were all experiencing similar fears, anxieties and
   hopes, and this created a sense of community that helped us cope with
   the emotional burden. We shared our stories, gave each other advice on
   how to support our children and tried to be a support for each other
   during those dark times.

   When we received the offer to write about our experiences in a book,
   we all felt it was a challenge. We were confronted once again with the
   pain, the memories of the beginning of the war and our confusion. The
   need to work as a team and to find words and images was not a trivial
   task in our circumstances. So, this proposal raised many questions, both
   about the work of the group and about our current ability to contain
   and symbolize our experiences. We felt that after two years of full-scale
   Russian invasion of Ukraine, we could do more than we could at the
   beginning, and we dared to try.

KATERYNA ALPATOVA: For me personally, trying to write about my experience
   immediately confronts me with the problem of finding the words to convey
   this difficult experience. When speaking, I can see my interlocutor, allow
   myself pauses and intonations, and hope for a non-verbal transmission of
   emotions. Writing a chapter for a book, however, is an extremely challenging task. I return to the question raised in one of our groups – how do I

convey pain in writing when there are no words? Should I leave a blank page? Hope that the reader will see in it exactly what we want to convey?

Almost three years have passed since the full-scale invasion began, and my son is now two and a half. We have been through a lot with the group. I had the desire (and perhaps a little strength) to try to find words and symbolic forms to express my thoughts and feelings. The first thing I would like to say to the reader is that I sincerely hope that the experience of living through the war will touch you only in empathetic sympathy with the authors of our section. For me, who is currently in this situation, it sometimes happens that I can distance myself so far from the external and internal reality that I can fantasize about peaceful times. Sometimes it feels as if I can physically touch the separation I am resorting to in order to preserve my own psyche. My experience of being in a group of mothers who gave birth during the war is also prone to splitting and fragmentation as well as painful integration. The challenges of mother-hood are superimposed on the difficult conditions of the surrounding reality. I doubt that this experience can be fully lived without resorting to strong defenses that keep us afloat. Group work is a broad field where everyone expresses their unique position and moves the others. I hope that the form we have found for writing this chapter, in the form of a discussion, will allow us to share our experience more fully and give the reader space to explore it.

ANASTASIA SVYNARCHUK: In the group we all experienced the trauma of war in such a similar way, and at the same time there was an understanding that there was no single response to everything that was happening. We all had our own inner worlds and attachment styles. The difficulties of writing this chapter brought the participants back into contact with heavy and unbearable feelings that we unconsciously wanted to avoid.

Recognizing my trauma and its impact on my child felt like an irreparable loss and created a sense of inner disorientation. During this period of the group's work, aggression arose, possibly as a result of identification with the traumatic event and replaced thought and understanding with a sequence of specific events that we could not process. In my view, this created a threat within the group and undermined the ability to think symbolically and reflect on what was being experienced in the group.

Writing the chapter stimulated reflection on collective and individual experiences. Then we experienced personal breakdowns and disruptions in interpersonal relationships. We felt that good was being turned into bad, which we experienced as catastrophic. We temporarily lost our basic trust and hope that the good object, the group, would endure and prevail. The group experienced the destructive consequences of trauma by losing our mental capacity to think rather than just react during this difficult period, and the symbolic functioning of the group was disrupted. It was as if there had been a regression to a state where words had no meaning.

The emotional states of the group, described by Bion as "beta elements," took a destructive form and provoked hostility among the participants. It seemed that such hostility, in an exciting and numbing way, became a shelter that protected us from the suffering of the primary trauma of war. However, the supportive function of the group was rediscovered by participants who experienced hope in what we had created. This hope and love gave the group the strength to carry on.

Experiences of Pregnancy, Childbirth, and Childcare

ANASTASIA SVYNARCHUK: The desire to have a child has been a long journey of waiting. After treatments, hopes and despair, my long-awaited pregnancy began in November 2021. The feeling of happiness was indescribable. I dreamed and fantasized about becoming a mother and awaited the birth of my baby. But two days after I felt my son's first kick, I was confronted with the new reality of explosions and the start of war. At that moment it felt as if there was no distinction between what was actually happening and my fear and anxiety about what might happen.

I could not fully comprehend the danger I was facing on the doorstep of my country. I thought I could withstand all the challenges that came with the birth of new life, but I quickly began to understand that the external threat would become an internal threat within me, potentially leading to loss. We fought so hard for this birth that I had no doubt about the need to protect myself and my unborn child.

KATERYNA ALPATOVA: My experience of waiting for pregnancy is similar to Anastasia's – I went through a lot to make my dreams of pregnancy come true. Unfortunately, the opportunity to enjoy pregnancy was stolen from me. The invasion on 24 February 2022 caught me at the beginning of my second trimester. I had just begun to settle down after a stressful first trimester, and I had begun to allow myself to experience being a pregnant woman and to plan accordingly.

I was afraid to go on a long trip because of medical advice and the fear of being without my husband, who could not leave the country. I stayed in central Ukraine, where I could maintain a sense of attachment to my home, loved ones and familiar surroundings. There is no third trimester to remember, as I went into premature labor at twenty-nine weeks. This cut short my ability to experience myself as a pregnant woman with enough time and space for feelings and reverie about myself and my future child. This is what makes me acutely sad – the loss of carefree walks and fantasies about my future child. Instead, I was faced with a terrible choice – to cut off a part of myself and my war-related fears, or to fill our shared space with images of the external dangerous reality.

I felt insane when I thought only of the baby and distanced myself from the war, just as I felt mad when I immersed myself in the reality of the war and left my baby out of my thoughts.

KATERYNA ABASHKINA: It is very traumatic to try to reconcile my fantasies and ideas about the birth of a desired child with the reality of war. I will need a considerable amount of therapeutic time to digest and mourn the loss of hopes and expectations about the desired conditions and circumstances of my daughter's birth. There is a need to mourn the person who I was with all my plans and dreams before the war. As Caroline Garland (1998) aptly put it in "Thinking About Trauma," some of that mourning, as has often been pointed out, must be for oneself – for his own lost world, for his own pre-trauma life and identity.

The war caught me at the beginning of my eighth month of pregnancy. I refused to even think about giving birth in a cellar or a bomb shelter. Photographs of the first women giving birth in underground and metro shelters horrified me. I could not believe in this reality or that I would have to face it directly. Unlike my previous pregnancy, this one was very favorable, both physically and emotionally, and I was happy.

I dreamed that my child would be born in my hometown, accompanied by a doctor I trusted and in partnership with my husband. I imagined that we would return home with our daughter to our son, and that we would continue our lives as usual, gradually getting used to our new family member. I felt that all the conditions and prerequisites were in place to give this child a basic sense of security and safety in the world I would be bringing her into.

A woman enters the maternity ward with one identity and leaves with another. What happens there, I think, is of great significance. The maternity ward is ideally designed to contain the anxieties associated with "creating a new form of object relationship – motherhood, which can only begin when the child separates from the maternal body and enters the object world" (Pines 2010). But in the harsh reality of war, with constant sirens, and being in a bomb shelter with a newborn in my arms, this space was transformed into something damaged, further increasing my regression and fears.

I was forced to arrive at the maternity ward early and stay in the bomb shelter waiting for labor with other women who were as frightened as I was. I was lucky. Our first meeting with my daughter was in the delivery room, not the bomb shelter, and we managed to enjoy the silence in each other's strong embrace for the first hour and a half. But soon the chaos began and the siren went off. Shortly afterwards we heard that two cruise missiles had fallen near the maternity hospital. That day, my daughter and I spent time in the corridor and in the bomb shelter. Tears streamed uncontrollably down my face. Inside me and around me echoed the question, "When will this nightmare end?"

TETIANA STASIUK: In the first days of the war, when I was full of despair and fear, my six-month-old son did not seem to react at all to the events around him – he remained as cheerful and lively as ever. One moment stands out in my memory. Almost at midnight, a powerful bombardment began and everything around us exploded loudly. We took the children to the stairwell next to the lift and just stood there, trembling from the sound of the deadly shells. I held the baby in my arms and tried to talk to him, and he babbled back, looked into my eyes and smiled. At that moment, a wild fear for my children's lives overwhelmed me, and sweat ran down my back. At the same time, I was relieved that my baby was fine and did not seem frightened at all.

Almost a year passed after that event – a year we spent in a safe place, in a country without war. One evening, just before we went to bed, the neighbors started setting off fireworks. Suddenly my son started screaming, grabbed his clothes and pulled me down the stairs. It struck me that he seemed to be experiencing the same thing I had a year ago when I held him to my heart to the sound of cannon fire. This happened several times, and it seemed that he needed to be in similar conditions over and over again to work through the traumatic event. Did he really remember the terrible explosions, or did he perhaps sense my condition at that moment?

This also raised the question: how much is this related to the sleep difficulties we were experiencing? Was my reluctance to put him in a separate bed, despite the discomfort of co-sleeping, a factor? When he slept, my son would cling to my neck as if to prevent me from falling if I suddenly had to run somewhere. I also felt much calmer having him close. In this way I felt I could protect him better, shielding him with my body in case of danger.

The experience of war and the many losses we have suffered have complicated the natural process of my child's separation, even though we were physically safe outside our warring country. This situation makes me reflect on how deeply the war has impacted our souls and how important it is to find a way to restore our emotional connection in order to move forward.

Motherhood during the War

We are going through a difficult time of war, which has become an inseparable part of our experience of motherhood. This period brought many challenges. We faced feelings of helplessness and fear as danger loomed around us. We are filled with loss and pain as we are forced to mourn that we cannot provide the support our children deserve in these difficult times.

ANASTASIA SVYNARCHUK: At the moment I was experiencing the trauma of war, I was going through the process of forming a bond with my child. However, it seemed that the identity of my maternal role as a protective shield for my child was weakened. I often longed for a protective embrace for my child, our vulnerability, my helplessness and the horror of the events around me. I would read or watch the news about people's tragedies and imagine, with desperate horror, that my child unconsciously felt this in my arms, where I longed to protect him. My dreams of fleeing from missiles flying overhead, or the desire to hide my child "on me," gave me insight into the power of fear and the sense of vulnerability I was experiencing. At the same time, I was trying to live a life where there was room for joy and happy moments with my loved ones and my little son.

KATERYNA ALPATOVA: The feeling that I had to care for and protect my child in a situation of physical danger almost shattered my soul. I was outraged that I could not give my child the necessary maternal sensitivity and be in tune with him. Feeling vulnerable and able to sense my child's emotions was at odds with my need to rely on coarser psychological defenses to help me survive in a state of chronic stress. It seems to me that the experience of motherhood in war is full of losses (both actual and psychological) and requires a constant process of mourning.

KATERYNA ABASHKINA: Before the war started, I lived comfortably in my country and had a private psychotherapy practice. For years, I worked with children and adults in a cozy office next to my dear colleagues, and I gradually discovered the psychoanalytic universe. I continued my studies with interest and planned my professional development.

Together with my beloved husband, we were raising our son, had bought a house and were planning to renovate it beautifully in preparation for the arrival of our daughter. I never planned to emigrate when I was eight months pregnant. I could not even imagine in my worst nightmares that I would give birth under the explosions of enemy rockets and that our first days with our daughter would take place in a bomb shelter. I could not imagine that my fear of death would become so real that it would manifest itself in a desire to hold my daughter constantly, even when there was no need or threat. The ruthless and cruel war in the frontline town where I lived was getting too close. With almost no alternative, we made the difficult decision to move to another country.

Now I am a forced migrant, a refugee. What is left of my former life? What have I lost forever? What can I do? Who am I here? What's next? These and many other questions swirl around in my mind. I often feel emotionally overwhelmed, anxious and exhausted. It is well known that constant contact with the immature psyche of an infant captivates the psyche of an adult and causes regression through introjection and projection. This process reactivates personal internal parent-child conflicts, making it difficult to establish contact and interaction with the child. But in the context of additional traumatic circumstances, such as war and forced migration, it seems incredibly difficult to bear the burden on the psyche. Sometimes it is barely possible.

TETIANA STASIUK: Motherhood has always been a great challenge for me, and in the context of war this experience has become even more complicated. War not only destroys physical spaces, but also deeply affects my psycho-emotional state. I was forced to leave my home during the conflict and this experience became a real test of my identity and my ability to care for my children. Under intense stress, I found myself emotionally unavailable to my children. My internal container was overflowing with despair and fear, and at that moment it seemed there was no room for the fears of a small child. I was tormented by the worry of how my internal state would affect my child. I was afraid that he would begin to feel my fear, helplessness and despair and that this would harm his development.

To Leave or to Stay: Support and its Absence (during Pregnancy and Motherhood)

We found ourselves in a situation that changed our lives forever, and each of us experienced this in our own way. The war took us by surprise, forcing us to leave our homes and loved ones. Each of us sought a way forward in this new reality. Some tried to find a safe place for their families, while others struggled with internal conflicts, trying to remain true to themselves and their values.

ANASTASIA SVYNARCHUK: I travelled to another country, not knowing for how long. I left my husband and grown-up son and felt an unbearable desire to return home and embrace them. I spent two months there, recovering and wondering what to do next. I dreamed of giving birth at my husband's side and returning with the baby to our home that was waiting for us. I was afraid of returning to a country at war, but I was even more afraid of giving birth in a foreign country without the tender and warm embrace of my husband and my son. I gave birth to my son, reassured by the physical presence of my husband at my side, and then descended into the basement eight hours after the caesarean. I reminded myself of the need to be strong for my baby, who was met with hostility from the world. I nursed him in the basement when he started to cry, and I cried inside, watching mothers who were also trying to soothe themselves and their newborns. I was overwhelmed with anger at my own helplessness and the need to protect myself and my tiny child. I prayed and believed that I would return home to introduce my son to his new home. In my dreams, our home would become a fortress for him, where he could feel safe and grow up surrounded by loving people.

TETIANA STASIUK: The war took us by surprise, even though I was already familiar with the first powerful explosions I heard at five in the morning on 24 February 2024. My heart was pounding, and tears were streaming down my face. I felt a sticky, hot déjà vu from the summer of 2014 in Donetsk. For a moment I was immersed in memories of a wonderful family and peaceful life – my husband, children, friends, our beloved summer cottage for weekends. Then, with the Russian invasion, I lost everything at once. I could not believe that the situation was repeating itself. My heart cried out in denial, but reality broke into my consciousness with terrible explosions from all sides.

I had fled Donetsk under fire with two children, and now I had a newborn with winter outside. For the first few days I was in a state of utter shock and helplessness, with an almost animal fear for my children's lives. Somewhere around the fifth day, after another terrifying night in a cold basement with a screaming baby, I realized that I had to secure a safe place for all of us. If something happened to me, who would look after them?

My emotions ranged from helplessness to inner strength and determination. The fear of danger on the road for my baby was extremely heavy for me, but staying was much more dangerous. We were very lucky to reach the station, board a train to Lviv and finally leave the country. The arduous journey took more than four days. Due to all the circumstances and the intense stress, we fell seriously ill, especially the little one. The day after we arrived, my child began to have difficulty breathing at night and his body was covered with terrible eczema. But the most important thing was that we were finally safe!

KATERYNA ALPATOVA: My personal decision to stay in Ukraine is largely based on the feeling that my child and I needed a supportive environment. I had the opportunity to stay in a safe place, close to my family, and I was able to stay in my country, where I feel at home. I see my son developing, having relationships not only with his mother and father, but also with his grandmothers and grandfathers. He goes to nursery school where he communicates in his mother tongue. This setting is a bulb that envelops him and allows him to grow. This supportive containment also nourishes my state of mind as a mother.

It stems from the beginning of my motherhood, when I became acutely aware of the importance of the environment. I had an emergency caesarean section, and my son was in the incubator. I felt a heavy sense of guilt and a sense that I had failed. I remember my first glimpses of my child through that heavy state. And I remember my surprise when I began to hear my husband's words of gratitude for our son and the words of support from family members who were able to come and see my son in the incubator. Their words brought me back to the joy of meeting my child for whom I had fought for so long. It was this support – family nearby, friends who could come, my analyst and the medical staff at the perinatal center – that helped me to overcome my depressive feelings. I shudder to think what would have happened if my eyes had continued to see my child through my problems, burdening him and hindering his development. How different my experience of motherhood and my child would have been if my guilt had not been pushed into the background by the support of those around me.

ANASTASIA SVYNARCHUK: There was a moment at the beginning of the war when I had no support from an important person in my life. I was filled with pain, despair, resentment and disappointment. I immediately wanted to destroy this unreliable connection, which I found so hostile and distant. During this period of inner suffering and reflection, I found support from people I did not expect to feel so much care and acceptance from. Sometimes these were complete strangers to me who seemed to be saving and supporting me. I then opened up to so much faith in my ability to go on living without destroying anything, and felt able to accept what I received.

Today

KATERYNA ALPATOVA: Today it is almost 3 years since the war started. Unfortunately, it has become a part of life. We live with it, go about our daily business and look after our children. I don't know how the war will affect my son, but I feel that I am slowly beginning to accept how the war has changed my life – not only the future, but also my past. My memories of the house where I spent my childhood are marred by a Russian missile. The war is leading to a re-evaluation of my life. I have a lot of psychic work to do to mourn my life, my motherhood and the life of my son.

Since the Second World War, there has been a category for a vulnerable population in Ukraine known as "children of war." These were children born during the war who carry this status throughout their lives and receive state support. I hope that this burden will not be too heavy for my child and that I will have enough strength to help him see the world not only through the lens of war.

ANASTASIA SVYNARCHUK: In the lives of ordinary people, events occur that make each of us wonder: how does a person survive after such a trauma? What happens in the psyche to accept the consequences and the knowledge of what has happened? How can such a massive loss be understood? How does a person recover, and what does recovery mean for each individual? Understanding trauma requires immense psychological work, and the main part of this work is mourning a series of losses – literal losses as well as the loss of a dreamed life and lost circumstances and opportunities. I understand that motherhood during war is a challenge and a limitation, but at the same time I have a lot of hope and faith in a better future for myself and my family.

It has been more than two years since the war began and much has changed in that time; we are raising a wonderful little son who is full of strength, energy and resilience. We enjoy watching him grow and invest a lot of effort in creating a space for him to live and explore his world. We are watching him find new meaning in life and new understandings of life.

TETIANA STASIUK: There have been 904 days of war in Ukraine. During this time the country has suffered countless terrible destructions and hundreds of thousands of its sons and daughters, including innocent children, have been killed by Russian aggressors. How many more will it take? This thought fills me with fear and hopelessness. But life goes on.

The other day I was reading a bedtime story to my son. It was a new book. The illustration for the Ukrainian folk tale "Ivasik-Telesyk" showed a multi-headed, fearsome dragon trying to reach the boy Ivasik, who was hiding in a tree. The dragon had almost gnawed through the trunk and was about to reach the boy. When Sasha saw this, his eyes sparkled with determination. He immediately held out his hand, shouted "Sasha is a tank!" and began to imitate the sound of gunfire. In his childlike simplicity, he was trying to save Ivasik from destruction.

This reaction touched me deeply. I had never seen my son, who has no toy guns, react emotionally to a situation like that. He simply adores technology – various machines, buses, excavators, tractors, combines. His favorite game is to imitate an excavator. He hides one hand behind his back and with the other imitates the movements of the bucket, making characteristic noises as if he were "digging" for something important. The digger is exploring new depths, building something new – perhaps new meanings? The tank protects those closest to it from intruders. These two figures that appear in his imagination can coexist and manifest depending on the circumstances.

I also feel that alongside the all-encompassing, unspeakable emptiness that exists within me as a result of the trauma of war, new shoots of hope and strength are emerging to continue fighting for the future. I try to find light in the simple moments of joy that a child's imagination provides. And this light, though fragile, becomes my support, my inspiration.

KATERYNA ABASHKINA: Our little daughter is now 2 years and 8 months old. How quickly time flies. It seems like only yesterday that I was thinking about when to introduce solid food, how to get her to sleep at night, or how to wean her from breastfeeding. Now she makes me an omelet for breakfast, chooses her own clothes for a walk, dances and sings and says she doesn't want to sleep but wants to go out with her friend. She has so much tenderness, love, energy and curiosity. It is a real miracle. And it is a true joy to watch my daughter grow, to learn new skills almost every day, and to enjoy our bond.

Much has been done to gradually alleviate the strong fears and anxieties from the beginning of the war and the early stages of our interaction. I was very worried that my daughter would be suspicious of this world and people, as she was at the beginning of her acquaintance with the outside world. But now I see her interest in the things around her, her desire to communicate and her persistent attempts to connect with other children and adults. This is incredibly inspiring and gives me hope and confidence in the restorative power of the psyche.

Now I have no women nearby to share the physical care of the children with, and this is sometimes very difficult. Fortunately, I have the opportunity to meet online twice a week with women, my colleagues, who also gave birth during the war. This is a treasure that cannot be overestimated. Communicating with other women who, like me, are currently facing the challenges of motherhood during war is incredibly valuable and helpful.

We are all different, but what unites us is a thirst for life, resilience and love. This is a space where our children are full participants in our group and recognize each other every time they appear on the screen. As it turns out, our children made online friends before they learned to use the potty.

We don't always understand each other, and we don't know how long this group will last. Nevertheless, it is such a feminine space that has a containing function and allows us to transform our unbearable feelings. We share the joys of motherhood and the successes and struggles of our children's development. Our older, experienced facilitator, Emanuela, shared her subtle and deep insights "from the outside." At times she shared her reflections through poems, lullabies, films or papers and we felt our creativity and symbolic thinking restored.

References

Garland, Caroline. 1998. Thinking about Trauma. In Caroline Garland (Ed.), *Understanding Trauma: A Psychoanalytical Approach*. New York: Routledge.

Pines, Dinora. 2010. *A Woman's Unconscious Use of Her Body: A Psychoanalytical Perspective*. New York: Routledge.

Chapter 5

Children in a World at War

Mónica Cardenal

Introduction

A child in Ukraine represents all the children in the world. This thought reverberates within me, evoking the Middle East, Africa or Central America, just to mention some of the regions of the planet where children and adolescents suffer extreme violence in different forms. They experience both openly declared wars and less visible ones.

I would therefore like to invite you to imagine the emotional world of a child who might have lost everything in the fire of violence. We can also ask ourselves what our role as analysts is in the face of such an emotional catastrophe, where there is much to do. At the same time, the task is delicate and complex. It is not only about wanting to help at any cost.

When cruelty emerges in a massive way, it is very difficult for us not to succumb to its impact and to continue thinking in a way that keeps us emotionally close to those who are suffering, especially children. In this context, rescuing individualities is crucial, a task which is inherent to psychoanalysis. I therefore remember Melanie Klein (1945) and Richard – their evacuation out of London, the bombings, the improvised consultation room. All the fantasies that Richard displayed in the relationship with Klein. Both patient and analyst, are bravely building a relationship in the midst of a line of fire. Today there are many brave people too. Analysts continue to work in the worst circumstances with our method.

Even though the war in Ukraine is geographically far from me, I am in close conversation with colleagues from that country and from countries assisting refugees. I am close to the discomfort and to the urgent need for assistance for Ukrainian children. Some of these children are displaced on their own; some are orphans. These are children who have seen the worst and have lost everything.

I try to understand what the emotional world of these children, and what our task is. I do not want to become attached to any theoretical idea on trauma. I believe that might prove to be a bit sterile. I prefer to think about the children closely, and focus on the capacities of the mind to be able to

DOI: 10.4324/9781003520757-6

symbolize, as the only way to elaborate violence internally. When I write, I am inspired by my experience with refugee children and adolescents in Latin America in the last few years, through an organization in Mexico. They are children who flee from wars in their own countries in Central America. They are impacted by drugs, gangs, and extreme violence. Some are abducted, indoctrinated, and forced to become child soldiers. Most of the time these children flee alone. Sometimes they leave with their mothers, who might die on the way. Second-skin phenomena emerge as Esther Bick (1968) described, when the mind remains without an internal container object capable of sustaining and unifying the Self. In the presence of catastrophic events such container objects can be destroyed or fragmented, leaving the Self without integration and support. This is experienced concretely at body level; somatic symptoms, false independence, reckless and impulsive conducts predominate. The Self defends itself by fantasizing that it does not need anybody on whom to depend lovingly. We might wonder in which ways traumatic events suffered throughout these children's short lives test the capacity of their most mature aspects to sustain new experiences in which pain might be tolerated, in order to grow from them. Experience has shown me that whenever we can intervene psychoanalytically, children and young people have hope of leading a full emotional lives. We now have new tools for assisting in humanitarian crises, such as applied observations with pediatricians, brief emergency interventions and Work Discussion Groups (WDGs) with volunteers or professionals who assist. Some of these interventions are carried out by the IPA Committee dedicated to the international Psychoanalytic Assistance in Crises and Emergencies (PACE). Being close to the children means being able to help them be themselves and help them develop their receptive and introjective capacities of the mind, where a range of emotions – including the most negative ones – can be contained.

I will focus on a photo taken by the Argentine photographer Rodrigo Abd in Bucha, Ukraine, published in *La Nacion* newspaper on April 9, 2022. A 6-year-old boy stands in front of his mother's grave, very near his home. She died of depression, the article says, after spending too much time in a shelter; she refused to eat. The child and his 10-year-old brother started to take food and drink to the grave, which had been improvised close to what had been their home. They also played near it. My feeling was not primarily one of sadness; I thought of the resources that the mind has to carry on. The idea was theirs, since their mother refused to eat at the shelter and they probably wanted to remember her "eating" in order to have her internally.

How can one find within oneself a comforting presence that can be an internal support in the face of the pain caused by such a traumatic loss? Maybe some of the first good experiences with that object can be one of the bases to investigate and work on for the recovery of the mind in situations of extreme violence and loss. As I have described, the possibility of having within oneself a containing object, which has been capable of giving the Self the emotional experience of being lodged within a receptive container during the first years of life, might be the only possibility the mind has to support itself in the face of urgent demand and pain. It may also aid in the child's capacity to accept the help that another provides externally.

If we could be close to the emotional experience of a child, close to their way of constructing the world without imposing our theories, I believe we could be better analysts for them. Child psychoanalysis works to facilitate an emotional experience of learning about oneself, about bonds and the world, even when times are critical due to a pandemic, a war or other catastrophes. I like to think that our task is to help children help themselves, offering them the opportunity to acknowledge their own emotions as they arise, even in the most dramatic circumstances.

This paper attempts to conceptualize these processes displayed within the framework of PACE's work with groups of analysts and psychotherapists assisting the most vulnerable ones: babies, children and young people who have experienced devastating losses in the midst of a humanitarian emergency. I will explore the introjective identification processes through the presentation of brief clinical vignettes of material related to work with children.

Through crises of different types which predominate in the world of today, psychoanalysis has the possibility to help investigate the early mental states which are key for the personality on its way towards maturity, as Matha Harris (1987) argued. Times of intense social destructiveness in different forms force us to develop our knowledge of high levels of very primitive anxieties and their possible evolution, as well as the defenses that the mind sets in motion in situations when cruelty is extreme.

We could ask ourselves: how can terror be transformed into symbols and emotional experiences? My presentation will attempt to transmit the work we have been carrying out, where identifications turn out to be vital mental processes.

Psychoanalytic Task in the Midst of the War: History and Particular Characteristics of the Work Discussion Groups

Through our work at PACE we offer a possible vertex of comprehension about the emotional states and possible defenses that can be displayed in critical situations of intense terror. This is only one possible perspective of work which tries to be close to those who assist at different levels, opening paths for discovering inner landscapes of those who are suffering loss and extreme violence, considering the unconscious effects of these processes on the assistance task itself and on the professionals who provided the assistance, a kind of knowledge that only psychoanalysis can offer. Within the WDG framework the unconscious identification processes are detected and understood, and work is focused on them.

The psychoanalyst Martha Harris had the idea of expanding the reach of the benefits that psychoanalytic thinking offers " beyond the walls of our consulting rooms"(Klauber 1999). This inspiration led her to articulate and apply infant observation to the social and assistance areas, helping different professionals to understand the unconscious processes involved in tasks carried out with small children, adolescents and families. The final aim was to improve the strategies of intervention. Motivated by this, she implemented a seminar at the Tavistock Clinic that was initially called Work Study (Klauber 1999), and later Work Discussion. Its main objective is centered on making a detailed observation of the task performed by the professional, so as to be able to later discuss what occurred between this professional and the person being assisted, as well as about everything that has happened in the work setting. Esther Bick's method of infant observation is instrumental in helping the professional to grasp and appreciate the unconscious emotional factors of the interrelation with the other that is implied in his task. Exposure to her method broadens his understanding about his own mental state and that of the person whom he is trying to help. Through this psychoanalytic work frame, a space of discovery emerges for all those involved in the task (Cardenal 2021).

Following Martha Harris (1987), I suggest that those of us who are in a position of responsibility for the welfare of others must necessarily be engaged in thinking about the conditions that allow individuals to develop and to share the fruits of that development with others, even in the most critical circumstances. From working in groups, an emotional experience emerges.

The Case of Clara: Children in a World in War

A 10-year-old girl has had to emigrate to a European country because of the war in Ukraine, together with her mother and three younger brothers. Her parents are separated and have a very bad relationship. The father has also emigrated, but to another country. The girl behaved very badly towards her mother. She insulted her, hit her, and had even run away from home. The mother said she cannot deal with her any longer. Clara was isolated, without friends. She insisted that she wanted to return to her country in spite of the war. One of the therapists from the group brought this material and commented that she did not know if she would be able to help her. She saw Clara online as she was in Ukraine. In the few sessions they had had, the child played with her computer in spite of being connected by telephone with the analyst. From the point of view of the psychotherapist, the girl seemed uninterested in her help and that attitude irritated her. The analyst saw a good deal of hostility in Clara, who often dropped the telephone by which they communicate.

In the third session with her patient something interesting came up which we were able to think about as a group. The therapist consulted us on how to continue, and whether it was possible to do so. The girl had begun to talk about Halloween, her own dress, and how strange her classmates looked with their colored hair and their "sharp faces." She described them as "strange." She said that they had gone trick-or-treating in English. The psychotherapist observed that Clara was watching cartoons in English, the language of the country she was in, on her phone. They spoke about school. She said she always arrived late to class. She suddenly stood up and cried: "there is nothing to do here," "we are traveling all the time" (they had moved to different countries), "I want to go back to my city." She spoke about a friend who was in Russia and the analyst asked her if this angered her. Clara said it didn't and added: "what does she have to do with any of this?" She shouted even louder and said: "Don't speak to me about this, in 2024 the war will be over." After 50 minutes in the session, the analyst said she would speak to her mother about arranging another session, and asked the girl if she would help her with this. She agreed. In the group, we felt that it was very important that the girl had a stable framework of sessions. This was a way of offering her a safe container.

The first thing we had to understand is the enormous effort this girl had to make to adapt to a new country. She had lost everything: her father, her home, her friends, her extended family. This comprehension also gave us a dimension of the anxieties in a pubescent girl who is in her own process of development and the effects of the war on this. She was physically safe with part of her family, but there were infantile parts of her which were not.

Her use of projective identification and its intrusive effects through her screaming and hatred towards her mother and towards therapy was a prominent defense in the face of her vulnerability and loss. The girl explained it very well. She also explained how much the identification processes, which were key for her growth, were affecting her, alongside war and losses, and the "strangeness" that invaded her at times. Our work in the group involved illuminating the effects of those processes for the analyst so that she did not lose hope with this girl. Instead of a desperate escape having nowhere to go, what better place to be than the inside of the analyst's mind? The analyst could also understand that the girl's infantile parts were interested in learning another language "the way children like to do it," through cartoons. Surely if her infantile aspects could be understood in the other language, the one related to analysis, they would also be appreciated, and the introjective processes which seemed to be in place could be consolidated.

Young Children: A Mind to Contain Terror and Hatred

A psychotherapist working with a group of small children, in a city affected by war described the following dream after a situation with the group which moved her deeply: The first thing she said was that they were very dark days, since there had been a power cut. "I dreamed that I was returning home and couldn't find the garbage bin which is usually in the street. I felt it was the end for me, that everything was ending" She was very worried about one of the 4- to 5-year-olds in the group. He clearly manifested his hatred towards the enemies. He said: "I want to transform into a [he used a very hostile adjective to refer to enemies] in order to know how to kill them." During those days, the child talked with a fellow member of the group about the noise of the missiles falling the previous night. They were the only two children who referred to it, in spite of the fact that everyone there had gone through the same experience. It was a striking exchange for the therapist due to the degree of repetition of phrases in the dialogue. Subsequently, in one of his games, the child built a robot and said: "the robot has bitten me until I started bleeding." Then he added that he wanted to get rid of the robot, and that he would throw it in the garbage so that it wouldn't hurt anyone. "I want to throw away the robot so that it explodes in the trash." The level of terror behind his hatred was clear, the way the children concretely experience it. From the boy's game material, we gained a better understanding of the therapist's dream. She felt that death had arrived when she couldn't find the "garbage bins." I found her dream very elaborative, as she was able to bring it to our work group and in this way it could also be "dreamed" by us. She wondered in our work group how such a young child could have so much hatred, and how she could help him. Then, she had that dream. An emotional experience arose and surprised her. In order to comprehend it, she needed a container able to receive the hatred and fear that the boy projected on her. Her unconscious received the projection and transformed it into a dream.

The Case of Peter: Identification with the Outsider and the Return to the Inner Family

This experience arose in a WDG with a group of psychotherapists who are working in Kiev during the war. A therapist in the WDG was running a group of five 9-year-old children, who attend after school for an hour and a half twice a week. The therapist, Rose, supervised the material that arose in the group as the children expressed their emotions and fears.

On this occasion she was very worried about one of the children, who is part of a large family of four brothers, one of whom was a newborn. Rose commented that the boy entered the group that day and said: " I have a secret." He took out a 25 cm-long knife, which he kept close to his body and observed from time to time, taking it out of its sheath. Rose invited the group to talk and the children started asking Peter how he had come to be in possession of the knife. Peter commented that after school he used to go to the shops in town to steal with a classmate. They stole knives, earphones and speakers. He added that "the plan" was to take the knife home in his school bag, hide it behind the piano, and later take it to "the base," a hideout where they hid the things they stole. He told the other children that he was very excited about the hideout and its treasures, so he would not allow anyone to know its location. He then asked Rose for some sheets of paper to test the sharpness of the knife. He tried to cut them but he couldn't. His classmates laughed at this, so he asked Rose to hold up one taut sheet of paper. He stuck the knife in the middle. After this Rose once again invited them to talk about what was happening, and told them that it was very dangerous to play with knives and steal. She offered him options in exchange for the knife, even suggesting that he could hand it over to his parents or return it to the shop. Peter told her that he would stick to his plan to hide it behind the piano at home and take it to "the base" to play with. He could not speak to his father about this because he would hit him, nor could he tell his mother because she would inform the father. Rose was very worried about the danger of the situation and because she thought she was facing a "young gangster." The WDG dedicated itself to understanding Peter's mental state in order to be able to help him, and help the parents contain him.

It seemed to us that we were with a "real gangster" (Meltzer 1973) with a set criminal "plan" who was very convinced that he would succeed. We tried to think about what this oedipal mental state meant, in which the inner family made up of parents, siblings and the baby is abandoned with the aim of organizing another type of group, one that stole and preferred knives. In this kind of mental state there are no parents. The children are confused and believe they are adults, which is why they can carry guns. Possibly, all of this was exacerbated by the actual war.

We tried to understand what the arrival of the baby meant for Peter and its effect on the bond with his mother. He was surely very affected by this and defended himself, avoiding that pain by unconsciously iden-tifying himself with the outsider, the one who is outside the inner family, and somehow works against its loving creative values. He and his friend had their own stolen "treasures." In his phantasy these were stolen from the parents who were capable of producing the most envied treasures: babies. It was interesting that Peter hid the knife at home behind his mother's piano.

In view of this analysis of Peter's emotional world and Rose's comprehension of what was happening to him, we suggested that she should invite the children to play in the group in the construction of two spaces: a house and a "hideout." These were spaces they could enter and leave, while being understood and contained by Rose. This was a very different group from the one from the street and a different hideout from the "base" where the loot was hidden. We also suggested that Rose should meet with Peter and his mother to try to transmit Peter's emotional state and his way of building bonds at that moment. She could think with them about this without giving indications or reporting him.

I can only add that the boy handed Rose the knife, a good sign. Some of his confusions seemed to have been clarified. In this way, we are working in our group so that the "gangster" can return to his family.

Acknowledgements

My deepest thanks to the group of psychotherapists from APPU, coordinated by Taras Levin, for their work with so many children and young people in the midst of war together for sharing their clinical experience with the IPA PACE team. Thank you very much to Tetiana Chernokovtseva and Liudmyla Uspenska for sharing your work with PACE.

References

Bick, E. 1968. The Experience of the Skin in Early Object Relations. *International Journal of Psychoanalysis*, 49, 484.

Cardenal, M. 2021. Special Time: Working with Street Children. *Psychoanalytic Dialogues*, 31(4).

Harris, M. 1987. Towards Learning from Experience in Infancy and Childhood. In Meg Harris Williams (Ed.), *Collected Papers of Martha Harris and Esther Bick*. Karnac.

Klauber, T. 1999. Observation at Work. The Application of Infant Observation and its Teaching to Seminar Work Discussion. *The International Journal of Infant Observation*, 2(3).

Klein, M. 1945. The Oedipus Complex in the Light of Early Anxieties. *International Journal of Psychoanalysis*, 26, 11–33.

Meltzer, D. 1973. *Sexual States of Mind*. Clunie Press.

Chapter 6

There's a Hole in Daddy's Arm
Making Contact with the Opioid Epidemic in Clinical Practice

Benjamin Fife

I.

Amalia[1] is a working class Latina mother who came into treatment because of the stress of parenting a son with special needs and his younger, typically developing sibling. She arrives for a session considerably late and quite agitated. With barely a breath after a sharp apology for being late, she tells me that her brother is in jail. This, his second arrest in 6 months, both is and is not a shock. The previous arrest happened when he drove high on oxycodone on expired plates to try to score more pain pills in another town. While his prior arrest was deeply concerning and she struggled to set boundaries around her involvement, her current distress feels much more intense and complicated.

Amalia describes the circumstances of the arrest – In response to a confrontation about his substance abuse Amalia's brother has beaten up his wife. His wife, who provides Amalia with child care when she is at work, has been in Amalia's estimation, the most reliable person to marry into her traumatized and often highly conflictual family. When she told Amalia about the violence Amalia advised her sister-in-law to make a police report. The police were responsive. Her brother was arrested.

Amalia's speech from the moment of her arrival is fast and forceful. She tells me that her family is furious with her sister-in-law and is pressuring Amalia to post her brother's bail. She is refusing and wracked with a feeling of "wrongness," for having the thought that an arrest may be what her brother needs to get his act together. She also laments that if she did post bail only she and her husband would have the space to take her brother in and that would mean giving up on the childcare her sister-in-law provides. She feels guilty for having seen her brother's signs of untreated severe mental illness for many years and never successfully getting him help. She has trouble squaring this with memories of her periodically very competent brother, who was loved at work during long periods not abusing opiates. When he could hold jobs, he was genuinely proud of his ability to help others. She is furious with her own mother, nieces and nephews who are claiming that her sister-in-law is

DOI: 10.4324/9781003520757-7

responsible for the violence. Amalia is especially worried about how her own distress and distraction have left her emotionally absent when interacting with her children.

Amalia's rapid-fire speech evokes in me a sense of something that is both past and can never be in the past; something that keeps happening and bears the urgency of a constant present with no room for looking back. Searching for some way to make sense of all this information I think after the session about her family history. I imagine that an intergenerational experience of flight from war has linked up with this crisis to fuel a sense of being trapped in something timeless and repetitive. Amalia was born in the US to parents who fled a violent, repressive dictatorship. When, during Amalia's childhood, civil war erupted in her parent's country of origin, the family home in the US became a refuge for family and friends fleeing violence. As a young child she was sexually abused by the older son of one such family. Physical symptoms that emerged following the abuse were misdiagnosed by doctors, and Amalia did not receive treatment related to the sexual abuse until after she graduated from high school and sought out her own therapy. The abuse was never discussed in her home, yet her mother is often preoccupied with an idea that Amalia's son's developmental delay could have somehow been caused by sexual abuse at school, which Amalia sees no evidence for. Her father developed a drinking problem during her childhood that he overcame with the help of AA, but he later became addicted to pain medication following a workplace injury. He committed suicide when Amalia was in her early twenties. She understands his suicide as a quick decision he made when faced with a reduction in opiate prescription when a new doctor was unexpectedly uncomfortable with the amount of pain medication he was prescribed.

Over the next weeks the crisis of the brother's arrest, with its boundary violations, denial, victim blaming, danger, and substance dependence, ripples throughout Amalia's world creating internal and external conflicts for her. She starts to feel she is the only person in the family who can look out for her children. Her trust in her husband begins to deteriorate sharply over the next month after he makes playdates for their younger child with a neighbor she does not think they have adequately vetted as safe. The fight over this escalates to the point where her husband leaves the home for several days, which is something that has not happened before in their relationship.

I have a private thought during one session that Amalia's need to be hypercompetent may serve to fill with affect and action that which her brother's opiate abuse empties. She must be alert and responsive to all things, while he medicates the pain of their childhood and their family's losses. His opioid use seems to take him further and further outside of family life, while also linking him to their father. In responding to his increased abjection, she feels an inordinate sense of guilt for his increasingly desperate state. This guilt in turn threatens her capacity to be present in her own children's lives, leaving her distracted when out in the world with them. The threat of repetition of intergenerational trauma looms large.

As I struggle to think in a way that can help me feel, understand and relate to Amalia's pressured speech, overwhelm, and implicit expectations of self and others, I find myself returning to her lateness. Frequent lateness has been a figure in the treatment for a long time, and, until the crisis, I've distracted myself away from entertaining the lateness as meaningful. I've allowed myself a somewhat paternalistic pseudo-understanding of the lateness as part and parcel of work with an overburdened working class parent for whom responsibilities outside of the therapy often interrupt. This time I am able to entertain a thought about the function of lateness. It reinforces Amalia's need to squeeze every word possible into the time we have together and pushes to the margins any opportunity I might find to usefully respond, verbally or nonverbally. Late start sessions begin with an apology from Amalia that now makes me think of a starting gun. Words rush and race out after the "sorry," right up to the end of the hour.

I ask myself what would it mean to start regarding "late," as one signifier in a chain of signifiers pointing to one or more objects. I start to think about things that come late:

- My thinking about the lateness feels like it has come to me late. Until this crisis I think I mobilized a defensive and flattening pseudo-sociological understanding, meeting the lateness as something already known about instead of being curious or wanting to know what the lateness was doing.
- The friends and family who stayed in Amalia's home when she was a child, including the one who abused her were "late arrivals," coming after her family immigrated and depending on her family's hospitality and resources.
- Priority was put on responding to friends and family fleeing war. Attention to and acknowledgment of her childhood experiences of abuse by family members came late and was something she had to figure out without family help.
- Treatment for her abuse came late, when she was already almost an adult.
- Her brother's opiate abuse might represent another late treatment for psychic pain, the only thing left that can offer him some comfort for injuries left untreated.
- Her brother's opiate use evokes their "late" father. Does her wish to have helped her brother point to a relationship to a dead object, a parent who while living was dead in the sense of not being emotionally alive?
- Another patient with a history of opiate abuse following many years in foster care described the feeling of taking oxycontin as like a, "hug I didn't get the whole time I was away from my mother." Opiates for so many are experienced as a late arriving, finally available comfort that was needed long ago.
- The United States, my country and the country of Amalia's family's refuge, was instrumental in creating during the cold war the conditions that Amalia's family fled. Acknowledgement in the US of this role was denied for many years and can be said to be arriving late, when it arrives at all.

- Amalia worries that attention to the needs of her children are being delayed because of the focus on the crisis with her brother. Her children's needs seem fated to be addressed late because of the space taken up in her calendar and in her mind by the crisis.

I find some relief when a new thought arrives that Amalia and I are both stuck between two ideas about lateness – pulled between the comfort of "better late than never," and the accusation of being "too late."

This transferential and countertransferential tension between being "too late." and believing in "better late than never," is a frequent feature of opioid adjacent work with adults, children, and families. In Amalia's world, opioids have shown up, as they often do, in a vulnerable, traumatized family whose members are trying to find ways to manage immense psychic pain. In this instance the opioid abuse is linked to state violence as well as to family violence. The brother's reckless driving and arrest, and his violence towards his wife raise family ghosts related to the experience of flight from war and her father's suicide. The opiate use brings to the foreground the risks inherent to acts of protection, the possibility of future state intrusion and further loss. The arrest is like so many things in Amalia's life, arriving too late to help and causing a chain of too late responses to the needs of her own children. Without the crisis however it feels to her that there would be no hope of anything helpful arriving at all.

The therapy is, to some degree, just such a dangerous late arriving object. In the therapy are possibilities for containment, intergenerational change, and mourning. This exists alongside the risk of witnessing from a position of relative helplessness the continued unfolding of traumatogenic elements. Such a position can induce a feeling of certainty that one understands this is "just how life is," and the therapist can unknowingly cut off opportunities for new understandings to emerge. Pressured presences, confusing absences, and unsettling and intertwined experiences of loving and hating come up over and over again in work with patients impacted by the opioid epidemic. Clinicians are faced with complicated work when it comes to holding, containing, interpreting, and working through.

II.

Within the meaning-making matrices of Amalia's family as well as the broader social world she and her children inhabit, drugs, and in particular opiates, can be understood as a signifier that condenses many meanings. Link theory provides one model for seeing how this operates. In this model a given sign's unconscious meanings and ways of pointing towards various objects can be understood by looking at two axes. A vertical axis links the sign to meanings in the social world. The horizontal axis contends with the meaning across generations of a given sign (Scharff 2011).

The signifier "drugs" is part of a network of horizontal links that determine the meaning of this particular family crisis in the social world that Amalia belongs to and vertical links that connect the event backwards and forwards across generations (Scharff 2011). Horizontal links include the meaning of drug use as represented in mass media, legal discourses, and medical discourses. Unconscious meaning is produced around the signifier "drugs," as Amalia's relationship with her brother forces her into contact with spoken and unspoken questions regarding social rules, norms, and expectations. Her relationship to her brother and to those she identifies as like her brother (her children, aspects of her self-image) becomes impacted by social meanings of drug use ranging from the legal consequences of drug use to the limitations that identification as a drug user puts on job prospects and partner choice.

Vertical links include Amalia's understanding and feelings about who in each generation is using or has used drugs and why; who in each generation has supplied drugs to whom, and why; and who in the younger generation might be impacted by current use or be vulnerable to drug use in the future. I find the metaphor of a knit fabric useful for understanding this process of a chain of signification where key signifiers like "drugs" or opiates take up a space in a discursive cloth that has a history and is actively being constructed in the discourse of the therapy. In knitting, a cloth is formed from a single thread that links both to a previous row and to the stitch or stitches next to it. This is something akin to how meaning is formed in both individual and social discourses – signs and signifiers link to one another according to the meaning established in each epoch or generation, and also to the meanings established earlier, in prior epochs, imagined and real.

Drugs come to stand for psychic objects while at the same time pointing to aspects of the external world. In Amalia's understanding of her family, opioid drugs are an object that soothes, but should not be used. Opiates represent that which takes away, makes violent, and separates. They slow people down and speed people up at the same time. Understanding what they mean in her world will help me to listen for her expectations related to her own capacities to soothe and be soothed, and for her ways of understanding and sometimes judging the provision of comfort itself. Opiate as signifier is also intertwined with the ways she relates to her own capacities for violence and pain, the meaning of pacing, and of getting help with pacing.

As a therapist working in San Francisco, a city of about 870,000 that reported 806 overdose deaths in 2023, opioid drugs are unavoidably an important signifier in my own social and psychic worlds (Rodda 2024). I, like any other therapist, will bring all sorts of meanings to the signifiers "drugs," and "opioids." Some are from my own idiosyncratic experiences and some are from the meaning invested in drugs and communicated to me by others- parents, partners, school curricula, mass culture, the news. Drugs in American life represent unseen, suppressed and hidden conflicts at the level of the social as much as at the level of the intrapsychic. As analytic therapists we can better attune to the individual meanings the signifier "drug" holds for our patients if we can also hold key aspects of the meaning given to drugs and to drug use in our larger social context.

Like treating patients during the COVID pandemic, or in a country wracked by war or natural disaster, providing treatment during the opioid epidemic puts strain on any clinician's capacity for binocular vision – that Bionian ideal of being able to be inside a phenomenon with a patient while also looking at it from another place, from an outside with some claim to objective reality (Bion 1969). Our shared condition along with our patients as participants in a society actively making and contesting the conscious and unconscious meanings of the overdose crisis risks privileging one form of seeing over the other. We may be drawn to see opiate use as reflecting only repetitions of psychic or relational trauma and neglect the profound socio-historical meaning that also structures our patient's experience. Alternatively, we may be seduced into seeing only a socio-political problem and neglect how patients and their families unconsciously and uniquely create for themselves their own version of the charged signifier we call "drug use." The knit of the fabric that links up questions of war, colonization, and legality to opiate drugs starts during a period of colonial conquest temporally and physically distant from life in contemporary Northern California, but it contains constitutive links to my current clinical encounters with this population.

In his 2022 book *Quick Fixes*, Benjamin Fong argues that popular notions of the drug user have their roots in the late-eighteenth-century and early-nineteenth-century expansions of the English empire. In the late 1700s the early British industrial workforce was, to an appreciable degree, fueled by the "English cuppa," with the massive expansion in caffeinated tea consumption enabling a chemically stimulated, alert workforce to staff English factories. Massive amounts of British tea were being imported from China. China was insisting in the 1790s on payment for that tea being made in silver. At this time opium use was very widespread in British controlled India. The British East India company made possession of opium illegal in India, except as an export to foreign markets, and it flooded China with opium produced in India. This created a demand for the drug that effectively stopped the outflow of British silver to China. Over the next seventy years the British would fight two wars in China preventing the Chinese authorities from banning the sale and possession of opium and ensuring that the drug remained legal there. A nineteenth-century British ambassador would state in parliament that "we forced the Chinese Government to enter a treaty to allow their subjects to take opium" (Fong 2024, 79). Meanwhile in the industrializing West throughout the 1800s, opium would, in Fong's words, become the "opiate of the people" (Fong 2024, 80). Throughout much of Europe and the United States laudanum and other opium derivatives were a main ingredient in patent medicines marketed to cure everything from headaches to cramps to colic.

As Chinese immigration to the US became a source of class and racial anxiety, the image of the "opium eater," originally created to justify British colonial wars, gained popularity in the United States. Chinese immigrants were vilified in much the same language we currently see used to represent drug users, as slovenly, lazy and out of their minds. They were also represented as a threat to white women and girls who were cast as potential victims of enslavement in opium dens. The craze for opiates in patent medicines gave way to a regulatory regime in the early twentieth century focused on the control of opiates and a split between "good" supervised medical use and "bad" illicit use. The images of the drug user as predatory, slovenly, racially other, corrupting, and hidden, which formerly described a foreign other, now represented a part of the body politic that was in need of particular kinds of discipline and control. Ideas about the drug user as an enemy internal to the state were reified into policies that contributed to creation of a feared object in the minds of the citizenry. Drug policy became structured on protecting others from the predations of users, inhibiting curiosity into the reasons for mass drug use and effectively shutting down inquiry into the social and psychological roots of substance abuse problems.

Illegal opiate use, especially heroin use, and later the use of stronger medically manufactured opiates like oxycodone and fentanyl, shows up as a problem that tends to be statistically concentrated among populations with the highest rates of trauma subtended by poverty. The legal responses to the use and trade in these drugs push users into highly dangerous settings. This effectively recreates in successive generations populations of drug users and fears of drug users that fit the description first seen in the propaganda for colonial wars.

Fong argues that in the US forfeiture laws developed in the 1970s and still in place today demonstrate exactly how abject one becomes when one is suspected of being associated with drug use. Suspicion of illegal possession of salable amounts of opiates is an adequate legal basis for any possessions believed to be linked to the sale of those drugs to be seized and held by the police. This seizure of personal property does not have to involve a determination of guilt by the police, with radical implications for the legal subjectivity of people associated with drug use.

> Based on little more than their own "reasonable belief," police today can legally enter your home, take whatever they want (including your home), and not give it back – *ever*. Most Americans take consolation from the idea that while you might not be guaranteed a decent wage, healthcare, any safety net, et cetera, at least the right to property that you might have is sacred and inviolable. By the early 80s the generals of the War on Drugs had eviscerated this core tenet of American capitalism.

(Fong 2024, 92)

Whatever our understanding of the psychic damage associated with drug abuse, we must also hold in mind that proximity to drug use puts complete legal abjection and a potential loss of all that has been gained, on the horizons of our drug-adjacent patients.

III.

In my own practice one out of three of my clinical hours involve work with patients whose difficulties I have come to think of as "opioid-adjacent." I see children who have lost parents to fentanyl overdoses, children with parents in recovery from opioids, young adults who have lost siblings to overdoses, teenagers anxious about who will be carrying Narcan if they go out dancing with friends, parents struggling to manage relationships with extended family members in the throws of addiction, and parents struggling to reduce dependency on opioid pain medications themselves.

I find I can only stay in contact with patients' experiences of the opioid epidemic by developing ways of thinking about absence both at the intrapsychic level and at the level of larger social forces. Take for example the social and psychological experiences that leave children vulnerable to loss in working class families where there is an opiate using father. Case and Deaton (2020), propose that opiate use and other factors contributing to the increase in "deaths of despair," are tied to new absences in the social contract. Working class white and Latino men without college educations face an absence of jobs that pay enough to support families. In many instances these men are for the first time in generations making less than the generation of their parents, and are encountering an absence of the life they expected to live and to provide. In turn they are increasingly unlikely to be seen or to see themselves as economically viable partners and they become more likely to be absent from family life and to be more stressed when they are part of it. If they turn to opiate or alcohol abuse they become psychologically absent if they are high at home and are more likely to be physically absent longer than they promise or intend if they use outside of the home. The experiences of absence become clinically important phenomena not only if and when these fathers come in for treatment, but also in the experiences that children and other family members have of these vulnerable adults.

Two different spatial metaphors for psychic life have helped me to organize my thinking around work with this population. The idea from Link Theory (described earlier in this chapter) of horizontal and vertical links that give signs like "drug users" meaning by connecting them to chains of signification, and a spatial metaphor developed by André Green to describe different types of patients who come for analytic work.

In 1982 André Green proposed a model – also using horizontal and vertical axes – for thinking about the tracks along which the unconscious can develop. The first developmental line he says following Freud, is repression, a vertical structure for the development of unconscious life. A present caregiver is wanted body and soul by an infant. The infant develops a sense that what he or she wants is too much and could harm or change the caregiver. The infant feels guilt in relation to his or her desire. This conflict between desire and guilt (primary process) is covered over by conscious thought (secondary process) that denies or obscures the presence of the desire. The awareness of some communication from below, in the form of symptoms, slips of the tongue, dreams, forms a bridge between the unconscious guilt and desire and conscious thought and action. This constitutes a tertiary process by which the unconscious can come to be, to some degree, known.

Green argues in the same paper that when early development is characterized by the absence of an available other, the unconscious develops a radically different, horizontal structure, oriented around splitting. There are the wishes to have the (absent) object and there is the infant's need to prematurely accommodate a reality where the object cannot be reached. These psychic states of need on the one hand and lack on the other, exist side by side, in fragments that seem to be incommunicado with one another. What is unconscious here is not the conflict, but the fact of the split between these states of being. Feelings towards the object of love and hate are also split. It is as if each thought a person has about what is happening is made up of different pieces that do not fit together as a whole.

Repression, the unconscious developed along a vertical axis, whereby desire is kept unknown "underneath" conscious thought represents for Green a "logic of hope" (Green 1986, 24). Repressed unconscious life constellates around desire for an object that is experienced as undoubtedly having reciprocal feelings for the subject, that would love (or hate) the subject completely. In unconscious life formed along a horizontal axis where splitting is the dominant unconscious process,

> the object is in the forefront, not the wish, not the prohibition. If the happy union is experienced as being impossible, it is because the subject cannot feel loved by the object or cannot love the object. It is a different logic, over the conflict between the wish and the prohibition, because the conflict between the ego and the object about love and hate prevails.
>
> (Green 1986, 24)

Importantly there is no tertiary process available for communication between these different sides of psychological experience. The analyst or analytic therapist is called upon to use his or her own capacity to communicate between conscious and unconscious thought in service of the treatment. The impact on the therapeutic process is profound. When working with people whose unconscious life is organized horizontally around a split,

It is as though the analysand only hears the analyst's interpretations with one ear. The other ear continues to let itself be rocked and cradled by the instinctual impulse mermaid song, completely ignoring the message received by the other ear. The two logics are in contradiction with one another. There is a refusal to choose any of the items. Prior to the discovery of the ego's unconsciousness, unconscious repression had to disguise itself in order to express itself. Behind the conscious "No" we can reveal the unconscious "Yes": In the present case the ego's strategy changes. It says "Yes" and "No" at the same time. What is important is not so much the double game the ego plays by splitting, it is that the splitting is unconscious.

(Green 1986, 26)

As much as the link theory model helps me to think in the face of immense pressures like the ones Amalia brought to the session described earlier, Green's model has helped me find ways to be with the often disturbing sense of fragmentation and incompatible truths expressed by people impacted by the opioid epidemic. It gives me a way, in the here and now, to understand the psychological impact of the encounter between need and absence.

In the imaginary play, for example, of a child whose father died from a fentanyl overdose, figures that occupy the position of a paternal role, enforcers of rules and order, are often bureaucratized, punitive, and distant. If in the play I am assigned the role of a father or other holder of order (teacher, police officer, doctor, railway conductor) I can expect to be kept at a distance. I am either given a set of scripted roles meting out harsh punishment for minor infractions, or I am put in different disguises and assigned to watch the child's character secretly. Should I express concern in the play in the form of discomfort with meting out a harsh punishment or attempt to make contact with a feeling, there will be a great crash or disaster. I will be found guilty of causing it, and I will find my character hated for any act of care or concern. I will be punished violently and severely, possibly for the remainder of the hour.

Containment in therapy is always partial. When our opioid-adjacent patients bring forward their own needs for recognition in the world and encounter absences in the capacity of those they depend on (within and outside of families) to hold their experiences, further splitting and confusion around contact, care, love, and hatred emerge. Sometimes there are explicit needs for dynamic psychotherapists and psychoanalysts to consult with other people in a child's world in ways that protect the privacy of therapy while opening up some space within the other for increased understanding of the child. However, given the scale of the issues at hand here, I believe our field needs to find ways to reduce the splitting that is so prevalent in our contemporary social world when encountering those impacted by the opioid epidemic.

For Green, the encounter between need and absence that mattered most was a very early one that would lay the groundwork for unconscious dynamics that remain with the subject for the rest of his or her life. I have no sense of certainty about whether what creates these dynamics is primarily an infantile relationship that repeats or whether splitting dynamics are the product of societal processes where basic needs for contact and care are being denied on a mass scale. I am not convinced that for clinical practice a precise etiology of the symptoms we encounter matters all that much. I think that what matters here is the psychic reality of our patients, which is that needs that encounter nothing but an absence of contact create splits and confusion. That can happen early, creating a lens on future experience. It can also happen later. It is re-enacted in our offices, and outside our offices.

The generation of analysts and analytic psychotherapists I belong to has made what many are calling a "turn to the social," pushing training institutions to consider a social unconscious as a shaping force in psychic life. This turn reflects, I believe, a wish for greater involvement of psychoanalysis in the world. One, perhaps undervalued, way that we as analytic communities can be more involved in the world is to try to bring our non-analytic colleagues, physicians, teachers, politicians, paraprofessional caregivers and parents into safe contact with the emotional landscapes of children and families whose lives are shaped by loss and absence. These diverse adults are also coming into contact with more of the fragmenting pain and distress caused by large-scale traumas like the opioid epidemic. They need ways to understand the complicated feelings evoked by this contact, including the confusional mixes of love and hate. I think that this involvement with non-analytic colleagues can take many forms – consultation, teaching, writing op-eds, and getting involved in political projects that impact policy in useful ways. If we as an analytic community, with what we know about the short and long term impacts of loss, cannot be present and attempt to help those in need understand these experiences, we risk becoming one more absent object in a time of need.

Acknowledgement

The title of this chapter is taken from John Prine's 1971 song, "Sam Stone."

Note

1 The case material that follows is a composite of material from a number of cases I have worked with in clinical settings in the last 10 years. Composite material is used throughout this chapter to ensure confidentiality and because of the notable patterns clinicians see when treating patients with opioid adjacent concerns.

References

Bion, W. R. (1969). Introduction. In W. R. Bion, *Experiences in groups and other papers*. Routledge.

Case, A., & Deaton, A. (2020). *Deaths of despair and the future of capitalism*. Princeton University Press.

Fong, B. Y. (2024). *Quick fixes: Drugs in America from prohibition to the 21st century binge*. Tantor Media.

Green, A. (1986). Psychoanalysis and ordinary modes of thought. In *On private madness* (pp. 17–29). Karnac.

Rodda, L. (2024). Preliminary accidental drug overdose data report for January 2023 through December 2023. https://media.api.sf.gov/documents/2023_OCME_Over dose_Report.pdf.

Sharff, D. (2011). The concept of the link in psychoanalytic therapy. *Couple and Family Psychoanalysis*, 1, 34–48.

Chapter 7

Forbidden Games
Anti-War Manifesto

Ana Belchior Melícias

> The opposite of play is not what is serious but what is real.
>
> (Freud, 1908)

Produced only shortly after the end of the Second World War, and still remarkably close to the memory of its horror and devastation, *Forbidden Games* (1952), directed by René Clément, is a film inspired by the novel *Les jeux inconnus* (1947) by François Boyer, which "manages to be incredibly deep and emotionally complex, despite its seeming (and deceptive) simplicity …" (Eu Sou Cinema 2014).

What is the meaning of *interdits* or *inconnus* after all? The death drive in the lethal/legitimate adult games which superimpose death on life? The innocent-secret-forbidden children's games, which challenge the limits of the law tethered to both Church and Father?

During the Nazi occupation, a convoy of civilians in an exodus from Paris to the countryside is bombed on a bridge. While trying to save her puppy, the five-year-old Paulette (Brigitte Fossey) ends up witnessing the death of her parents, absorbing the blame and guilt of her attempt, in vain, to save the dog. She touches her dead mother's face and then her own. She touches her puppy's stiffened snout and then her own face again. Tustin considers warmth and cold, the call and absence of response, the first differentiation between animate and inanimate. Orphaned, she begins to wander in an absurd and brutal state of non-reality, holding the body of "Jock-her-dead-part" in her arms.

Is it possible to mourn what cannot be represented? What can be done both externally and internally about trauma, death and violence? About the mechanism of traumatogenesis, Ferenczi says: "first, there is the complete paralysis of all spontaneity, then of all thought work, including the occurrence of physical states similar to those of shock, or even coma, and then the establishment of a new – displaced – situation of equilibrium"(Ferenczi 1992). Without representation, the traumatic becomes "Sisyphean," returning and repeating itself incessantly.

Freud warned of the insidious installation of melancholy when "the shadow of the object falls upon the ego." The alternative, in intrasubjective terms, is

DOI: 10.4324/9781003520757-8

the creative and transformative power of thought often through analysis. But in intersubjective terms, what truly ends up prevailing over many generations, decades and even centuries, is the trans-generational transmission and the unavoidable propagation of the frozen historical trauma, which becomes part of the collective identity of the group/nation that suffered catastrophic loss, helplessness, and humiliation at enemy hands (Volkan 2020). To which is added, in the long run, and as a feature of this destructive lineage, the inflammatory, ideological use of the "chosen trauma" (Volkan 2020) by the leader(s) of the large group/nation.

René Clément's film is a true anti-war manifesto, which reflects the vulgar but symbolic "warfare" between neighbours Dollé and Gouard; the traumatic consequences of nameless dread, especially in childhood; the impossibility of fulfilling the farewell rituals crucial to the work of elaborative mourning; and the de-humanisation and violence leading to exodus, broken families, orphaned children, and millions of refugees all over the world, as we have witnessed, with horror and shock, in the Russian invasion of the Ukraine and the return of war to Europe and in the war in Gaza.

Paulette, the angelic Parisian orphan, meets the ten-year-old Michel (Georges Poujouly) and is taken in by his rural family (Dollé). The film denounces the asymmetry of resources in the countryside and city, while demonstrating in psychic parallel, that it is not enough to welcome and feed; it is necessary to also transform the beta into alpha, so they become elements capable of being digested, transformed and integrated.

Michel defies the limits of both Church and Father, giving himself fully over to primary maternal preoccupation of Paulette, this small, charming little treasure. In doing so, he spares no effort to protect her from any frustration or pain. A Ferenczian wise-baby, Michel protects Paulette, thereby protecting himself from his own helplessness.

For the two children, play becomes the potential intermediary space (Winnicott) for re-connecting to life. Bion shows us that the third element which opens up the binary matrix of Love (play) and Hate (war) is Knowledge, embodied in the owl in the film, is a symbol of wisdom and memory, responsible for retaining history and the possibility of thinking.

Paulette cries out for the rituals necessary to the mourning process, which she was unable to partake in following her parent's deaths. These are duly fulfilled – prayers, mass, funeral, burial, flowers – with Michel's brother.

United in their complicity and tenderness, they engender an illusory world to cope with tragedy and to represent the traumatic, moving from the passively lived to the actively played (fort–da). They indefatigably pursue the meaning of death through the creation of a cemetery-artwork, interweaving aesthetics and ethics in a final humanising act, foreclosed by war. "The dead should not be without company," says Michel. Paulette asks, "Are they buried so they don't get wet when it rains?"

A series of forbidden episodes, animated in equal measure by the strength of secrecy and infantile cruelty and sadism, give shape to the construction-elaboration-transformation of a microcosm: in a crescendo of evolutionary complexity (phylogenesis), the children imagine the living beings that they would bury – worms, snakes, lizards, cats, dogs, cows, horses, people – while they go on burying small animals. They steal crosses, the religious symbol par excellence, desecrating the real cemetery, and distribute fourteen crosses (stations of the Passion of Christ) throughout their cemetery-artwork, according to the size of the buried animals, amplifying the collective death of war.

"May God welcome you in paradise" is the parroted phrase at each burial, ritualising hope. The children personalise the graves, moving from the mineral (stones), to the vegetable (flowers) to the animal world (snails) and finally to the realm of the symbolic-word (crosses and signs with names written on them), choreographing a true ritual of farewell. The cemetery is finished and the mythical apple signalling the exit from the paradise of childhood is here, offered by Michel and refused by Paulette.

The tenderness and passion interwoven throughout the narrative, do not generate the usual "confusion of tongues": the tenderness of the Hail Mary and Our Father prayers, which Michel teaches Paulette, are juxtaposed with the passion in the courtship, *à la* Romeo and Juliet, between the older sister and her neighbour sweetheart. Paulette wonders about the meaning of "womb" (developing curiosity about the inside of the maternal body) while Michel's sister asks to be absolved for her sexual life.

Must every crime have a punishment? The end of innocence, ambivalence, and the beginning of reality are internally imposed, and the tribute for the forbidden games arrives through betrayal: "Paulette is constantly torn away from her attachment figures, from her emotional landmarks, which are absolutely essential for her growth. She is torn away from her parents, then from her dog, and finally from her adoptive family" (Avis-Gallu 2021).

The film ends abruptly, placing us counter-transferentially in touch with the horrors of war and its transgenerational and traumatic destructiveness.

A couple, her parents, dies on the bridge. Another meets again in the final scene at the train station. Only then does Paulette verbalise her orphanhood, calling out, "Mama!" We are left profoundly shaken by the real pain of her helplessness and orphanhood, wishing that Paulette – like all refugee children in the world today – might be able to remain in the childlike situation of protection and support so essential to children. As the poet Adília Lopes says, "there is no consolation for very sad things, only revolt" (Lopes 2022: 14).

And the beginnings connect to the ends...

In the beginning Paulette calls out: "Michel ... Michel ... Michel ... I'm afraid of the dark." At the end, in the train station and on her way to the orphanage, in the dark caused by this new, brutal severing, she calls out: "Michel ... Michel ... Michel ..." She gives herself his surname, Dollé, and thus leaves anonymity, naming the hope of a future reunion and taking with her a living object to search for.

It is also up to us psychoanalysts to build that most difficult of bridges, connecting despair with hope through culture and civilisation, as Freud tells us in his reply to Einstein:

> And how long shall we have to wait before the rest of mankind become pacifists too? There is no telling. But it may not be Utopian to hope that these two factors, the cultural attitude and the justified dread of the consequences of a future war, may result within a measurable time in putting an end to the waging of war. By what paths or by what side-tracks this will come about we cannot guess. But one thing we can say: whatever fosters the growth of civilisation works at the same time against war.
>
> (Freud 1933)

References

Avis-Gallu. 2021. Opinião sobre o filme Jeux interdits (1952) – A cada um sua própria cruz por gallu. *SensCritique*, 18 (35).

Eu Sou Cinema. 2014. Jeux interdits. http://eusoucinemapt.blogspot.com/2014/10/jeux-interdits.html.

Ferenczi, S. 1992. Análise de crianças com adultos. In S. Ferenczi, *Obras completas, Psicanálise 4*, 69–83. Martins Fontes.

Freud, S. 1908. Escritores Criativos e Devaneios. In *Edicao standard brasileira das obras psicologicas completes de S. Freud*, vol. 9, p. 149. Imago Editora.

Freud, S. 1933. Why War? In J. Strachey (Ed.), *The Standard Edition of the Complete Psychological Works of Sigmund Freud, Volume XXII (1932–1936): New Introductory Lectures on Psycho-Analysis and Other Works*, 195–216. Hogarth Press.

Lopes, A. 2022. *Pardais*. Assirio & Alvim.

Melícias, A. B. 2021. O enigma da morte – Brincadeiras Proibidas (1952). https://cine mapsicanalise.pt/2021/02/21/o-enigma-da-morte-brincadeiras-proibidas-1957/.

Volkan, V. 2020. *Large-Group Psychology: Racism, Societal Divisions, Narcissistic Leaders and Who We Are Now*. Phoenix.

Chapter 8

Totalitarian Regimes and a Child's Mind

Cría Cuervos

Mary T. Brady, Adriana Prengler, Ana Belchior Melícias and Virginia Ungar

Spanish director Carlos Saura's (1976) masterpiece *Cría Cuervos* (which Saura also wrote and produced) is used here to describe the way violence and repression, enforced by a totalitarian regime, affect the mind – not only for those of the immediate generation, but also of the next. Faimberg (2005) writes that transmission between generations (both political and personal) is an "invisible" object in psychoanalyses that needs to be constructed, or here made visible in film. We will argue here, that through art Saura makes the enforced silence as well as the violence of totalitarian regimes visible as it is transmitted into the film's characters and partially transformed in Ana, the central character of this film.

Eight-year-old Ana (played by then ten-year-old Ana Torrent) believes she killed her dead father (played by Héctor Alterio) and is frequently visited by hallucinations of her dead mother (portrayed by Geraldine Chaplin). Saura vividly depicts the way children's fragile psyches are frozen in time by trauma, here particularly the personal historical of the excruciating illness and death of Ana's mother. *Cría* thus also explores the interpenetration of the past and the present when time has been fractured by a traumatic loss.

The interpenetration of reality and fantasy is brilliantly played out in the opening sequence of *Cría Cuervos*. In a white nightgown, Ana descends a dark staircase. As the camera focuses on her pale, expressionless face, urgently whispered adult words – "I love you"; "I can't breathe" – are heard from behind a closed door. A half-dressed woman runs from the room. On entering the now silent room, Ana finds her father in bed, apparently dead. Impassive, she takes a glass to the kitchen and washes it in the sink. As she opens the refrigerator, her mother comes into the shot and addresses her tenderly. Only later do we learn her mother had died some time ago.

The psychological and the political are inextricable in *Cría Cuervos*. The title refers to a Spanish proverb: "*Cría cuervo y te sacarán los ojos*" ("raise ravens and they will tear your eyes out"). Ana's father was a Fascist military officer, so the film's title implies a legacy of political and personal violence. Thus, Saura foregrounds in choice of title the violence that is generated by one generation that will be vengefully returned by the following generation.

DOI: 10.4324/9781003520757-9

Saura conveys what Bollas (1992) has discussed psychoanalytically – that fascism is a pathological state of mind as well as a political phenomenon.

Saura shot *Cría Cuervos* in the summer of 1975, as Spanish dictator Francisco Franco lay dying. The film premiered in Madrid in 1976, forty years after the beginning of the Spanish Civil War and received the Special Jury Prize at the 1976 Cannes Film Festival.

Saura's early films were heavily censored by the Franco regime and contributed to his development of a metaphorical style. Ironically, the censors' insistence on cutting Saura's films, "of specific social, political and historical references left his characters in a vacuum that, with powerful irony, was wholly representative of Saura's view of contemporary Spain as a place of fear, denial and falsehood" (Stone 2002, 69).

Ana seems a wise child (Ferenczi 1955 [1933]) who has witnessed personal, familial, and cultural disasters, grasping the violence and beauty of life, (albeit infused by a child's magical and omnipotent thinking). Ana could be seen, in part, as motivated by the epistemophilic instinct, or in other words, the desire to understand. In Bion's view the desire to understand fundamental emotional truths is more central than the pleasure principle.[1] For instance, Ana is willing to face her Aunt Paulina's disapproval when she refuses to kiss her father in his coffin – reflecting her insistence on emotional truth.

Ana is also motivated by revenge: she has tried to kill her father, whom she holds responsible for the death of her mother due to his cruelty and infidelity. Ana did not actually cause his death, nor is her father literally responsible for her mother's cancer, yet in Ana's child mind both are true. Her actions exceed typical childish murderous wishes. Ana mixes a powder that she thinks is poisonous into father's drink and then clearly thinks she has been successful in killing him. She later kindly offers this powder to her grandmother whom she thinks wants to die. She later tries to murder Aunt Paulina with the powder and this time is surprised to find that she is unsuccessful. Thus, Ana has imbibed the murderousness of her father and his Francoist regime, although unlike him she is multi-dimensional and capable of great sensitivity as well. In this sense she is not overall in a fascist state of mind with a central preoccupation on violence and power, but an element of her father's murderousness is present.

Sabbadini's (2014) commentary on this film emphasizes Ana's preoccupation with death.[2] Child analysts are aware of the way children are often occupied with the great mysteries of life, including death, in ways many adults do not take seriously. Ana's preoccupation with mortality has been intensified by witnessing the deterioration and death of her mother, as well as by her own murderous wishes.

Cría's Ana is a child of great intuition, who takes in her personal and cultural surround at a depth and judges it unsparingly. Saura said:

> *Cría Cuervos* is a sad film, yes. But that's part of my belief that childhood is one of the most terrible parts in the life of a human being. What I'm

trying to say is that at that age you've no idea where it is you are going, only that people are taking you somewhere, leading you, pulling you and you are frightened. You don't know where you're going or who you are or what you are going to do.

(quoted by Stone 2001, 102)

Children *are* absorbed with understanding some of the darkest mysteries of life and are dependent on those around them to protect them from terrible experiences of helplessness, but Saura's childhood was also darkened by growing up in Madrid during the Spanish Civil War. Stone (2001, 98) cites Saura's biographer, Enrique Brasó who quotes Saura:

I suppose that for Proust his childhood was a series of details more or less poetic of his family and surroundings; for me those memories are much more violent: it's a bomb that falls on my school and a little girl bloodied with shards of glass in her face. And that's no literary invention, it's a fact.

(Brasó 1974, 23)

Saura was born in 1932 in the Republican controlled capital of Madrid, which was under siege by Franco from 1936–1939. During the Spanish Civil War, Madrid became a symbol of anti-fascist resistance, enduring a two-and-a-half-year siege, fighting against the rebel forces led by Franco. Madrid was the first major city in history to experience aerial bombardments of its residential neighborhoods and civilians, which was ordered by Franco as punishment. The bombardment was aided by German and Italian aircraft to extinguish the Republican resistance (Bordes & De Sobrón 2021). Madrid fell to Nationalist forces on March 28, 1939. On April 1 the Republican Army surrendered in all of Spain. After the fall of Madrid, Franco conducted mass punishment and purging of the defeated, and poverty was a punishment, "handed out to the vanquished zone" (Stone 2001, 61).

The great Spanish directors of this period, Luis Buñuel, Victor Erice, and Saura, have created narratives that allow us to reflect and represent the silence and violence of the Franco era. Buñuel denounced fascist power that felt totally justified and approved, destroying freedom. He likewise aimed his criticism at the Catholic Church in Spain, for instance in his 1961 *Viridiana*. Buñuel managed to largely evade the Spanish censors in this satirical and absurdist film, which when shown provoked outrage from the Catholic Church and was then banned in Spain until after Franco's death.

Erice's masterpiece *The Spirit of the Beehive* (1973), set in Spain in 1940, chronicles Spain's isolation and silence under fascism, as well as the marginalization of intellectuals. Like Saura's *Cría*, this film also evaded the Spanish censors by using a metaphorical style but vividly conveyed the silence and violence of its fascist period.[3]

Just as psyches are froze in time by trauma, one might see *Cría* as Saura's commentary on the potential for a social structure to be traumatized and frozen. Analysts of children, adolescents and adults must struggle to think in analytic fields dominated by non-thinking states engendered by trauma and splintered by dissociation and splitting. Trauma (including societally, economically, and culturally induced trauma) overwhelms the psyche, while both psychoanalysis and artistic creations such as film allow the psyche to make meaning of trauma. Children and adolescents are often the group most affected by cultural changes and catastrophes. They are like the canaries sent into coal mines to signal the presence of gases, imbibing cultural, societal, and economic changes in a rapid and powerful way.

We will use the axes of characters, spaces and settings, as well as memory, to think psychoanalytically of the effects of personal trauma, as well as a totalitarian regime's political violence and enforcement of repressive silence, on the current and succeeding generations.

The Axis of Characters

In addition to eight-year-old Ana, the other characters in the film are her mother Maria, her father Anselmo, her eleven-year-old sister Irene, her five-year-old sister Maite and their grandmother. Rosa the maid and Paulina the aunt (who comes to take care of the girls after both their parents die) complete the household. The final characters are a couple – Amelia (with whom the father is making love when he dies) and her officer husband Nicolas (who becomes the lover of Aunt Paulina).

The viewer is immersed in the darkness of Ana's family, metaphoric of the external socio-political situation. Entrenched authoritarianism is not only relevant to government: it can shape society, mold the family, and impose itself on the structure of the mind.

Anselmo, the military father, can be thought of as representing the Franco regime. He is the powerful dictator in his home, symbolic of oppression, abuse, and intolerance of the other. He follows his own desire, without considering anyone else's wishes. He submits others to his will. As a dictator, he needs to remain in power, to command others without limits, to exclude the other, devalue them, and subjugate them into dependence. Anselmo mirrors Franco at the socio-political level. The others in the family cannot escape and have no choice but to submit to father's authoritarian will.

Maria, the mother suffers from a terminal illness. She is weak, powerless, and trapped. She can be seen as representing the sick, mistreated, and dying Spain and its oppressed, humiliated people. She is "Mother Land," without the attention or love she needs, until she dies. Submitted and powerless before her husband, she begs for care and love, as the submitted Spanish society does under the Franco regime. Authoritarianism not only submits the people to dictatorial rule, but also justifies their persecution by blaming them. Similarly,

Anselmo blames his wife for being annoying and not letting him live happily (in his view without any restrictions or responsibilities), thus justifying his actions. This is consonant with Bollas's assertion that: "the Fascist mind transforms a human other into a disposable non-entity" (Bollas 1992, 203).

As mentioned above, Ana blames her father for the suffering of her mother, and this awakens her patricidal desire. Her patricidal impulse, rather than an intrapsychic oedipal conflict, seems an effort to avenge the mistreatment her mother received from her husband Anselmo. It is also an assertion of Ana's own desire for freedom. Ana decides to kill her father with a substance she thinks is poisonous but is really baking soda. Ana can be seen as a raven in relation to the titular proverb. Her attempt to murder her father can also be seen as an effort to be rid of him in order to free herself. We can see her attempted murder of her father as symbolic of the desire of the humiliated Spanish people to eliminate the tyrant who oppresses them, often with an ineffective weapon.

Ana silently witnesses her father's infidelity and his subsequent death. As his lover Amelia runs from the bedroom, Ana's gaze meets Amelia's eyes, and they stare at each other, not as woman and girl but as if they were two women cognizant of the truth about a man. In this moment, Ana loses her childish innocence forever. Surely this scene confirms something that Ana already knew about Amelia. But from that moment on, Ana and Amelia share a secret. Eventually, Ana (through her sister Irene) dares to reveal to her Aunt Paulina that her father was with Amelia when he died, possibly in an attempt to break the unspoken alliance with Amelia. This can also be seen as the way truth is not accepted in a repressive regime but may eventually come to light.

Paulina silences her, forcing her to deny what she has seen with her own eyes: "You did not see anything." Paulina silences the truth and forces Ana to go against her own perception. This can be seen as a mandate to submit to authority without thinking, without speaking, forcing Ana into a disavowal of the truth.

Ana's internal world suffered the same vicissitudes as the external world under Franco's dictatorship. If you tell the truth, you will be punished or killed. In a totalitarian regime there is no place for the truth, for freedom of expression, for sharing, for healthy alliances.

The three generations, Ana, her mother, and her grandmother, are forced into the fate that has befallen them. They are not to speak. This is epitomized by the grandmother, who literally *does not speak*. She only dreams while listening to the music of her past and looking at the photos that remind her of a happier time long, long ago. The three generations suffer fear, uncertainty, threat, confusion, loneliness, helplessness, and hopelessness, at home and societally. Once again Ana must submit to silence and an imposition, now from her aunt. Ana's wish to kill her aunt may reflect her rage at her aunt for replacing her mother, but also her belief that it is better to kill her aunt and attain some freedom. Both father and aunt, in different ways, represent the abuse of power and the disconnection of affections.

In one compelling flashback, Ana sees her mother screaming in pain on her deathbed. This scene conveys Ana's impotence and sadness. It is analogous to the impotence and the pain in Spanish society, which was traumatized and frozen in fear, as is Ana's gaze in this scene. Ana is wordless as she sees her mother die. She looks off into space, paralyzed, and expressionless. She does not feel entitled to hug her mother or talk to her on her deathbed. How will she be able to mourn this terrible loss? The scene can be thought of as conveying how difficult it is to grieve for someone one hasn't been able to say goodbye to, as in the cases of those who have been disappeared by a repressive regime. As in a frightening dictatorship, there is no one to talk to and no one to trust.

Ana feels guilty for having "killed" her father, and for wanting to kill her aunt with her poisonous baking soda. The poison conveys the rage that Ana feels and yet has no right to express openly. In one scene, Ana tells her doll, "You're bad!" projecting onto the doll her guilt for having in fantasy murdered her father and having tried to kill her aunt. In a subsequent scene, Ana plays a version of hide and seek with her sisters, in which those found must fall to the ground "dead." They play at dying and being resurrected, perhaps with the wish to resurrect their mother. One can see the attempt to make representable what is unrepresentable about death. One can also see the link that Ana makes between killing and dying, so characteristic of childhood games. At the time of filming in 1975, Spain's freedom had long been dying, but there was hope that it would soon be resurrected.

At one point in the film Ana asks her grandmother, "Do you want me to help you die?" Ana also in part seems to have wished to die. There is a repetitive, bizarre image in the film of a plate of chicken feet seen when Ana opens the refrigerator. When made into soup chicken feet are a common home remedy for many ailments. Chicken feet can also be an amulet for protection and good luck, counteracting bad energy. They stay fresh in the refrigerator, ready to be used, as if they are the source of magical hope. The film seems to leave in question how much Ana can meaningfully, psychically recover and survive (e.g., with the help of the chicken feet) or whether this is just a magical idea in the face of overwhelmingly harsh realities.

Ana is a victim of a terrible family situation in which she is submerged. She knows that she should hide herself, hide the poison, hide the truth – not only the material poison, but the poison of pain and anger. She watches and "kills" without any expression. Ana hides her "crimes," as dictatorial regimes do, trying to keep up appearances, just like her father Anselmo who goes to church and then unsuccessfully hides his relation to his lover.

Near the end of the film, Irene, the older sister, tells Ana her nightmare, in which kidnappers threatened to kill her. Irene finishes her story with the following words: "they put a gun to my temple, and when they were going to kill me ... I woke up." In her dream, she was tied up, abused, not able to talk, submissive and scared. Of the three sisters, Irene is the one who tries to adapt to the tyrant, to be obedient, to try not to get in trouble. Her dream seems to express her underlying experience of defenselessness, her uncertainty about her survival, and her hope that she won't be killed.

Psychoanalysis looks for the "truth" of the patient, with all the consequences that it may bring. What is unconscious is made conscious. In a dictatorial system the opposite occurs. The regime tries to deny truth and disavow reality. Being raised in a dictatorial system shapes the mind. The possibility of being severely punished, rejected, or killed makes it difficult to differ from publicly mandated "truths." It takes years or perhaps generations to trust that thinking differently from leadership will not be punished by death or exclusion.

Ana suffers experiences that will take decades to repair – if she succeeds at all. There is a similar challenge for a society casting off the cloak of oppression. A denied trauma is never forgotten because it remains ready to be repeated. As Freud would say, we need to remember trauma to be able to forget it and avoid repeating it. It seems that the compulsion to repeat occurs not only individually, but also at the level of societies in wars, dictatorships, killings, and discrimination. Internal and external realities are imbricated.

The Axis of Spaces and the Axis of Memory

In this section we will discuss another element of the film – that of spaces and settings. External spaces are intimately interconnected with internal spaces in *Cría*, similarly to how collective history contains the individual. Another axis of the film is memory, as a potential space for transformation. Saura, in resistance to dictatorship, makes full use of fantasy. He creates films between the psychological and the political, between the feminine and the masculine, childhood, adulthood and old age, vulnerability and violence, freedom, and oppression. Ana personifies remaining connected to psychic reality – both to her process of mourning and loss, as well as to the positive and loving relations that she insists on remembering. She rejects narratives imposed on her that she knows are not true. Daydreaming and playing allow her to resist, actively acting out what she suffers passively (*fort–da*).

The mansion (in which most of the film is set), is enclosed by high walls, like a bunker, a non-home. It could be thought of as representing a totalitarian space of mind – a space of confinement and dictatorship, a space of negative links and psychotic parts of the mind (Bion 1984 [1962]) such as hypocrisy, betrayal, violence, and arrogance. The mansion is in contrast to Bachelard's concept of the house: "the house shelters daydreaming, the house protects the dreamer, the house allows one to dream in peace" (Bachelard 1989 [1969], 26). The interior spaces of the mansion are suffused with darkness and shadow, reflecting anguish and pain. Ana dreams of a liberating flight from the silent and gloomy mansion-dictatorship to the vitalized, almost deafening noise of Madrid. In one scene she stands on the roof of the mansion, but she does not fall; she imagines herself flying, as often happens in children's dreams.

The paralysis and dependency represented by grandmother's wheelchair conveys involuntary imprisonment: lack of freedom of movement and speech, as well as silencing of communication and dialogue. Conversely, Ana has the most intimate and empathetic relationship with her paralyzed grandmother, who is dependent and unable to speak. Ana remains internally connected to loving links – her guinea pig, her sisters, her grandmother, Rosa the maid – remembering the warm embrace of her mother(land), through piano music and bedtime stories.

Ana's true and deep gaze entails a "double movement of looking outward and seeing inward. A wise child, Ana speaks little but observes and listens a lot, vigilant to external noises as well as internal ones, in search of knowledge" (Melícias 2021).

Ana lovingly cares for her guinea pig Roni, who is in a cage space with a wheel. Her lovingly feeding Roni lettuce can be seen as in contrast to being in a cage on a repetitive autocratic wheel-mind preventing creativity and relationships.

The parents' room/bed, instead of representing a primal loving scene space, is linked to the painful memory of mother's agonizing illness. The room/bed is also linked to witnessing father's betrayal and to father's death – desired and magically achieved in fantasy.

On the other hand, the children's room presents itself as a vitalized and potential space (Winnicott 1975 [1971]) for experience. In one scene, the three children playfully act out the arguments they have heard between the parents. In one scene the children imitate interchanges overheard from their parents. Their healthy sororal complicity allows their staging of the (non)love of their parents. Femininity also blossoms as they investigate and play with women's magazines, makeup, dances, and love songs, preparing for adolescence. The sororal dimension is decisive in many traumatic situations as well as in resistance groups and organized struggle. But the children's room also houses fears: of the night, of darkness, nightmares, and mourning so painful to process. Winnicott warns us that "there is a limited value to internal freedom ... if it is consciously experienced only in persecutory circumstances" (Winnicott 1989 [1969], 185).

The setting of the unsettlingly empty swimming pool in the backyard implies the absence of the uterinely aquatic womb and instead can be seen as tomblike. This space without content becomes occupied by Ana, in which she builds a hut. She keeps alive the capacity for "make believe" that is necessary for the elaboration of reality, from passively experienced to actively imagined. Through playing with the dolls/infants Ana elaborates internal conflicts, processing ambivalence toward a good and bad mother. But the authoritarian aggressor can be seen here as well in her superegoic games of punishment and guilt, negating vulnerability and difference. Ana's play and creativity fills the empty pool, transforming the claustrophobic emptiness and dehumanization of authoritarian regimes.

Contrasting with this negative space where she conveys her "capacity to be alone" (Winnicott 1998 [1958]), Ana feels accompanied by Rosa in the kitchen, an ancestral-oral space of nourishment. The kitchen is a primary space of concreteness but also a space of curiosity about the mysteries of origin and birth. With Rosa, she learns prosaically and without taboos about men, love, betrayal, disease, death, and birth. With Rosa, she quenches her curiosity about her own birth. She also asks to see Rosa's breast and satisfies her voyeurism, surprising herself when she sees Rosa's big breast – both erotic and nourishing.

We now turn to the bathroom – space of secrets and taboos. It is in the bathroom that Ana reveals her secret to Aunt Paulina that father died while having sex with Amelia. Aunt Paulina then insists it isn't true, teaching denial and hypocrisy. The bathroom setting is the space of anality *par excellence*, of control and authoritarian power exercised by Aunt Paulina trying to force Ana into a lie.

Perhaps the wall of photos that grandmother and Ana view together iconically represent the entire film. Viewing the photo wall together, Ana empathetically offers to help her grandmother – either to die if she so desires or to live by traveling through her good memories. Sharing memories allows us to surpass the limits of space and time. Memory is also a place to "go outside inside." *Cría* begins with Ana's memory album – "the day I was born" – and is a narrative dialogue between Ana as a child and the memories of adult Ana shown in flash forward scenes. Memory, a generator of thought, is a potential space for freedom, creativity, reparation and hope.

Memory entails a kind of language of emotion that fosters mourning (breast, mother, father, childhood, countries, relationships, etc.) from which new narratives can be dynamically articulated. New narratives can arise from the use of negative capability (Bion 2019 [1977]) including the toleration of doubt, uncertainty, and mystery. Memory can also be a space which can resist manipulative and misleading discourse utilized by totalitarian regimes. Memory, by allowing us to create a new experience, transforms the tragedy of catastrophe into benign catastrophic changes (Bion 2014 [1966]) and transforms the fatality of fate into creative opportunity.

As mentioned earlier, just before the final scene of the film, Irene tells Ana a nightmare as they eat breakfast, "They put a gun to my head, and when they were going to kill me ... I woke up." Spain and the children wake up from a long nightmare of fascism. In the final scene of the film, the girls emerge from the mansion on their own for their first day back at school. This scene embodies relief and hope. It becomes possible to go outside, away from Francoism and the claustrum-mansion. In the wide space of the freedom of the city street, we follow Irene, Ana, and Maite. They are released from the walls, on their way to school, an exogamous space of expansion, knowledge, culture, and enriching affective exchanges.

Conclusion

Man-made humanitarian catastrophes such as the Holocaust, wars, political and ethnic persecutions use dehumanization and destruction of subjectivity to annihilate the historical and social existence of individuals and groups. Psychoanalytic work attempts to integrate these traumatic experiences into a narrative context. Collective traumas also require a social discourse on the historical truth of the traumatic events, as well as on their defensive denial.

Psychoanalysis works in the search for truth. If there is censorship in a country, if there is no freedom of expression, if the law does not prevail, and if justice becomes a non-functioning structure, the most profound and essential part of a child's development is threatened. The traumatic effects of lack of freedom appear in the impairment of subjectivity that can be observed analytically. *Cría*'s characters convey the dynamic processes of subjectivization, which occur in the children's interaction with their time.

We live in a world with the ghosts of the past. Today's wars, as well as violence in its various forms, racism, xenophobia, and fanatical leaders, are eloquent proof that the tendency towards human self-destruction is still alive. Films play a key role in creating a reflective space for societal traumas, especially when so much of our lives are spent in virtual reality, and cinema has managed to reach the masses through online platforms. Films can function by intervening, modifying, and interacting with the society that produces them, that is, as agents of history. They set in motion the memories and direct experiences we have of different events. History is superimposed by the images of films, which in this way give additional content to our own images of the past.

The psychoanalytic approach to art and aesthetics has a long tradition. Some approaches focus on the analysis of literary works as a key to biographical studies of their authors. Some study the work of art itself. In relation to the former, *Cría* is a journey through Saura's childhood. The film was released in 1976, at the time that Franco was slowly dying, inaugurating what was called the Transition period in Spain. This historical interval is reflected in the transformation processes of *Cría*'s protagonist Ana. Saura masterfully used a narrative technique disoriented with flashbacks and flashforwards. Present and past situations are shown in a dreamlike manner, leaving unclear what is memory, hallucination, or part of the current narrative. There is a flow of superimposed memories and images, similar to how thoughts and emotions crowd our minds.

Saura created a narrative of great beauty, even within societal and familial horror, in which he manages to blur the boundaries that separate life from death and childhood from adulthood. He did it with a delicate vision, attentive to detail, always seen through Ana's hypnotic gaze.

The creative union of cinema and psychoanalytic thinking is an important agent in the process of construction and support of collective memory. We have hoped to convey here, how societal and domestic violence, lack of free-dom, fear, sadness, and hopelessness deeply affect development and shape a child's vision of the world. *Cría* conveys a sense that Ana will wrestle with these issues for a lifetime, as Saura acknowledges he did himself.

Ana and the sisters (representing childhood) inhabit virtually every scene in the film. They share a similar view of the events that build their lives and yet have unique ways of relating to these events. The spectator is one more child in this same confusing and depressing reality. However, the sisters also share the creativity of play – they kill and resurrect, disguise themselves, dance and imagine – thus allowing a potential space necessary for life itself.

Acknowledgements

This chapter was based on a film panel entitled "Totalitarian Regimes and a Child's Mind: *Cría Cuervos,*" presented at the 53rd International Association Meetings in Cartagena Colombia on July 29, 2023. The chair was Mary T. Brady, and the panelists were Virginia Ungar, Adriana Prengler, and Ana Belchior Melícias. These conference papers were published in 2023 as "Règims totalitaris i una ment infantil: Cría cuervos," *Revista Catalana de Psicoanàlisi*, XL/2, 69–87.

Notes

1 Ambrosio Garcia discusses cinema in Bion's terms as a "thinking space ... which drives the subject in the search for truth even if it is painful" (Ambrosio Garcia 2017, 50).
2 Sabbadini also asserts that Saura, (which he sees as rare for male directors), has a deep understanding of female internal worlds. Sabbadini comments that all the central characters in *Cría* are female, and the male characters only marginal.
3 For a recent paper on *The Spirit of the Beehive*, see Gerhardt and Slobin (2024).

References

Ambrosio Garcia, C 2017. *Bion in film theory and analysis: the retreat in film*. New York: Routledge.
Bachelard, G 1989 [1969]. *The poetics of space*, trans. M Jolas. Boston: Beacon Press.
Bion, WR 1984 [1962]. *Learning from experience*. London: Karnac.
Bion, WR 2014 [1966]. Catastrophic change. In C Mawson (Ed.), *The complete works of W. R. Bion*, vol. VI. New York: Routledge.
Bion, WR 2019 [1977]. *Capacidade negativa*. In WR Bion, *Capacidade negativa: Um caminho em busca de luz* (Trans. A Stürmer). Zagodoni.
Bollas, C 1992. The fascist state of mind. In C Bollas, *Being a character: Psychoanalysis and self experience*. New York: Routledge.
Bordes, E & De Sobrón, L 2021. *Madrid bombardeado: Cartograia de la destrucción, 1936–1939*. Cátedra.
Brasó, E 1974. *Carlos Saura*. Madrid: Taller de Ediciones Josefina Betantor.

Faimberg, H 2005. *The telescoping of generations.* Hove: Routledge.

Ferenczi, S 1955. Confusion of tongues between adults and the child. The language of tenderness and of passion. In S Ferenczi, *Final contributions to the problems and methods of psycho-analysis* (pp. 156–167). London: Hogarth.

Gerhardt, J & Slobin, D 2024. Victor Erice's film *The Spirit of the Beehive:* A coded warning against submission in the aftermath of the Spanish Civil War. *Psan Dialogues,* 34 (2).

Melícias, AB 2021. Blog cinema & psicanálise – totality is the non-true. https://cine mapsicanalise.pt/2021/10/06/a-totalidade-e-a-nao-verdade-cria-corvos-1976.

Sabbadini, A 2014. *Moving images: psychoanalytic reflections on film.* New York: Routledge.

Stone, R 2001. *Spanish cinema.* New York: Routledge.

Winnicott, DW 1998 [1958]. A capacidade para estar só. In DW Winnicott, *O ambiente e os processos de maturação.* Porto Alegre: Artes Médicas.

Winnicott, DW 1989 [1969]. A liberdade. In DW Winnicott, *Tudo começa em casa.* São Paulo: Martins Fontes.

Winnicott, DW 1975 [1971]. O brincar: uma exposição teórica. In: *O brincar e a realidade.* Rio de Janeiro: Imago.

Chapter 9

The Need of Truth for Healthy Psychic Development

Antònia Grimalt

Introduction: The Emperor's Clothes

Hans Christian Andersen astutely analyzed certain weaknesses of the individual and society. "The emperor's clothes" has become a phrase often used to describe pomposity, the falsity of which is not discovered or denounced because the person who perceives them is afraid of being called insensitive or rude and unable to appreciate the truth of what he observes. Through his story, Andersen describes complicities and organizations within society and within individuals that pose serious obstacles to truth:

> The vain emperor was enslaved by two brigands who, in exchange for a reward, offered to weave a wonderful cloth that, apart from giving great splendor to whoever wore it, would allow the misdeeds of officials who mismanaged the country to be discovered, thanks to a quality as simple as it is extraordinary: the stupid, the wicked and the swindlers would not know how to see it. The brigands installed themselves in the royal workshops and pretended to work day and night. When the emperor went to visit them, he saw nothing on the loom. Suddenly he remembered everything he had promised, but he hadn't done or had done half-heartedly, and he didn't dare open his mouth to say he didn't see the fabric; on the contrary, so that his traps were not discovered, he praised the beauties of what he did not see. The same thing happened to all the State officials who went to admire the fabric, in fact non-existent. The more anxious they were to be discovered as deficient and reprehensible officials, the more they tried to praise the non-existent work of the loom. Finally, in a procession, in which the emperor "showed off" in the midst of his flattering entourage the dress made of that very subtle fabric, a child exclaimed: "*But the emperor does not wear a dress!*"

Andersen reminds us that a fraudulent system allows individuals to be surreptitiously used. We know too well that the power of tyrannies depends less on military force than on the corrupting cunning that

DOI: 10.4324/9781003520757-10

ensnares citizens. Such power allows them to benefit from any kind of administrative injustice and keep citizens silent. It goes without saying how difficult freedom of expression is, when to the censorship imposed by the governing institutions there is added even more severe censorship by the individual, who does not want and cannot take charge of his own weaknesses.

The story is a good point of reference to describe psychic organizations that hinder contact with internal and external reality. Without some clarity of both realities, progress becomes difficult, if not impossible. Psychoanalytic research in recent years has been applied to the study of resistance against change that does not depend on the difficulty of the unconscious becoming conscious (Freud 1925) nor on the inability of the ego to perceive and think about what is perceived. The currently studied resistance depends much more on the complicated nature of the individual's defensive organization. One of these organizations is well embodied in our story. The entire organization of the emperor was leading him and his entourage to an increasingly catastrophic and irreversible situation, with a growing dependence on the lies. In this situation, the possibility of change that happens through the discovery of the great fallacy on which the organization has been based, is experienced as increasingly terrifying (Eskelinen de Folch 1983).

Wilfred Bion (1962) described two main ways of dealing with a reality that the individual feels is flawed and frustrating: either he takes charge of the deficiency and tries to modify it, or he hides the perception and thus escapes the reality. The first way of proceeding favors development, the second leads to psychic organizations that we call psychotic, perverse or neurotic depending on the quality of the anxieties and defenses they involve.

Naming, knowing, inventing lies, and finding thoughts are recurring themes throughout Bion's later writing. I wish to suggest that they are as fertile a ground for exploration in the field of the group as they are in that of the individual. More than that, I also believe that these two fields provide, in Bion's phrase, a "binocular vision" for exploring and understanding the ground of human knowing and unknowing, becoming and being, without which we are prisoners of our fears and terrors in our private and public life.

I will proceed to describe the dramatic tragedy of a little child who, instead of colluding with the group lies, fought to not deny her own perceptions and doubts. Her struggle provoked intense anxieties and feelings of not being "normal". This impressive, dramatic and insightful experience from a then 10-year-old shows the intensity of guilt and disorientation when the truth is substituted by lies. This poignant vignette allows us to think about the emotional toxicity of ongoing lies and malignant manipulation of the truth:

A childhood is painted by a palette of illusions – that the world is safe, the adults are fair, and the future is bright.

On April 26, 1986, when Chernobyl's reactor No. 4 exploded, I was a 10-year-old living 60 miles away, in the Soviet Ukrainian city of Kyiv. It was a sunny Saturday, and I had spent most of the day outside, playing with other kids from our apartment building ...

Our next-door neighbour Olena, a researcher at the Kyiv Institute of Nuclear Physics, came over one day. Without the usual niceties, she drew my mother into our room and closed the door behind them. She told my mother that there had been an explosion at a nuclear power plant, and that radiation was escaping the reactor in Chernobyl, reaching dangerous levels in Kyiv. She said we should keep our windows closed, and that I must stay home instead of going to school.

I wondered if Olena could be right and the government wrong. It didn't seem possible. How could one person know more than the whole government, especially the government in Moscow, where they had the best specialists in everything? What Olena said about radiation sounded like a scary fairy tale: You couldn't see it or smell it, you couldn't get rid of it by sifting or boiling water, and yet it could kill you. I wiped my sweaty palms on my skirt.

A heated discussion ensued, the result of which was a unanimous conclusion that Olena was exaggerating a minor problem to flaunt her expertise. The three women, matriarchs of the families with whom we shared the communal apartment, nodded at one another and pursed their lips. They rolled their eyes at Olena's attention-seeking. I exhaled. Everything would be fine, it seemed.

Radiation was spreading through the air and through the rain. Buses brought refugees from Chernobyl into Kyiv, carrying additional radiation on the refugees' bodies and on their possessions. I was unaware of all of it. The explosion at Chernobyl blotted out my childhood. The Soviet way of dealing with problems was to soldier through with no whining or self-pity, and so I built a sarcophagus over the pain of my experience.

They knew what they were doing ... It took me a while to watch the HBO miniseries Chernobyl.

I was watching the reactions I'd yearned to see when I was 10 years old. Someone back then should have pounded on the table, gaped at the government lies, yelled at the hypocrites. Because nobody did, my own emotions seemed capricious. Because nobody ever showed remorse, my grievances seemed unjustified. Watching the series felt like receiving a diagnosis for a subtle but devastating malady, one that's hard for those not afflicted to appreciate, or even believe: It felt validating.

(Moskalenko 2019)

"Ukrainian War Is Not a War": The Toxicity for the Mind

The sincere communication of this child witness allows us to think about the *nuclear and emotional toxicity* this catastrophe represented, which continues with the permanent lies and use of confusion of some malignant minds around the war in Ukraine. According to the perpetrator of the invasion of Ukraine "this is not a war"! Putin's propaganda has portrayed the invasion of Ukraine as if he is a savior and not an invader.

The Russo-Ukrainian war began in February 2014. Following Ukrainians Revolution for dignity, Russia Occupied and annexed Crimea from Ukraine and supported pro-Russian separatists fighting the Ukrainian military in the Donbas war. Putin said the operation was "to protect the people" of the Russian controlled-breakaway republic.

The best evidence we have now suggests that Putin believes that an independent, democratic Ukraine is a threat to him and to Russia. Unfortunately, he's decided that the only way to deal with that threat is through the use of overwhelming military force over a long period of time.

Lies We Tell in the Face of the Unbearable

Sophia's Moskalenko living witness at 10-year-old is an impressive illustration of lies parents may tell to shield children from the terrors of war. We cannot forget the genocide in Gaza and the lies that we are all impelled to tell ourselves to live with the unbearable realities that so many children are living in war zones. UNICEF just declared that 2024 was the worst year for children and war in its history. Almost 1 in 5 children are living in conflict zones. As we struggle to acknowledge the lies we all might tell ourselves (i.e., there's nothing we can do, governments will eventually sort it out, someone will look out for these children, countries must defend themselves, etc.) to go on with everyday life, Bion's sensitive and detailed theories of the psychic necessity for and resistance to truth can be valuable and timely. How do such everyday lies corrupt our humanity and our own capacity to apprehend the reality that children are being brutalized? Do these lies not represent a retreat from this painful reality?

Early Infantile Emotional States

The idea of proto-sensoriality as something that first becomes an image and later a symbol is what Bion expresses through the concept of *transformation from beta to alpha*. He deduced and described how early infantile emotional states, pleasurable as well as painful, are experienced concretely and as such are not available for mental growth. These states cannot be thought about, imagined, dreamt of, or remembered (as opposed to being repeated), until they have been transformed into abstract experiences. An infant cannot acquire this capacity for transforming its primitive experiences from beta (a proto sensation, something that has not yet become thought) to alpha elements except by identification with an object capable of performing this fundamental function.

Such identification is achieved in healthy development via the use of projective identification as a normal mechanism. In this situation an infant evacuates an unmanageable, indigestible, conglomeration of good and bad experiences into the care-taking part object. This receptive part-object offers a realization of the infant's inborn expectation, its preconception, that there is a somewhere in which the unmanageable can be made manageable, the unbearable bearable and the unthinkable thinkable. The primary part-object, the breast in Kleinian terminology, acts through alpha-function, upon the projected beta elements and render them into thinkable, storable, dreamable alpha-elements. These are projected into the baby and introjected by it. The result is an identification with a part-object capable of performing alpha function. Perhaps one should speak of a trace identification because the word "identification" can seem to describe so formal and final an activity. It can seem to imply that one contact with one manifestation of alpha function in an object and all is well. It is likely that the baby needs a million or more occasions of such a transformation. The nature of a baby's need for repetitive experience is worryingly and increasingly misunderstood (Isaacs 1983).

A normal baby's capacity to tolerate frustration is very small. When a neonate experiences frustration, it is in a terrible state of suffering, which feels endless to the infant. This sense of urgency that an infinite agony must be stopped right away is projected into the mother. She must often, and normally can, react both appropriately and speedily. Reverie, or alpha function, can occur almost instantaneously. Such responses, meeting the baby's preconceptions with appropriate realizations, must be repeated on innumerable occasions, and over a long period of time, if normal mental growth is to be built on a firm foundation. Yet, reverie also means prolonged concentration with patches of musing and intense attention that the patient, pre-occupied mother accesses with her infant or an analyst with his patient. For appropriate growth in sensitivity and depth in any relationship depends on the object learning from experience as well as the subject.

Bion's idea of the breast as the prototype of the container, the recipient of the projected contents known by him as the contained, is not as static as the symbol suggests. Throughout life people use each other, their environment in this active way and are in search of an active response. Among situations perceived by all infants as damaging to the object and to the baby is the disappearance of the external object. For babies "gone, not there" equals "bad, broken" and "broken" to some extent always equals "will break me." The absent object gives no evidence to the contrary and such evidence is essential. Furthermore, the absence, the unavailability, of an essential object and the anxiety resulting from its disappearance both result in angry, greedy and envious responses and so leads to further attacks on whatever good internal representation of the object may have been achieved. In fact, thinking does not come from the mere absence of the object, but from his return after an absence that was bearable.

Bion hypothesizes that the mother's capacity to create bonds and represent them can facilitate the child's capacity to develop his or her own alpha function and, ultimately, to know emotional truth. The containment function allows the child to develop the dreaming function (i.e., the first step in the construction of the apparatus for thinking).

The Vital Importance of Emotional Truth

Healthy mental growth seems to depend on emotional truth as the living organism depends on food (Bion 1965). Bion argues that emotions are the significant core of experience, which seeks a symbolic form and representation to be thought of. Thought is driven by the human need to know the reality of who one is and what is happening in one's life. Truth-seeking as human beings is the core of research in the development of thought.

Grotstein (2005) thinks that the child may be endowed very early on with a function that is capable of attributing an initial personal meaning to the emotional fact that impacts him. In relation with another mind capable of reverie, α elements can continue their transformative path toward elements of the dream, the contact barrier, and memory. In the opposite situation, facts are denied by the mind and degraded into β elements and therefore remain impersonal in a space devoid of meaning. Personalization, then, is initially guided by another mind that enters into contact with that of the child and, since the transformative system depends on a relationship, it is always in a state of precarious equilibrium.

The baby's and mother's sensations constitute a first step towards sensory integration, a first "common relational sense, shared relational meaning" a "first sense of truth" in each subject and between them. Later this first integration will be able to evolve into a "truth-functional-statement," a second level of mental functioning but is more verbal and representational.

In Bion's work the impetus for knowledge is the desire to grasp what, in fact, is unknowable-the truth. The ultimate truth exists, whether it is discovered or not; the awareness of its existence is one of the elementary experiences of the mind. Truth is autonomous and part of the Faith of man. As Bion says, truth is essential for psychic health. Yet there is an awe that separates man from truth. Truth requires from the thinker modesty to accept something outside his powers. If the thinker takes an omnipotent attitude, truth is no longer what he must accept, but what he has created: truth becomes dependent on the questioner.

It is only by virtue of reverie and the mother's capacity for holding and handling, that a newborn succeeds in emerging from a state of non-integration and in coordinating his senses in such a way that they confirm each other. It is the mother who makes up for the infant's immaturity with her own mind. What is true for the infant is perception, in the sense of "common to the senses." Touch tells sight: "Received loud and clear; what I am feeling corresponds to what you are seeing," and vice versa. This correspondence stems from an infinite set of micro-experiences of emotional confirmation, unison, and consensuality. These minimal emotional experiences (of sense) interweave to form the stuff of thought. The mother helps the child to confer order on the chaotic flow of stimuli in which he is immersed and so to speak passes on the method of so doing to him.

At the beginning of life in particular, the drive towards integration and the construction of a psychic space can obviously be fulfilled only on the basis of emotional unison, and not yet by intellectual agreement, because, if any type of categorization is already present, it is preverbal and prereflective, semiotic, affective and not yet truly conceptual in nature. At this stage, the emotions directly express the principle of value essential for survival, and they continue to do so throughout life, even when accompanied by logic-rational thought. There will always be reasons of the heart or body which the mind cannot represent to itself but can only feel (Civitarese 2016).

So, the truth drive is found to consist in the search for this special attunement with the other. Bion's conception is social and relational, and it demonstrates better than any other term that the factor which promotes life and makes for development of the mind is the thirst for sociality.

The value of the concept (reformulated as the urge for emotional attunement) is not confined to the sphere of analysis. It helps us to see that any type of truth recognized by a community – at the limit, including, too, the truths of science – is a particular case of communication, and also has its roots in emotional accord, in being-in-unison. Given the centuries-old tradition of contrasting the passions with reason and dreaming with waking, the idea that judgement concerning factual reality is ultimately rooted in emotional and "oneiric" reality may seem bizarre, but, given the premises of Bion's theory, it is not. In moving on from "somatic emotional categorization" (Grotstein 2007, p. 276f.) – i.e. semiotic categorization – to the concept, all we are doing is getting to know reality by simplifying it and reducing it to a system of relationships.

In his work on "basic assumptions" Bion (1961) develops the concept of unconscious fantasy within a group (and of course he considers the analytic couple as a group). Individuals are endowed with "valencies," that is, the spontaneous and instinctive (unconscious, automatic and inevitable) capacity for setting up an emotional bond between one another in order to share a basic assumption and to act on that basis. The basic assumptions (which may be one of pairing, fight-or-flight, or dependency) gives rise to "mental activities that have in common the attribute of powerful emotional drives. It is the cement that binds the group" (López-Corvo 2006).

Valencies and basic assumptions express aspects of the individual's psychological functioning dictated by the "proto-mental system." Bion coins this concept, which is among his most speculative in that it "transcends experience" (Bion 1961, p. 101), to explain the tenacity of emotional links which consolidate groups, joining the members in a common psychological situation, and to indicate a psychic dimension in which basic assumptions may be located during those moments when they are inactive. The proto-mental system is one in which physical and psychological or mental are undifferentiated. It is a *matrix* from which spring the phenomena which at first appear – on a psychological level and in the light of psychological investigation – to be discrete feelings only loosely associated with one another. It is from this *matrix* that emotions proper to the basic assumption flow to *reinforce, pervade*, and, on occasion, to *dominate* the mental life of the group. Since it is a level in which physical and mental are undifferentiated, it stands to reason that, when distress from this source manifests itself, it can manifest itself just as well in physical forms as in psychological (Civitarese 2016).

A particularly interesting aspect of Bion's reflections on the relationship between truth and lies is that he places the emphasis, from the beginning of his observations, on the importance of the relationship between two minds, and this relational root is what distinguishes psychoanalysis from other scientific disciplines (Bion 1970). Freud had already anticipated the idea of the importance of truth and that the push in the direction of truth is inscribed in the psyche. He thought that unconscious, repressed desire is in itself a form of a lie but maintains a pressure toward the truth. Indeed, Freud writes that the unconscious, "which is ordinarily our opponent, comes to our help, since it has a natural 'upward drive' and desires nothing better than to press forward across its settled frontiers into the ego and so to consciousness" (Freud 1938, p. 179). We could say that there is an inclination toward truth and communication in the repressed unconscious.

Bion recognizes that man's capacity to survive depends also on his capacity for illusion – broadly, for lying to himself and lying in general: "It is difficult to accept real life because frustration is an essential feature of real life. In an extreme position it obstructs the development of thought" (Bion 1977, p. 28). And returning to the challenge of facing the truth of disaster- the self-deception that arises in the face of such a task:

> man owes his health, and his capacity for continued health, to his ability to shield himself during his growth as an individual by repeating in his personal life the history of the race's capacity for self-deception against truth that his mind is not fitted to receive without disaster.
>
> (Bion 1992, p. 192)

War and Lies: Basic Assumptions that Aid in the Dehumanization of the Other Including Children

The basic assumptions in a group allow for dehumanization of the other and this Othering of the enemy profoundly affects children. What is the impact on children who do not go along with dehumanizing basic assumptions?

The *Boy in the Striped Pajamas* (directed by Mark Herman) is a powerful fictional story that offers a unique perspective on how prejudice, hatred and violence affect innocent people, particularly children, during wartime. A German boy whose father is a Nazi befriends a Jewish boy in a concentration camp. He doesn't understand that the boy is a prisoner and gradually comes to realize what's happening and the horrifying basic assumptions of his collective. He stays rooted in his own emotional truth.

Through the eyes of an 8-year-old boy largely shielded from the reality of the Second World War, we witness a forbidden friendship that forms between Bruno, the son of Nazi commandant, and Shmuel, a Jewish boy held captive in a concentration camp. Though the two are separated physically by a barbed wire fence, their lives become inescapably intertwined. The imagined story of Bruno and Shmuel sheds light on the brutality, senselessness and devastating consequences of war from an unusual point of view. Together, their tragic journey helps recall the millions of innocent victims of the Holocaust. It also reflects a capacity, in the face of tremendous psychic pressure, to resist basic assumptions that enable dehumanization.

References

Bion, W. R. (1961). *Experiences in groups*. London: Tavistock.
Bion, W. R. (1962). *Learning from experience*. London: Heinemann.
Bion, W. R. (1965). *Transformation*. London: Heinemann.
Bion, W. R. (1970). *Attention and interpretation*. New York: Basic Books.
Bion, W. R. (1977). *Two papers: The grid and the caesura*. London: Karnac.
Bion, W. R. (1992). *Cogitations*. London: Karnac.
Civitarese, G (2016). *Truth and the unconscious in psychoanalysis*. London: Karnac.
Eskelinen de Folch. T. (1983). We versus I and you. *International Journal of Psychoanalysis*, 64: 309–320.
Freud. S. (1925). The resistances to psycho-analysis. In J. Strachey (Ed. and Trans.), *The standard edition of the complete psychological works of Sigmund Freud*, vol. 19. Richmond: Hogarth Press.
Freud. S.(1938). Some elementary lessons in psychoanalysis. In J. Strachey (Ed. and Trans.), *The standard edition of the complete psychological works of Sigmund Freud*, vol. 23. Richmond: Hogarth Press.
Grotstein, J. S. (2005). Projective transidentification. *International Journal of Psychoanalysis*, 86: 1051–1069.
Grotstein, J. S. (2007). *A beam of intense darkness: Wilfred Bion's legacy to psychoanalysis*. London: Karnac.
Isaacs, S (1983). Bion and babies. In J. S. Grotstein (Ed.), *Do I dare disturb the universe*. London: Karnac.
López-Corvo, R. (2006). *Wild thoughts searching for a thinker*. London: Karnac.
Moskalenko, S (2019). I was a child of Chernobyl. www.vox.com/the-highlight/2019/6/25/18716117/chernobyl-evacuated-true-story.

Part II

Pandemic

Kristin Fiorella

Introduction

> Because certain conditions of life and living are laid bare by the circulation of
> the virus, we now have a chance to grasp our relations to the earth and to each
> other in sustaining ways, to understand ourselves less as separated entities
> driven by self-interest than as complexly bound together in a living world that
> requires our collective resolve to struggle against its destruction.
>
> – Judith Butler (2022)

Like the climate crisis, the pandemic makes clear that, no matter where we
are on this planet, we are interdependent and entangled with one another and
also with the more-than-human world. Both the pandemic and our ecological
crisis upend the sense that we are discrete, bounded individuals. Both illuminate
also the material dimensions of this entanglement. Our bodies are inextricably
porous to one another and also to everything that is *not* us.

While both crises reveal a common world that is, in Latour's (2017) use of
the word, Earthbound, they simultaneously illuminate all the ways in which
we do not share a common world. Because of systemic and racist inequalities
rooted in violent colonial histories, for some collectives the pandemic intensified poverty and precarity. In the U.S., Black and brown people were three
times as likely to be infected and twice as likely to die. In Brazil, as Alicia
Lisandro describes in Chapter Thirteen, systemic inequalities and poverty,
coupled with governmental corruption and mishandling of the pandemic,
resulted in over 700,000 deaths.

During the pandemic, we all lived with death. As these chapters demonstrate,
this reality profoundly impacted children and adolescents, even though they were
less likely to die of Covid. The pandemic also put on full display inequities in
access to healthcare and how those inequities define certain populations as, in
essence, dispensable. As Padrón's and Lisandro's chapters describe, children and
adolescents whose lives were not recognized as valuable during the pandemic
were robbed of subjectivity and agency. When they suffered systemic attacks on
their bodies and minds, they struggled to think freely and to anchor meaning.

DOI: 10.4324/9781003520757-11

In the U.S., in the context of a health care system brutal to Black and brown people, there was the ever-present drum beat of Black people being murdered by police. It was a moment where both of these realities intersected and made increasingly visible the "somatic sense of dispensability" that Black Americans have lived with for so long (Butler 2022). Against this background, George Flyod was murdered, and the Black Lives Matter movement mobilized a global demand for the dismantling of white supremacy, cessation of violence against Black bodies, and safeguarding of the dignity and value of Black lives.

Such decolonial movements are vital efforts to de- and re-construct the very structure of our social world. Postcolonial scholar Achille Mbembe expressed that reconstructing an interdependent, shared world is "by definition a planetary enterprise, a radical openness of and to the world, a deep breathing for the world as opposed to insulation," (Nilsen, quoted in Butler 2022). The pandemic, along with the climate crisis, exposed the disavowal of a world held in common. Both these phenomena have also mobilized efforts towards a radical reworking of interdependency and the creation of new social imaginaries. As chapters in both this section and the next illustrate, many children and adolescents are highly attuned to and involved in these efforts towards building a common world.

The pandemic radically disrupted the lives of children and adolescents, as well as their parents. The continuity of their social and academic worlds was upended. This dislocation of the world they knew sometimes had devastating impacts on their psychic lives, as anyone who worked with children and adolescents during the pandemic knows. It was also a time in which children and adolescents reached for new modes of inhabiting their worlds. Technology, already vital to the lives of many children and adolescents, became even more central.

During the pandemic, technology became the medium by which we all maintained social connections and also through which psychoanalysts maintained connections with patients. Several of the papers in this section take up the challenges and possibilities that accompanied this move to virtual treatments. Kathleen del mar Miller's chapter beautifully demonstrates the need for a rethinking of core psychoanalytic concepts in the light of our entanglement with technology, particularly since the pandemic. She argues, along with several other chapters in the climate section, that psychoanalysis must expand to include the more-than-human environment, of which technology is a part. Such an expansion disrupts certain psychoanalytic concepts of subjectivity. By contrast, in Chapter Twelve Fernando M. Gómez highlights some of the potential risks of the virtual world for children and adolescents.

Overview of Chapters

Part II begins with Carlos Padrón's chapter demonstrating how colonial legacies and systemic racism, in the context of the pandemic, had a nefarious impact on the material, symbolic, and psychic life of poor and working class children and adolescents of color (Chapter 10). Left outside the symbolic, material and legal order, these individuals suffer systemic attacks on their bodies and minds. This fact, in tandem with the devastating impacts of the pandemic, produced what the author calls an experience of "the end of the world." Three main consequences of all these configurations are discussed: (1) failed identifications with whiteness; (2) loss of play; and (3) "confusion of tongues." The need for new social lullabies, ones that invigorate our social capacity to dream the (colonial) state of affair as being otherwise and that create communal solidarity, is proposed.

Then, in Chapter 11, Kathleen Del Mar Miller cautions that if analysts are to continue working virtually in the pandemic's aftermath, then psycho-analytic theory and technique must be reconsidered. Miller describes also her profound experience of transitioning from working in-person to working virtually during the pandemic with Lila, a mixed race, trans adolescent. Although Miller experienced Lila as difficult to reach when working in person, the shift to working virtually enlivened the treatment and allowed Lila to explore her early trauma, as well as aspects of her mixed race identity for the first time. Drawing on her dream of a techno-baby alongside the work of Ferenczi, as well as more recent writings on technology and embodiment, Miller argues that psychoanalytic conceptions of subjectivity must be expanded to include the more-than-human environment.

Fernando M. Gómez, in Chapter 12, explores technology and its post pandemic centrality for adolescents from still another point of view – the unprecedented social complexity generated by the omnipresence of social media and the rapid assimilation of an AI. He questions whether adolescents possess sufficient ego resources to navigate the fast paced exposure to social ideals linked to a "perfect self," violence, bullying, and AI "other" who is always and unconditionally available. He also explores the post pandemic surge in depression and suicide in adolescents He examines the numerous domains of psychic work that characterize adolescent subjectivation in the context of our post pandemic world.

In Chapter 13, Alicia Beatriz Dorado de Lisondo looks at COVID's exacerbation of an already catastrophic social and economic situation in Brazil. Systemic inequalities and poverty, coupled with governmental corruption and contempt for science, resulted in over 700,000 deaths by the time S. O.S. Brasil established a psychoanalytic, multidisciplinary team approach serving babies, children, and adolescents. S.O.S Brasil provides short term interventions to sew the holes left by trauma. It is an expression of a psychoanalytic social responsiveness that Lisondo argues is urgently needed in the context of our current global situation.

In the final chapter of Part II, Christine Anzieu-Premmereur describes some of the impacts of the pandemic on young children (Chapter 14). The persistence of trauma during COVID gave rise to what André Green calls a disconnection between Eros and Thanatos. In this phenomenon, primitive defenses are activated and emotional contact is avoided. When parents suffer in this way, they are unable to provide adequate containment for young children. Some children, during the pandemic, were able to maintain auto-erotic fantasies, enabling them to play with persecutory anxieties and fantasies of sadistic intrusion of the virus. For other children, transitional space collapsed, leading to fetishistic activities, instead of symbolically elaborated ones. Anzieu-Premmereur offers several clinical vignettes to demonstrate the ways in which child analysts worked to create virtual psychic envelopes with their child patients in order to offer containment and awaken transitional space.

References

Butler, Judith. 2022. *What World Is This?: A Pandemic Phenomenology*. New York: Columbia University Press.
Latour, Bruno. 2017. *Facing Gaia*. Cambridge: Polity Press.
Latour, Bruno. 2022. *How To Inhabit the Earth*. Cambridge: Polity Press.
Nilsen, T. 2021. Thoughts on the planetary: An interview with Achille Mbembe. In Jaco Barnard-Naude (Ed.), *Decolonising the Neoliberal University*. London: Birkbeck Law Press.

Chapter 10

Other Lullabies

Attacks on Blackness, Confusion of Tongues, and the Loss of Play

Carlos Padrón

Duerme negrito

Duerme, duerme negrito
Que tu mama está en el campo, negrito
Duerme, duerme negrito
Que tu mama está en el campo, negrito

Te va a traer codornices para ti
Te va a traer muchas cosas para ti
Te va a traer carne de cerdo para ti
Te va a traer muchas cosas para ti

Y si negro no se duerme
Viene el diablo blanco
Y Zas, le come la patita

Yakapumba, yakapumba
Yakapumba, yakapumba
Apumba, yakapumba, yakapumba, yakapumba
Apumba, yakapumba, yakapumba, yakapumba

Duerme, duerme negrito
Que tu mamá está en el campo, negrito
Duerme, duerme negrito
Que tu mamá está en el campo, negrito

Trabajando
Trabajando duramente
Trabajando, sí
Trabajando y no le pagan

Trabajando, sí
Trabajando y va cosiendo
Trabajando, sí
Trabajando y va de luto

Trabajando, sí
Pal' negrito chiquitito

DOI: 10.4324/9781003520757-12

Trabajando, sí
Pal' negrito chiquitito

Trabajando, sí

No le pagan, sí
Duramente, sí
Va tosiendo, sí
Va de luto, sí
Duerme, duerme negrito
Que tu mama está en el campo
Duerme, duerme negrito
(Canción Anónima)

(Traditional)

English translation:

Sleep Little Black Boy

Sleep, sleep, little Black boy
That your mother is in the field, little Black boy
Sleep, sleep Black, little Black boy
That your mother is in the field, little Black boy

She's going to bring quail for you
She will bring many things for you
She's going to bring pork for you
She will bring many things for you

And if the little Black boy doesn't fall asleep
The white devil will come
And yes, he will eat the boy's little leg

Yakapumba, Yakapumba
Yakapumba, yakapumba
Apumba, Yakapumba, Yakapumba, Yakapumba
Apumba, yakapumba, yakapumba, yakapumba

Sleep, sleep, little Black boy
That your mother is in the field, little Black boy
Sleep, sleep, little Black boy
That your mother is in the field

Working
Working hard
Working, yeah
Working and not paid

Working, yeah
Working and sewing
Working, yeah
Working and in mourning

Working, yeah
For the little Black boy
Working, yeah
For the little Black boy

Working, yeah
They don't pay her, yeah
Hard working, yeah
She is coughing, yes
She's in mourning, yes
Sleep, sleep little Black boy
That your mother is in the field
Sleep, sleep little Black boy
(Anonymous Song)

A Lullaby from the Margins

Atahualpa Yupanqui, a well-known Argentine folk singer who archived and disseminated Latin American popular music all over the world, said that he first heard this anonymously created lullaby sung by a Black woman in the frontier between Colombia and Venezuela.[1] He says: "The song belongs to the Black people, and, like all lullabies, it has its feet on the ground but is also a little bit metaphysical."

I remember another lullaby, this time in Freud's *Civilization and Its Discontents,* one that comes after his description of a metaphysical battle between two cosmic giants. It is a lullaby that can be thought of as the obverse of "Sleep Little Black Boy." Freud references a "lullaby from Heaven" present in Heine's "Deutschland," a satirical poem critical of German nationalism yet earnestly devoted to the notion of a Fatherland:

> And now, I think, the meaning of the evolution of civilization is no longer obscure to us. It must present the struggle between Eros and Death, between the instinct of life and the instinct for destruction, as it works out in the human species. This struggle is what all life essentially consists of, and the evolution of civilization may therefore be simply described as the struggle for life of the human species. And it is this battle of the giants that our nurse-maids try to appease with their lullaby about Heaven.
> (Freud 1961, p. 122)

Freud critiques what he considers to be a cover for unconscious processes which construes the notion of an afterlife as a duel between good and evil, including an imagined reward for goodness and of God's omnipotent benevolence. The form of paternalism expressed by the "lullaby of Heaven" satisfies a need to be lulled and contained from the existential angst produced by living amidst a struggle between Life and Death where sometimes the latter takes the upper hand, as Freud had experienced during the Great War. Civilization can only evolve when Life, or Eros as a binding force, wins in the battle against Thanatos.

Freud is preoccupied here with European civilization; one whose constitution, at least from early Modernity, depended on colonization and extraction of free labor from non-white *racialized* others. While Freud doesn't address colonization head on, Beverly Stoute (2021) says that his theory of the duality between Eros and Thanatos can be used to begin gaining some understanding of racial hatred. And yet, she specifies, "Freud formulated a theory of the 'universal mind' detached from cultural and social context ... that made it difficult for psychoanalytic theory to subsequently recognize, formulate, and integrate the experiences of the oppressed" (Stoute 2021, p. 260). In tandem with a "theory of universal mind" is the assumption that European civilization is the site of the Universal. This fact suppresses the social and cultural context of colonization and racism.

Extending from Descartes's *cogito ergo sum* ("I think, therefore I am"), Argentine/Mexican philosopher Enrique Dussel relates to the *ego cogito* (the "I think" as foundation of the modern subject) as inseparable from the *ego conquiro*, that is, the "I conquer," the subject of colonial drive and empire (Dussel 1992). It is only upon the presumptuous discovery of the Americas that Europe anoints itself as the place of the Universal, the alleged peak of evolved Humanity, disavowing the history of barbarism and oppression that was necessary to establish this "peak." The lullaby "Sleep Little Black Boy" touches on this history that Freud's analysis of Heine's shrouds.

Coloniality of Race Relations in Children

The little Black boy's lullaby, following Atahualpa Yupanqui, depicts a metaphysics of colonial and post-colonial (object) relations. The Black boy is left in the care of a Black woman, a neighbor, who is not his mom because his mom needs to labor as a servant. I want to stress that this Black woman is a friend or neighbor and not a bourgeois nursemaid for hire as in Freud's lullaby. The mother, we intuit, is in poor health and works hard while mourning. What is she mourning? Perhaps the boy himself, or the death of the little boy's father, or his absence, or her home in Africa. These are all possibilities, but the song suggests an ongoing state of the blues, or the "blue devils," meaning melancholy and sadness.

The caretaker, trying to soothe the Black boy to sleep, promises that the mother will bring him things that belong to the realm of fantasy: quail, pork, "many things." She assuages the Black boy's imagination through the evocation of reward. But she also says that if he does not go to sleep the white devil, who is the slave owner, then the colonial master, will come and eat his feet. The colonial stage is set: the life-form of Black enslaved people is characterized by material deprivation, violence to the body (the white devil will eat your feet), unfulfilled desires and fantasies of satisfaction and identification (perhaps one day being like the Master, as Frantz Fanon would say), abandonment, loss of home, paternal absence or death, illness, interminable mourning. However, just offstage in the wings is a communal Black solidarity and extended family that we still find today in inner city slums both in North and South America – the work of Eros confronting Death.

The lullaby speaks to pervasive systemic racism and class divisions that continue post-chattel slavery. One in which the poor and working-class Black mother, for example, leaves home to work for the meager wages with which to feed her child, or where a disproportionate number of Black women work in motherly ways (as daycare nannies, and so on) for other (typically white) mothers with higher incomes to attend to.

The white devil, internalized as a colonizing internal *object*, is used to put the Black boy to sleep with the warning that he will punish the boy if he stays awake. The Black boy must learn how to control his aliveness, desire, rage, will, autonomy and power, under conditions of slavery, colonialism, and its legacies. The arms of the Black boy's caretaker represent the containing role which is both psychological and cultural. It is not that dissimilar from the talk that every Black parent still has with their kids about what to do if stopped by the police.

The white devil eating the boy's feet is not a fantasy of castration but a historical fact and real threat *vis-à-vis* lynchings in the United States and in areas of Latin America. I take castration anxiety in the broader sense of an anxiety around the body's integrity within a political and social system where the Black body becomes the site of physical and discursive violence and projective identifications of what whiteness considers *abject*. The absent Black father in the lullaby is substituted with menacing whiteness which maintains its domination through threat and death, whereas Blackness defies this sensibility through shared experiences of *jouissance* and vitality.

Brief Excursus on Whiteness

Whiteness is the hegemonic, universal, disembodied *position* from which the racialized other is judged, measured, construed, imagined.[2] In "Decentering Whiteness," Charley Flint and Jeff Hitchcock characterize the position of whiteness as the standard, the background, the normal, the undifferentiated, that which is defined by the other or the outsider, the obvious, the familiar, the impervious to contradiction, the morally correct, the dominant, the place of resources and wealth, the ordained by God; a self-centeredness that makes it better than its obverse: the "marginal" (Flint & Hitchcock 2015). This position is unconscious, invisible. The eye that can't see itself. Whiteness, as a racial position, is elaborated in opposition to Blackness, and Blackness, *in identification with whiteness*. Frantz Fanon clearly saw this in *Black Skin, White Masks*: "[T]here is no longer any doubt that the true Other for the white person is and remains the Black person" (Fanon 2008, p. 139).

Racial difference is a social placeholder for a split between the central and the marginal that seeks to perpetuate itself through racism.[3] Mass shootings, police brutality, and lynchings are violent incidents constitutive of this split created by whiteness. They are normally criticized by "well intentioned" white liberals, including white psychoanalysts, to unambiguously position themselves as allies, and, many times, unconsciously, as morally superior to the racist "bad whites." They are the "good whites." An analysis of the almost manic search amongst the "good whites" toward reparative action in order to tolerate their own inevitable connection to internal and external violence is beyond the scope of this chapter.[4]

A Game You Cannot Win

In the context of the protests ignited by the assassination of George Floyd, a Black, heterosexual, cisgender, late adolescent female from a working-class family starts a session with me by saying that she decided to let her hair curl naturally after years of straightening it through a procedure that would burn her scalp. She says she loves it and can now look beautiful in two different ways. She says that she went to sleep and woke up every two hours. The dream she wakes from is the following:

> I'm in a house from which I can't get out. The house is surrounded by men with rifles. The house is dark, I can't see anyone, but I know it is packed with people. There's an ominous feeling that I sense. Being in the house means being part of a game. People are trying to win a game, but I feel that whatever outcome, losing or winning, is not good. If you win, you become part of a group you don't want to be part of. Losing means, well, just losing. The game creates the illusion that you should win but winning is not good. I suddenly feel that my fate in the house involves being burnt or branded with something hot on my skin. I then wake up full of anxiety.

She tells me that in the house she is not under the spell of the others who want to win the game. She wants to yell this to the people but has no voice. She says that she senses that the game is the house and that the house is the game. All she wants to do is go to a *different place she cannot find*. She suddenly remembers that the house has a gate where she hears evil spirits whispering. These are spirits that say the only way out is to let your body be snatched by them. Due to the work we have done in relation to her conflicts over being Black yet white identified, she comes to the realization that the game is a game of whiteness, or of proximity to whiteness. She says that in this game, if you win, then you lose yourself. I asked her at some point about her fate of being branded. With surprise, she connects it to her hair being burnt in order to be straightened, as well as Black enslaved people being branded by their enslavers. She further connects this with her own identifications with whiteness: it is the hair straightened forcefully by her mother to look beautiful like a white girl; it is having grown up as a girl in a white school where she was bullied for being Black and hence wanting to be the most beautiful and desirable white girl in the school; it is having had a "beautiful" white Barbie as her favorite doll and rejecting an "ugly" Black Barbie; it is her Black grandmother saying to her as teenager not to bring any friend or boyfriend to the house who was darker than her; it is her own grandmother denying she is Black but instead Brown. The patient says, "whiteness is a game you cannot win: if you lose, then violence is exerted against you; if you win, then you lose yourself, your connection to your own Blackness."

Jay Frankel says that identifications can be life-creating in so far as they are the stuff of which the self is formed in the midst of human relationships (Frankel 2002, p. 160). Following Ferenczi's idea of "traumatic aloneness," he claims that such identifications are born out of a fear of staying alone and its related experiences: "isolation, loss, separateness, otherness, rejection, being unwanted, banishment" (Frankel 2002, p. 159). All these are forms of potential trauma that might produce "psychic death." I would add that, in the case of marginalized and oppressed individuals, such as poor and working-class Black people, identifications with the status quo also function as a way of preventing *social death* by connecting them to others. Imagine my patient growing up in a Black yet Black adverse family and in a predominantly white school where she was bullied on account of being Black. The patient had no other solution but to identify with the discourse of whiteness to avoid psychic social death, to bridge the gap between herself and the libidinally charged people in her life: to stay connected. This use of identification, to expand on what Frankel says, is in reaction to a specific form of what I would call *social annihilation anxiety*, a potential severing from the social order imposed by whiteness that would be experienced as catastrophic. In this way, especially in children, whiteness is equated with the "good" or the "beautiful" while Blackness is associated with the "bad" or the "ugly."[5]

Failed Identifications with Whiteness

The lessons revealed by the lullaby show how identification with whiteness also entails *identification with the aggressor*. The logic is as follows: to avoid the trauma of isolation, the Black child identifies with whiteness – seeking to preserve their parents as "good objects" and the white status quo as the model of the "good." In doing so, the child internalizes the very aggression they sought to avoid in the first place: the "good" that becomes part of the fabric of the self ends up attacking it from within. In other words, the child trades one external trauma for an internal one, from which there is no escape.

This is what I call a "failed identification with whiteness"; in the words of my patient: a *game you cannot win*. Frankel says, while not addressing race, that a trauma causing someone to feel "outcast from the human community" produces a more basic feeling of "badness." This "badness" dissociates many people from their own Blackness, their history, culture, and vitality. Instead, it is replaced by what Frantz Fanon calls a "psychic schema" reflected by the *white gaze* which dehumanizes Black people by stereotyping them as hyper-sexual, primitive, violent, ugly, and so on. Fanon says:

> And then we are given the occasion to confront the white gaze. An unusual weight descended on us. The real world robbed us of our share. In the white world, the person of color encounters difficulties in elaborating their body schema. The image of one's body is solely negating. It's an image in the third person.
>
> (Fanon 2008, p. 90)

The idea of the white gaze constituting the "body schema" of the Black person is in line with W. E. DuBois's notion of *double consciousness* (DuBois 1909). This notion captures how the Black person, in seeing themselves through the eyes of a racist, white society, is not only dehumanized, but becomes torn within their very being. Prior to analysis with me, my patient only dated white men without full awareness. On dating apps, she said she would only swipe "white," not "right," a slip she made once. She discovered her inner fragmentation with her last white boyfriend, feeling ugly and fat only when she saw herself in front of the mirror. When no mirror was present, there would be no such self-perception. It was only through the *third person perspective of the mirror* that she saw herself this way: "how all those white skinny bitches see me."

Where is the Lullaby for George Floyd?

There is a violent discord between poor and working-class Black people and their "holding environment." This discord is with the psychic, economic, social, and material environment, but also with what Orna Guralnik calls the *discursive environment*. This is why people need to be reminded that "Black Lives Matter" and that "We can't breathe" (in reference to the respiratory symptoms of COVID-19, the death of George Floyd asphyxiated by the police, and the psychosomatic effects of racism). The dominant narratives, which position whiteness as the standard, often leave Black people, especially those from poor and working-class backgrounds, outside the realms of full life, citizenship, and subjectivity. These social structures don't just marginalize Black people in a general sense but also treat poor and working-class Black individuals as less deserving of rights and recognition. The impact of racism is intensified by class.

As Orna Guaralnik reminds us: "discord with one's discursive environment is very different from problematic object relations" (Guralnik 2010, p. 408). What is negotiated through the introjection of one's object relations has to do with "patterns of attachment and emotional coloring" (Guralnik 2010, p. 408), whereas what's negotiated with our normative environment are ideological instructions as to what is human and speakable. The suffering of poor and working-class Black people is not only in the area of interpersonal relations but also reaches the register in which "one's personal sense of self, reality and world gets constituted" (Guralnik 2002, p. 408). Guralnik reminds us that "our emotional internal economy is only one feature of what makes events traumatic. A major aspect of what is traumatic [...] is being designated socially unintelligible and having to incorporate what society wishes to turn its back on" (Guralnik 2002, p. 405). The traumatized individual is holding what "other's wish to repel, that which has no place in collective reality, and which is situated outside the symbolic order (society's *excess* and *abject*)" (Guralnik 2010, p. 405). As Beverly Stoute, among others, observes, violence becomes embedded in these projective identifications, creating an inherently unsafe environment for Black people – an environment that was made even more dangerous by the COVID-19 pandemic (Stoute 2021). I remember George Floyd dying at the hands of a white devil, calling for his mama. This heartbreaking event deeply disturbed me, leaving me to wonder: where was the lullaby for George Floyd, for the child he once was, for the poor Black children of this country?

Did he read Maya Angelous's *Life Doesn't Frighten Me* (Angelou 2018, p. 2018) to his own children, which included vivid illustrations by American artist Jean-Michel Basquiat?

The End of the World

The intertwinement of the COVID-19 pandemic and racism brought up fantasies of the end of the world, as Freud called them, which produced apocalyptic anxiety. While some people occupying the position of whiteness discovered this phenomenon for the first time during the pandemic, poor and working-class Black communities have been long accustomed to anxieties of annihilation through their close proximity to violence. This trauma disrupts the ability to rely on memory, to think clearly and flexibly, to make links and to imagine alternatives. This cognitive *concreteness* proper to trauma of social origin collapses the possibility of meaning and different forms of symbolic expression. Alongside Winnicott, I would suggest that what collapses is the illusion of having some control over life. Many poor and working-class Black children and teenagers in big cities are not given the opportunity to develop this illusion, this sense of core omnipotence that makes life and meaning possible.

A 15-year-old Black female former patient of mine was dealing with the trauma of domestic violence and had always dreamt of escaping to college. During the COVID-19 pandemic, she began suffering from panic attacks for the first time in her life: heavy chest, constricted breathing, claustrophobia, and depersonalization. She felt a disconnect between her mind and her body.

I associated her labored breaths with the widely used signifier *I can't breathe* – a condensation of both COVID-19 symptoms and the psychosomatic effects of racism. The patient's presentation reminded me of Billie Holiday's song *Strange Fruit*, because her body – like the Black bodies "swingin' in the Southern breeze" – seemed severed from her suffocated soul, "Strange fruit hangin' from the poplar trees ..." She felt trapped at home but learned to suppress her true feelings, fearing violent punishment. She would compulsively watch violent crime shows on TV, once falling asleep during an episode and waking up crying. Her associations in the session were of a fear of being killed, especially shot. She also feared that her mother, a working-class woman who left very early to work, when it was still dark, and who did not stop working during the pandemic, could be killed in their dangerous neighborhood. She shared an experience of seeing someone shot near where she lived. She confessed suicidal intrusive thoughts, including a scenario with a knife in her stomach and uncaring bystanders. I asked about her fear of being killed and her imagining of killing herself. She was struck by my comment and replied: "When I hear it that way it feels weird." Then she said: "Maybe I am trying to take death in my hands." She felt some control in inflicting the violence she might be subject to in her own hands. Death wins over Eros here in her struggle for escape and survival.

Another former patient of mine, a 16-year-old Black teenager, had a dream after George Floyd's murder and the protests that ensued:

> My mother is about to give birth to my seventh brother. My father and all my family go with her to the hospital. Precautions are taken because of Coronavirus. Right after my baby brother is born the police raid the hospital and kill everyone in my family, including the new baby. I saved myself by escaping in a helicopter.

In his associations, he told me that he related the death of his family, and especially that of the newborn, with there being no future for Black people. In connection to him saving himself, he related it to his own anger towards his family for rendering him Black. I understood his anger in terms of what Beverly Stoute (2021) calls "Black rage" that is the result of trauma and a moral injury inflicted by systemic racism and oppression. She says, "Black Rage, having evolved in the African American cultural context, has primed the defensive operations, on an individual and cultural level, that confer psychic protection and resilience in response to the assault on dignity" (Stoute 2021, p. 284). The escaping in the helicopter can hence be seen as both a defensive operation against annihilation anxiety, but also as a sign of resilience and hope.

The Loss of Play

To COVID-19 we lost human lives, health, the rituals that wove our ordinary lives together, a sense of futurity, safety (personal and institutional), freedom, embodied connectedness, and the fantasy that we are not going to die, even if we workout regularly and eat healthy. The new regime enforced by COVID-19 also made us lose a clear demarcation between inside and outside, past and present, fantasy and reality, private and public, near and far, personal and professional, health and illness. Do we still remember this?

During the pandemic, home became strange to many of us because we lost some of the boundaries that distinguish it from other places. It became un-homely. In this sense, the coronavirus crisis created conditions for the uncanny. In the words of Schelling: "something which ought to have remained hidden and secret but has come to light" (Schelling 2007, p. 23). The internal and external rites that keep us oblivious to mortality and to the abject disparities of our society, were radically unsettled. Children and teen-agers were particularly vulnerable, having lost a sense of stability and con-nectedness that is essential to develop and expand their natural, spontaneous aliveness, as well as playing that is at the core of creativity.

In *Creativity and Its Origins*, Winnicott discusses the creative impulse: "Presumably it belongs to the aliveness of some animals as well as of human beings" (Winnicott 2002, p. 91). This aliveness is present in the artist who produces a work of art, but is also present when anyone "looks in a healthy way at anything or does anything deliberately, such as making a mess with faeces or prolonging the act of crying to enjoy a musical sound or enjoying breathing, etc." (Winnicott 2002, p. 91). People who are attacked, dominated or unseen at home, or who undergo a lifetime of persecution under a cruel political regime, Winnicott argues, have difficulty in remaining creative. As COVID-19 intensified the effects of systemic racism, many poor and working-class Black children faced the threat of losing or suppressing their creativity, hiding it behind a façade of a false social self imposed by double conscious-ness. These children risked becoming tamed, acquiescent, and well-behaved "citizens" of a system that oppresses them.

Confusion of Tongues

When boundaries are unsettled, they can create conditions for productive exchange and negotiation, or conversely, lead to defensive blockage, dis-turbance, and confusion. One boundary disrupted by COVID-19 was that between the world of adults and the world of children. The pandemic fostered a dangerous and impoverishing lack of symbolic exchange between these material and psychic realms.

The boundary in question is one that can be thought of as primarily separating different modes of language exchanged from one world to another. For Ferenczi, the entry of children into the world of adult language is *poten-tially* traumatic. In *Confusion of Tongues Between the Adults and Child* (Fer-enczi 1949), Ferenczi puts it in terms of a confusion between the child's language of tenderness and the adult's language of passion. I would like to expand his idea of a confusion of tongues, which mostly focusses on traumatic sexual impingements from adults towards children, to a broader confusion between the world of adults and of children, in which there is always the possibility of a traumatic entry into the language of adults – what Miguel Gutiérrez Peláez calls a "destructuring trauma through the silencing of the subject through the imposition of the Other's word" (Gutiérrez Peláez 2018, p. 3). Since the child is drawn towards the language of adults, the child's desire can get caught up in the language of adults and its logic of passion. The result is that the child's voice is silenced and, with it, the aspirations and longings of tenderness that ludic experiences give expression to. Ferenczi's warns that an adult is more likely to misconstrue their needs with that of a child "especially if they have been disturbed in their balance and self-control by some misfortune" (Ferenczi 1949, p. 227). COVID-19 intertwined with systemic racism precisely produced this disturbance in adults.

In addition, Ferenczi observes that other effects of trauma, stemming from the confusion of tongues, include the child losing trust in their own senses, which become "broken," as well as the development of precocious maturity or premature maturation (Ferenczi 1949, p. 229). The child, whose language of tenderness seeks to become part of the adult world, remains governed by what Ferenczi terms "oppressive love." In an effort to protect themselves from feelings of helplessness and to preserve the innocence of their caregiver and the status quo imposed by whiteness, poor and working-class Black children are vulnerable to identifying with the aggressor, and effectively becoming *hypnotized*, as Ferenczi would put it.

The result is internal fragmentation, where the child feels both disobedient and compliant, a precocious adult and an immature child, both victim and victimizer. Using Ferenczi's language, the child becomes an acquiescent automaton in the world of adults, rather than pursuing their true desire: to play. This aligns with the work of Kirkland Vaughans and Lisa Harris, who argue that underprivileged Black and Latino boys – criminalized and perceived as potentially violent adults – are forced to mature too quickly, while still remaining children at their core (Vaughans & Harris 2016). They miss the developmental experiences of childhood and adolescence, which are essential for healthy mentalization. The system denies them the freedom to play and to cultivate their creative vitality.

Winnicott says that the child's play can be interrupted by an overexcitation produced by the eruption of aggressive and erotic drives, both from the children themselves and the adults. This is the language of passion. That is why play requires a form of what I would call an *elastic spontaneity* where desire and aggression can be modulated and transformed in the sphere of internal illusion and the experience of control over the world. I relate this to Ferenczi's language of tenderness. Play requires a fluid interplay between fantasy and reality, where the two poles are not opposed but are complementary and elastic. It is not only "oppressive love" which hinders the capacity to play. It is also the constant "oppressive reality" that many poor and working-class Black children undergo, and that we all experienced during the COVID-19 pandemic *in very different ways* depending on class, gender, race, ethnicity, and many other factors.

"Oppressive love" and "oppressive reality" stultify the child's capacity to imagine the world beyond the more rigid symbolic structures of adulthood. The tendency toward *concreteness* in the language of adults produced by ongoing trauma during COVID-19, may have stultified the possibility of the child having a more fluid and imaginative experience of the pandemic. In an environment dominated by adult anxieties and immediate concerns – health risks and unequal access to care, economic instability, racial violence, and social isolation, for example – the imaginative realm, necessary for psychic elaboration, may have been overshadowed by the harsh demands of a survival-focused present. Undoubtedly, the above-mentioned scenario, was far more challenging for poor and working-class Black families.

The long-term effects of COVID-19 on children remain uncertain. Likewise, the psychological and political lessons drawn from this devastating epidemiological crisis, which laid bare and exacerbated deep-seated social and racial inequities, have yet to be fully realized. I believe it is the responsibility of psychoanalysis to engage with the psychological, social, emotional, and developmental challenges brought on by the lasting effects of COVID-19.

Coda: Other Lullabies

Michelangelo is famously quoted as saying that "every stone has a statue inside it" and that he had seen the "angel in the marble and carved until [he] set him free." The sculptor, he says, has "no conception" which the rock does not already "contain within its mass" as a potentiality to which the artist remains "obedient." The psychoanalyst, in sharp contrast with the suggestive or directive therapist, is like the sculptor: beholden less to a personal preconception and more to the image of the angel contained in the rock. Angels are conveyors of truth or its potential messengers and attendants to suffering.

Returning to the introductory lullaby of the Black boy, we find there an intimate rhythm wrestling with dire themes, a beat that resembles that of a human heart, and a reference to a broader community of love. It is this collective love and vitality that might heal what Ferenczi describes as the "traumatic aloneness which renders the attack traumatic, and that is causing the [social] psyche to crack." The lullaby is a promise –the promise that, despite all, *mama will return home* – which creates what Winnicott calls *potential space*, an open field where the future can unfold. The lullaby serves not just to lull the child to sleep, but invites the child to *dream*.

A significant part of psychoanalytic work in devastating crises like the pandemic, which revealed deep inequalities and caused profound upheaval in our lives, involves using words and playfulness to express desires and distress.

It attempts, much like the lullaby, to keep open a potential space of life and meaning. In *Infancy and History: An Essay on the Destruction of Experience*, the Italian philosopher Agamben wrote that "today we know that the destruction of experience no longer necessitates a catastrophe, that humdrum daily life in any city will suffice. For modern man's average day contains virtually nothing that can be translated into experience" (Agamben 1993, p. 13). It is my hope that our work, especially with children and adolescents, can always create imaginative and hopeful experiences that run contrary to the violent and numbing effects of modern racial capitalism.

The possibility of a compassionate, solidary world will not be possible until we are all, psychoanalysts included, active in committing ourselves to criticize in word and action systemic racism and class inequalities. There is a need for a new form of openness and *attentiveness towards the other's suffering* that surpasses mere "tolerance," couched as it is in the discourse of multiculturalism. We are talking about ways of being and living where we accept, as Freud said, that "we all speak a foreign language," the point of departure being difference and otherness. We need an abundance of social dreaming that can sustain poor and working-class Black communities, aligning ourselves with their process of reclaiming the aspects of their selfhood and histories that the existing social order deems abject. Social lullabies can invigorate our social capacity to dream, to imagine the *present as otherwise*, as well as creating communal solidarity. We must leave doom-laden tales and further privatizations of mind and body behind to generate stories that empower, inspire new forms of action, and that transcend the strictures of our existing conditions.

Notes

1 A beautiful and haunting rendition of this lullaby sung by Argentine protest folk singer Mercedes Sosa can be found at www.youtube.com/watch?v=gKgEBBUI6U4.
2 Whiteness, as a hegemonic position, obscures its non-monolithic nature, which is shaped by a heterogenous historical construction rooted in ever-evolving ideas of race, gender, and, crucially, social class. Two books come to mind in connection to this: *The History of White People* by Nell Irvin (2010) and *How the Irish Became White* by Noel Ignatiev (2009).
3 The split between the central and the marginal is also constitutive of class differences. Classism, directed towards the poor and the working-class, is expressed and sustained by a system of attitudes, policies, differences in access to resources, means of economic production, institutions, values, and practices that perpetuates different forms of discrimination and inequalities. Although all Black people encounter racism in various forms, poor and working-class Black individuals face marginalization and attacks that middle or upper-class Black people may not. This essay focusses on the experience of poor and working-class Black families and their kids.
4 I discuss this more in depth in Padrón (2020). In this essay I connect the manic reparative impulse to what James Baldwin calls "virtuous rage," a state that is less motivated by a true and ongoing concern for the other and more "by a panic of being hurled into the flames, of being caught in traffic with the devil."

5 These uses of identification can be extended to the experience of some poor immigrant children of color. In this case, racial trauma is intersected with the immigration experience of the child and their parents. This causes multiple conflicts that might lead to what I call *psychic limbo states* at the core of the child's identity formation. We are talking about a double identification with what are felt as two aggressors: whiteness (in being stigmatized as the other), and the culture of their parents (in so far as there is also the desire to become Americanized and *not be* seen as belonging to their country of origin). Limbo psychic states are felt as inner fragmentation, the self being neither from here nor there, a non-place that is a form of inner suspended animation, the child carrying the trauma of both their parents and the racist and classist society we live in.

References

Agamben, G. (1993). *Infancy and history: Essays on the destruction of experience.* New York: Verso.

Angelou, M. (2018). *Life doesn't frighten me.* New York: Abrams Books for Young Readers.

DuBois, W. E. (1909). *The souls of Black folk.* Chicago: A. C. McLurg and Co.

Dussel, E. (1992). El encubrimiento del otro: Hacia el origen del "mito de la Modernidad." Plural Editores, La Paz.

Fanon, F. (2008). *Black skin, white masks.* New York: Grove Press.

Ferenczi, S. (1949). Confusion of tongues between the adults and child. *International Journal of Psychoanalysis*, 30, 225–230. https://icpla.edu/wp-content/uploads/2012/11/Ferenczi-S-Confusion-of-Tongues-Intl-J-Psychoa.-vol.30-p.225-1949.pdf.

Flint, C., & Hitchcock, J. (2015). Decentering whiteness. www.euroamerican.org/public/decenteringwhiteness.pdf.

Frankel, J. (2002). Identification and "traumatic aloneness": Reply to commentaries by Berman and Bonomi. *Psychoanalytic Dialogues*, 12(1), 159–170. doi:10.1080/10481881209348660.

Freud, S. (1961). Civilization and its discontents. In J. Strachey (Ed., Trans.), *The standard edition of the complete works of Sigmund Freud*, vol. XX1. Richmond: Hogarth Press. (Original work published in 1930).

Guralnik, O. (2002). Depersonalization: Standing in the space between recognition and interpellation. *Psychoanalytic Dialogues*, 20(4), 159–170. doi:10.1080/10481885.2010.502501.

Guralnik, O. (2010).

Gutiérrez Peláez, M. (2018). *Confusion of tongues: A return to Sándor Ferenczi.* New York: Routledge.

Ignatiev, N. (2009). *How the Irish became white.* New York: Routledge.

Irvin, N. (2010). *The history of White people.* New York: W. W. Norton & Company.

Padrón, C. (2020). Six inconclusive notes on the whiteness of the "good white". *Division Review*, 22, 30–32.

Schelling, F. W. J. (2007). *Historical-critical introduction to the philosophy of mythology* (Trans. M. Richey & M. Zisselsberger). New York: State University of New York Press.

Stoute, B. (2021, October). Black rage: The psychic adaptation to the trauma of oppression. *Journal of the AmericanPsychoanalytic Association*, 69(5), 259–290. doi:10.1177/00030651211014207.

Vaughans, K. C., & Harris, L. (2016). The police, black and hispanic boys: A dangerous inability to mentalize. *Journal of Infant, Child, and Adolescent Psychotherapy*, 15(3), 171–178. doi:10.1080/15289168.2016.1214454.

Winnicott, D. (2002). Creativity and its origins. In D. Winnicott (Ed.), *Playing and reality*. New York: Routledge.

Chapter 11

Caring for Cryptids

Welcoming the More-than-Human into Psychoanalytic Treatment

Kathleen Del Mar Miller

> It matters what matters we use to think other matters with; it matters what stories we tell to tell other stories with; it matters what knots knot knots, what thoughts think thoughts, what descriptions describe descriptions, what ties tie ties. It matters what worlds make worlds, what worlds make stories.
> – Donna Haraway (2016), *Staying with the Trouble*

Composed as a series of dated fragments, this chapter is disordered within linear time as a way of attempting to score the "dense temporalities" (Gentile, 2021) of its contents and the nonlinear ways in which psychanalysis, like time itself, unfolds. The fragment as a form offers an important revision to psychoanalysis, and therefore psychoanalytic writing. Rather than moving towards containment, cohesion, and the integration of the singular psychoanalytic subject, the fragment queers relations between time and space, subject and object, driving us to imagine new psychic and material forms of living – new subjectivities – that honor our entanglements and multiplicities. The knowledge produced as a result is one that emphasizes "feeling, trying, experimenting, hesitating, advancing, retreating, [and] revising" (Soreanu, Staberg, & Willner, 2023). It is a knowledge that "weaves paths and consequences but not determinisms," what feminist theorist and post-humanist philosopher Donna Haraway refers to as "tentacular knowledge" (Haraway, 2016).

For the sake of providing some measure of continuity, however, and in the service of weaving with diverse clinical, theoretical and personal threads, I foreground the evolution of my analytic work with Lila, a mixed race, trans adolescent who was born prematurely, spending the first several weeks of life in the neonatal intensive care unit (NICU).

May 2020

Two months into the "first wave" of the COVID-19 pandemic, while working virtually with patients from my New York City apartment, I have the following dream: I am holding a techno-baby. Folded in at each end, the techno-baby's body is thin and porous – a tangle of finely meshed wires. I unfold the

DOI: 10.4324/9781003520757-13

baby carefully, awkwardly cradling it in my arms. It's long, perhaps a foot, and about the width of a ruler. This techno-baby is strange, yet somehow familiar. It has a face, though not a human face. Though its eyes are not human, I know that this baby is looking at me. I might love it? I feel a kinship with it, though I'm uncertain of our relation. I speak to it softly and its face lights up, glowing brighter in response. It's as if I'm welcoming it into being.

June 2023

Through the viral waves of isolation and connection, as well as the entangled social and ecological crises that this pandemic has brought to bear, my techno-baby dream has given welcome form to provisional half-thoughts, enigmatic intuitions, and open-ended questions about my virtual analytic work and the densely webbed constellation it illumines between technology, subjectivity and body.

Virtual analysis is neither a poor translation of, nor a disembodied substitute for, in-person treatment. Nestled within "the opacities and capacities of digital technology" (Clough et al., 2023), virtual analysis is an entirely different modality and medium, replete with its own grammar and potentialities. These "screen relations" offer "new modes of relating and providing treatment that…from a digital perspective rather than a traditional 'analog' one, allow entry into qualities of mind that emerge in the technologically mediated frame" (Hartman, 2017).

Rather than fixating on the so-called disembodied aspects of working virtually, what shifts in our thinking-feeling might occur when we think of virtuality as differently or diffusely or dispersively embodied? When we think of ourselves, both on- and offline, as *not only* embodied, but embedded across time – as "galactically entangled" (Dodge, 2020) plural subjects (Braidotti, 2013)? What affects, and what versions of self, might the virtual be uniquely capable of producing and circulating?

Encountering virtual analysis *on its own terms* requires that we as analysts acknowledge our profound entanglement with the more-than-human environment. Our "matter is never a settled matter" – as material beings, we are "always already radically open" (Barad, 2015). Given our radical openness (Barad, 2015; Miller, 2020, 2021), might virtual analysis require – in Ferenczi's words: "an elasticity of [our] technique" (Ferenczi, 1929), as a way of welcoming the more-than-human into our treatments? And if so, might this acknowledgment *also* require a re-imagining of our foundational concepts, such as the frame, transference/countertransference, our defenses – perhaps even, the unconscious itself (Clough, 2018; Hartman, 2017, 2019; Gentile & Miller, 2021)?

We must keep in mind, of course, that the digital technologies which virtual analysis relies upon are not neutral. These technologies are touched by the social. They are in-formed, with differing effects, by the biases and power structures of the societies that they operate co-extensively with. While there are rich histories of digital technologies connecting and empowering marginalized communities, there are, conversely, myriad and ongoing examples of digital technologies being weaponized as forms of surveillance and control (Clough, 2018). Not only this, but the carbon footprint of digital technologies (high-resolution streaming media, in particular) contributes significantly to global greenhouse emissions (Marks, Spohrer, & Wicke, 2021).

May 2018

My Monday afternoon sessions with Lila are filled with dread. After the school day has ended, and before homework and violin practice have begun, she slinks into my office, a baggy black sweatshirt and skinny black jeans accentuating her slight frame. Placing herself at the farthest corner of the couch, her brown eyes hidden under the brim of a faded baseball cap, Lila barely acknowledges my human presence, absorbed instead by the more-than-human environment of my office: a painting tipped slightly off-balance sends flashes of irritation across her face, while cinematic shadows cast by the afternoon sun seem to arouse surprise and delight.

"What's caught your eye?" I'll ask, occasionally eliciting small bursts of speech. More often, though, my words seem to hang midair, then evaporate. I can't reach her – I'm missing something.

November 2019

A look of contentment passes over Lila's face, followed by a faint smile. "Did something just come to mind?" Lila shakes her head, avoiding eye contact. "Nothing at all? Maybe an image or a feeling?" More head shaking. "You looked really peaceful for a moment," I reflect. Lila shrugs.

We resume our silence, but I'm unable to let this moment go. "Sometimes I have the sense that while we're sitting here, you're somewhere else, having a totally different experience. Maybe this is someplace you go a lot?" Lila startles, sitting straight up on the couch as if she's taking me in for the first time. "Oh," she looks me directly in the eyes. "I was just in outer space."

Projecting my own existential anxieties, I wonder aloud if "hanging out" in this vast, boundless space ever feels frightening. Does she worry that she'll float away, losing all recognizable coordinates? "What?! No way!" Lila exclaims, rolling her eyes. As we explore her effortless more-than-human relationality, Lila assures me that when in outer space, she doesn't get trapped by the planets' gravitational orbits, but weightlessly travels alongside them, perhaps as another planet, an entity, of her own. As she rattles on about gravitational force, describing the ways that objects are brought into relation with one another, the timbre and tone of her voice changes: it's softer, musical, more alive.

Lila then unzips her backpack, pulling out a stack of small watercolor portraits, painted late at night on her bedroom floor. They are impressively rendered close-ups of planets, stars and galaxies, sometimes in unusually vivid colors. "I'm a part of Earth and Earth's a part of me, but then we're also a part of the Universe and that's just way bigger. It goes on and on and on."

Lila's sense of planetary interconnectedness is palpable. Over the past year, I've become aware of her tenuous connections to both her nuclear family and proximal social structures – the relational forces through which psychoanalysis typically considers the individual. It's as if Lila's relation to the more-than-human environment, in this case the planetary, is primary, while her human family and social relations are felt as secondary. Rather than feeling held by these human relations, it is – paradoxically – within the infinity of outer space that Lila feels she can come into being. Similarly, it is through sharing this experience with me that she begins to emerge in our session(s).

"Outer space can really hold you. You can really feel it holding you," I reflect. "We think of it as being far away, as being outside of us, but we are also in it all the time. It's here, inside of us, too."

Her body relaxes briefly, then tenses. "You know what sucks, though? In outer space, you really see all the bad things that people are doing to Earth. Like places where the ozone layer gets super thin. Or like, big fires and storms. Oceans drying up and forests disappearing." Like a weather pattern, Lila's entire mood has shifted.

"It does suck, you're right," I respond. "It's scary. We've been doing a lot of damage to each other and to the planet. Maybe this is one thing that's painful about feeling so connected to outer space? That you're incredibly sensitive, more than most people, to the ways that Earth is being harmed?"

She nods fiercely, then ticks her head repetitively from side to side, which I take as a building up and then, discharging of anxiety. She shifts her weight on the couch and averts her gaze, falling quickly back into silence.

Previously, I had wondered if Lila's lack of connection to her social world, which included me, might have been one sign of what Ferenczi refers to as an "aversion to life" (Ferenczi, 1929). Because her verbal communication was so limited and carried so little affect, I worried that she may have been silently embattled with self-destructive forces. Hearing her speak with such vivacity about her relationship with outer space, with the cosmological, however, reminded me that an aversion, a turning away from, is also always *a turning towards*. If Lila had an "aversion to life," perhaps it was an aversion to *human* life, in which case, she seemed to have turned herself towards her more-than-human environment.

May 2020

Now meeting from her bedroom, rather than the overdetermined space of my office, Lila positions her laptop on her desk as she moves freely, picking up colored pencils and rearranging stuffed animals, while I sit patiently, peering inside the technological membrane of her private world.

Rather than dwelling solely on the loss of the in-person body, what happens when we open ourselves to the unanticipated intimacies of virtuality? Unlike the dread of our in-person meetings, I find myself relaxing into the quiet choreography of these sessions as Lila plays with the space-time between us, drawing herself nearer and then, farther away again. In the moments when Lila moves towards the skin of our screens, I notice that the rhythm of my own breath syncs up with hers. Boundaries momentarily dissolve. Watching Lila breathe and listening to her breath feel simultaneously like touching and being touched – a kind of more-than-human skin-to-skin, screen-to-screen contact (Miller, 2021). I notice an eyeliner smudge on Lila's cheek and imagine brushing it away with my hand. This imagined gesture is filled with a sisterly tenderness that surprises me.

August 1987/2021

Whereas the sonic environment of my office was one of quiet continuity, my sessions with Lila are now punctuated by eruptions and collisions. For summer vacation, Lila and her sisters are sent to visit extended family in Los Angeles, staying only a few minutes away from the house I grew up in. To be in the present with Lila in these sessions, I'm forced to re-live some of my earliest sonic memories – planes flying low over the water, lawnmowers and leaf blowers, ocean breeze through the palm trees, wind chimes – and the often-painful associations they stir.

What does it mean that virtual analysis not only creates the conditions for, but amplifies these sonic collisions? Instead of taking this experience at face value – the bizarre result of a professional life now lived online – what meaning does my affectively saturated experience carry *within* the context of my relationship with Lila, at *this* moment in our work together?

In one of these sessions, as I struggle to sit still – skin crawling, stomach in knots – my thoughts drift and settle on my youngest sister, Keri. In the summer of 1987, Keri was born with tetralogy of Fallot, a rare congenital heart defect. Shortly after birth, Keri began to suffer from "tet spells" – terrifying episodes where oxygenated blood flow is in short supply and babies turn blue as they struggle to breathe, going limp and, occasionally, unconscious. Keri's condition would require complex surgeries, drawing on technologies which had only recently been invented. As a newborn, her heart was temporarily repaired with artificial materials, then later replaced with bioprosthetic valves transplanted from the heart of a young pig.

Following this string of associations, I'm able to imagine Keri as my first techno-baby, dreamed into the present because of my virtually unfolding connection with Lila, once a techno-baby herself.

January 2021

Mid-session, as Lila is mid-sentence, our internet connection becomes unstable. The invisible seams of sound and image are ripped apart. Our faces freeze and unfreeze, while our voices speed up and slow down, our digital skin-screens awash in deconstructed color. Minutes later, as our connection stabilizes, Lila recounts how this experience produced feelings of disorientation and loss: "I think that my birth was a glitch."

How might we theorize and work clinically with these technological glitches that now populate our daily, sometimes hourly, analytic sessions? Rather than treating these glitches as extra-analytic – writing them off as inconvenient or frustrating technical malfunctions – how might their disruptive powers produce new associations, new languages, new metaphors for describing experience (Russell, 2020)?

Lila's profound association to the glitch of her birth makes me wonder if working virtually might be reactivating aspects of her early life. Although Lila's never spoken directly about her birth, I know from her parents that she was born premature, "at the edge of viability" (D. Halperin, 2022, personal correspondence), spending several weeks in the NICU, which placed considerable emotional and financial strain upon her ill-equipped parents.

The emergence of this link allows me to wonder: How might Lila's prematurity, along with her parents' limitations and the entangled complexities of her marginalized identities (gender, sexuality, and race, to name only a few), have created the unfortunate conditions for what Ferenczi (1929) referred to as "unwelcome"-ness? According to Ferenczi (1929), the "unwelcome child," having been received "in a harsh and disagreeable way," "easily and willingly" slips back into a state of "individual non-being."

Rather than dwelling in this state of unrelated "non-being" however, might it *instead* be possible that Lila was brought into intimate relation with the more-than-human environment of the NICU, with its regular and irregular rhythms, electronic blips and bloops from monitors, voices of parents and hospital staff, the softness of a tiny cap or mitten, the feel and smell of latex, the many tubes and wires assisting with her basic functions such as feeding and breathing, the controlled temperature, the quality of light shining through her incubator's plexiglass? Rather than viewing Lila's experience in the NICU as *only* traumatic, might it *also* be possible that this "consequential collision" (Soreanu, Staberg, & Willner, 2023), deeply and creatively connected her to the more-than-human (Miller, 2021)?

July 2022

Vitalized by our "screen relations" (Hartman, 2017), Lila begins to initiate more conversation in our sessions.

"I was basically raised by Google," she informs me matter-of-factly. "I first started using it on the computer at the library after school, then on my mom's laptop, and then my phone. I could ask about anything, whatever, whenever I wanted."

We've been talking about her English class, where the teacher begins each period by asking the students to write for several minutes in response to a chosen prompt. Today's prompt, Lila explains, asked the students to reflect on a time when they felt they had experienced an adolescent rite of passage.

"I'd ask for friend advice," Lila continues. "Or like, questions I had about puberty and trans stuff. Google taught me how to shave my legs!"

What does it mean to feel that you've been parented by a search engine? What is the process by which one, in this case, Lila, internalizes this cyber-object? And how does this object differ, or does it, from other, analog internal objects? What are the relational implications?

With animated flair, Lila goes on to read me an excerpt from her assignment in which she details the experience of shaving her legs for the first time: the anxiety and excitement of covertly buying the pink plastic razor, the fear of cutting herself, and finally, the ecstatic sensation of her smooth, shaved skin.

Hartman, writing on "the temporal qualities of analog and digital platforms," explores the distinctions between analog and digital internal objects, acknowledging that they may be "blurry at best" (Hartman, 2017). Nonetheless, digital cyberobjects possess unique contours, textures and temporalities that are worthy of our attention. Notably, Hartman theorizes the internalized digital cyberobject "as searchable rather than losable such that the transference becomes less about indexing the past in narrative than articulating novel forms of presence" (Hartman, 2017).

Working virtually opens questions which remain open: How do I exist for Lila as an object? Was I first an analog object that, after our shift to working virtually, became a digital cyberobject? If so, have these objects remained separate or have they been spliced together, taking on mixed characteristics of both? Or, alternatively, given Lila's initial lack of engagement, did I even register as an object for her prior to our work online?

February 2022

"I have a secret …" Lila whispers into the computer screen. She holds up a notebook, the word "CRYPTIDS" scrawled in black marker across its cover. "One page for each cryptid,[1] see? I'm researching popular ones like Sasquatch, the Loch Ness Monster, the Yeti, Chupacabra, and my favorite, the Mothman."

"I've never heard of the Mothman," I interrupt. "Tell me more."

"The Mothman is really cool. He's got huge wings, like a moth, and glowing red eyes."

Lila's excitement builds – it's as if she's bursting out of herself. "Here, let me share my screen with you." Using Zoom's "share screen" function, Lila opens a drawing app, clicking through the digital drawings she's been making of her favorite cryptids. The intimacy of having the contents of *her* screen now contained within the borders of my own is striking, as is the pace at which we are suddenly moving – a pace which Lila herself now gets to control.

For those like Lila who live with "a vicious question mark snaked around being" (Gill-Peterson, 2018), whose bodies continue to be relegated to the societal and therefore, psychoanalytic margins, what potential spaces might virtual analysis open that have otherwise been foreclosed? Might these spaces facilitate the very conditions which make play possible?

As we explore Lila's kinship with these liminal creatures, I begin to understand how important they are to her. "You really care about cryptids. You're making a whole online world for them." She shrugs shily in response. "Being a cryptid is probably like being a trans girl. Some people see me and know that I'm real, but other people get upset or think that I'm pretending. Sometimes, it's like I don't even get to exist." Although I don't think Lila is aware of it yet, there is a complex lineage of trans people both resisting and reclaiming transphobic narratives of monstrosity (Koch-Rein, 2014; Preciabo, 2021; Stryker, 1994).

"Do you think race might be mixed in here somewhere?" I ask Lila in a later session. "I know that we've been talking about being a trans girl, but when we talk about how people see you and how you see yourself, I wonder if we're also talking about race?"

"I never used to think about it," Lila replies. "But lately, when I look at myself on the screen, my features seem different somehow. It's like I can see the part of me that's Native. I've even changed how I do my eyeliner – see [she points to her eye makeup], because I started researching what women and two spirit people in my family's tribe looked like." Lila smiles. "I think I look like my grandmother."

June 2023

From within psychoanalysis's normative, "anenvironmental" (Kassouf, 2021) paradigm, Lila's profound relationship with her more-than-human environment might have been interpreted as defensive or even, pathological. As psychoanalysis's skin becomes increasingly permeable to transdisciplinary material, however, and as more analytic treatments are now (and will continue to be) conducted virtually, emergent forms of theory and praxis offer radically open, and therefore, more capacious readings (Clemente & Goodman, 2023; Frankel & Krebs, 2022; Gentile & Miller, 2021; Hartman 2017, 2019).[2] There are also psychoanalytic threads we can draw on, which have been written about extensively and elegantly elsewhere (Kassouf, 2017; Gentile & Miller, 2021), such as Freud's linking of anxiety to the Ice Age, Ferenczi's writing on the geologic unconscious, as well as the oceanic, and Searles's works on the nonhuman environment – elastically adapting them to our contemporary contexts.

Drawing on transdisciplinary material alongside the psychoanalytic allowed me to deeply listen to my pandemic dream of the techno-baby, and eventually, to Lila herself (Oliveros, 2022). Holding Lila in mind as an *unwelcome techno-baby* transformed our Zoom room into an incubator – a Zoom womb – not unlike the incubator where she spent the first several weeks of her life. I no longer felt like a distant observer, as I had when we were meeting in-person, but an alive presence witnessing an early experience of being that must not be impinged upon but rather, allowed to unfold in the loving presence of an/other.

As Lila began to emerge in our virtual treatment, her constellation of more-than-human relations – the planetary, the digital, the supernatural – came into focus. Perhaps paradoxically, it was through her computer screen that Lila was able to see herself *not only* as Native, but as connected to a member of her human family, her grandmother, for the first time. Rather than fostering a sense of disconnection and disembodiment – a "pale simulation" (Hartman, 2017) of our in-person work – the unforeseen shift to working virtually vitalized our connection, allowing Lila the space to exist as an embodied, embedded subject, held within a dense network of human and more-than-human relations.

Acknowledgements

Aspects of the form and content of this chapter were inspired by artist Harry Dodge's more-than-memoir, *My Meteorite: Or, Without the Random There Can Be No New Thing*, as well as composer Pauline Oliveros's *Quantum Listening*. I'm grateful to them, as well as to Kelly Merklin, Karen Weiser, Kerry Downey, Katie Gentile & Patricia Clough for the agile and imaginative conversations. Previous versions were presented in 2023 at Division 39's Annual Spring Conference in New York City and the Ferenczi 150th Anniversary International Conference in Budapest.

Notes

1 Cryptids are supernatural creatures which some people report having seen, but scientifically, have not been proven to exist.
2 Pellegrini and Saketopoulou (2023) deftly caution that as psychoanalysis engages further with other disciplines, which it must in order to thrive, there is real risk of "effectively cannibalizing" this material, thus "preserving the field's structural stability" rather than opening or even, transforming it. "Psychoanalysis," they potently write, "will need to let itself be screwed, and even to find pleasure in how it can get screwed up by such contact, so that it may become screwed differently, re-screwed through this encounter with other disciplinary domains" (Pellegrini & Saketopoulou, 2023).

References

Barad, K. (2015). Transmaterialities: Trans*/matter/realities and queer political imaginings. *GLQ: A Journal of Lesbian and Gay Studies*, 21 (2–3), 387–422. doi:10.1215/106426842843239.

Braidotti, R. (2013). *The posthuman*. Polity Press.

Clemente, M. & Goodman, D. (Eds.). (2023). *The Routledge international handbook of psychoanalysis, subjectivity, and technology*. Routledge.

Clough, P. (2018). *The user unconscious: on affect, media, and measure*. University of Minnesota Press.

Clough, P., et al. (2023). Mediating the subject of psychoanalysis: A conversation on bodies, temporality, and narrative. In M. Clemente, & D. Goodman (Eds.), *The Routledge international handbook of psychoanalysis, subjectivity, and technology* (pp. 17–33). Routledge.

Dodge, H. (2020). *My meteorite: Or, without the random there can be no new thing*. Penguin Books.

Ferenczi, S. (1929). The unwelcome child and his death-instinct. *International Journal of Psychoanalysis*, 10, 125–129.

Frankel, R. & Krebs, V. (2022). *Human virtuality and digital life: Philosophical and psychoanalytic investigations*. Routledge.

Gentile, K. (2021). Kittens in the clinical space: Expanding subjectivity through dense temporalities of interspecies transcorporeal becoming. *Psychoanalytic Dialogues*, 31 (2), 135–150. doi:10.1080/10481885.2021.1889334.

Gentile, K. & Miller, K. (2021). An end of the world-as-we-know-it: After DaSilva. *Room: A Sketchbook for Analytic Action*, 10 (21), 35–44.

Gill-Peterson, J. (2018). *Histories of the transgender child*. University of Minnesota Press.

Goodman, D. (Ed.) (2023). *The Routledge international handbook of psychoanalysis, subjectivity, and technology*. Routledge.

Haraway, D. (2016). *Staying with the trouble: Making kin in the Chthulucene*. Duke University Press.

Hartman, S. (2017). The poetic timestamp of digital erotic objects. *Psychoanalytic Perspectives*, 12 (2), 159–174.

Hartman, S. (2019). Hashtag mania, or misadventures in the #ultrapsychic. *Studies in Gender & Sexuality*, 20, 84–100.

Kassouf, S. (2017). Psychoanalysis and climate change: Revisiting Searles' the nonhuman environment, rediscovering Freud's phylogenetic fantasy, and imagining a future. *American Imago*, 74, 141–171.

Kassouf, S. (2021). A new thing under the sun. *ROOM: A Sketchbook for Analytic Action*, 6 (21). https://analytic-room.com/essays/a-new-thing-under-the-sun-by-susan-kassouf/.

Koch-Rein, A. (2014). Keywords. *TSQ*, 1 (1–2), 134–135. https://read.dukeupress.edu/tsq/article/1/1-2/134/91717/Monster.

Marks, L., Spohrer, R. & Wicke, J. (2021). Haptic entanglements, organs of touch: A mail conversation with Laura Marks. www.schauspielhaus.ch/en/journal/20831/haptic-entanglements-organs-of-touch-a-mail-conversation-with-laura-u-marks.

Miller, K. D. M. (2020). Working clinically with the skin's surface: Tattoos, scars and gendered embodiment. *Studies in Gender & Sexuality*, 21 (3), 143–154. doi:10.1080/15240657.2020.1798188.

Miller, K. D. M. (2021). A radically open analysis: writing as wrapping, video as skin. *Psychoanalytic Dialogues*, 31 (5), 10–18. doi:10.1080/10481885.2021.1944159.

Oliveros, P. (2022). *Quantum listening*. Ignota Press.

Pelligrini, A. & Saketopoulou, A. (2023). *Gender without identity*: The Unconscious in Translation.

Preciabo, P. (2021). *Can the monster speak?: Report to an academy of psychoanalysts*. Semiotext(e).

Russell, L. (2020). *Glitch feminism: A manifesto*. Verso Books.

Searles, H. (1960). *The nonhuman environment in normal development and in schizophrenia*. International Universities Press.

Soreanu, R., Staberg, J. & Willner, J. (2023). *Ferenczi dialogues: On trauma and catastrophe*. Leuven University Press.

Stryker, S. (1994). My words to Victor Frankenstein above the village of Chamounix. *GLQ*, 1, 237–254.

Chapter 12

Adolescents in the Line of Fire

Between Chaos, Ideals, and Psychic Reality Today in the Adolescent Subjectivation Process

Fernando M. Gómez

Introduction

The modern world is a complex web of challenges that seem to multiply at a dizzying speed. People's daily lives are marked by a constant confrontation with a rapidly changing external reality. Old certainties fade away, giving rise to new uncertainties. Among those most affected by this accelerated transformation are adolescents, who must face not only the challenges inherent to their age but also the immense complexities characteristic of the twenty-first century. Technological advances such as increasing virtuality, social networks, and the use of artificial intelligence (AI) often act as a double-edged sword. Rather than serving as empowering tools, they frequently impose unprecedented challenges for new generations. Global crises, wars, climate change, the pandemic and post-pandemic phenomena also have severe impacts on adolescents. By the end of 2019, COVID-19 triggered a radical change in the way people interact, work, and live. Although the war and climate changes sections of this book delve deeper into these phenomena, I want to highlight them along with others that are closely related to pandemic and post-pandemic adolescents' subjectivation process.

These circumstances are further framed by another major issue: the perception of time and the development of a phenomenon of globalization and immediacy driven by social networks, excessive screen time, and the internet – platforms that adolescents consume extensively. This raises significant questions: How will adolescents internally confront an external world shaped by the complexities of the twenty-first century? Do they possess sufficient ego resources in a world dominated by speed and immediacy? What will happen to their identificatory and disidentificatory processes? How will they transition from an omnipotent Ideal Ego to a coherent, mature Ego Ideal with achievable aspirations? What new ideals proposed by culture will today's adolescents align with?

As a psychoanalyst, I believe we have an ethical and professional responsibility to bring to light and analyze cultural phenomena and their impact on the psychic reality of human beings, just as Sigmund Freud did repeatedly

DOI: 10.4324/9781003520757-14

throughout his work. These phenomena are, after all, human creations, giving rise to a contemporary culture that places the psychic reality of many infants, children, adolescents, and adults directly in the line of fire.

External Reality Today: A Line of Fire on Multiple Fronts

Social Media and New Communication Technologies

These have brought about a significant phenomenon of globalization through an unprecedented level of interconnection among people. However, they have also generated and continue to create a growing vulnerability to global-scale phenomena. In the political, social, and economic spheres, today's adolescents live under constant pressure to remain informed and up to date about events occurring anywhere in the world. Through social media, news travels faster than ever, and young people's emotions are deeply impacted by global events. The migration crisis, armed conflicts, social protests, and political tensions are just some of the issues that adolescents must now process in "real time," a temporality that contrasts sharply with the temporality/atemporality of psychic reality.

Moreover, we must not forget that social media often becomes fertile ground for anxiety among many adolescents, who tend to constantly compare themselves to others. The notion of "becoming someone" in such a competitive and globalized digital world creates immense pressure. At the same time, the virtual spaces fostered by social media have frequently been accompanied by the rise of phenomena such as racism, xenophobia, cyberbullying, grooming, and sexting – diverse forms of harassment that constitute new manifestations of violence and abuse. These forms of aggression are unique in that they transcend physical boundaries, remain present in the digital realm over time, and can reach individuals at any moment and in any place, thereby exacerbating their impact on the psyche of many adolescents.

The speed and immediacy that social media present, the "non-time" (Augé, 2009, 2014), is reflected in how today's adolescents navigate a multitude of virtual platforms, moving from one to another in an undifferentiated manner, with limited capacity for their symbolization. These platforms while enabling multiple communication channels, often function merely as spaces for circulation and immediacy, but also have the capacity to impose certain cultural ideals that invite strong comparisons, aspirations and demands, which often have a considerable impact on the self-esteem of adolescents, leading sometimes to extreme dependencies, consumption, and addictions: WhatsApp, Facebook, Instagram, Snapchat, Discord, TikTok, Twitter, Tinder, among others. Some communication platforms (Zoom, Google Meet, Teams, among others) while also facilitating communication, remote psychotherapeutic, educational and different kinds of supports, we know that these interactions lack many of the emotional and non-verbal nuances present in face-to-face conversations. The absence of physical contact, gestures, subtle body language, and emotional exchange can create feelings of emptiness in some people. To this list, we can now add artificial intelligence assistants such as Siri, Alexa, ChatGPT, and Google Bard, which, while providing efficiency and accessibility to information and many tasks, also pose significant challenges to the mental health of the individual.

Even more concerning is the exponential growth in the use of artificial intelligence (AI) by this age group. These tools lead some individuals to believe in the existence of an "other" who is unconditionally available – a disembodied voice at the service of immediacy and absolute reliability, supposedly knowing all that the individual needs, with the illusory promise of never causing frustration or making them wait.

Wars, Poverty, and Natural Disasters: Challenges that Shape a Generation

Conflicts, such as those in Ukraine, Gaza, Sudan, Afghanistan, Myanmar, Haiti and other parts of the world, create an environment of profound anguish and uncertainty. Many adolescents have been, and continue to be, forced to abandon their homes, separate from their families, and endure extreme living conditions as refugees. According to United Nations High Commissioner for Refugees – UNHCR, more than 4 million people have fled Ukraine, with a significant percentage being children and adolescents. In some cities, children study in underground shelters to avoid frequent and disruptive air raids, so these kids don't have natural light or a playground to play (ACNUR, 2024). In Sudan, since the crisis began, more than 11.5 million people have been displaced, more than 3 million women and girls were at risk of gender-based violence, including intimate-partner violence, and 19 million school-age children having no access to formal education (UNHCR, 2024a). By the end of 2023, there were 6 million Palestine refugees under the mandate of the United Nations Relief and Works Agency for Palestine Refugees (UNRWA) mandate, 1.6 million of whom were in the Gaza Strip (UNHCR, 2024b). Nearly 10.9 million Afghans remained displaced, almost all within their country or in neighboring countries; and more than 1.3 million people have been displaced within Myanmar in 2023 (UNHCR, 2024b). In Haiti nearly half of the country's 11.4 million people require humanitarian assistance (UNHCR, 2024b).

Thus, all these people who must flee from the war, must face not only material losses but also emotional and cultural ones, such as separation from their communities, the loss of family members, and the rupture of their cultural and linguistic roots. In babies, children and adolescents, this collective trauma becomes a significant challenge to their development, marking their lives with feelings of emptiness, hopelessness, and rootlessness.

Poverty is another grave threat in today's world. The economic impact of the pandemic exacerbated inequality across many regions. According to UN data, millions of people fell into extreme poverty, directly affecting adolescents from vulnerable families. In many cases, these young individuals had to take on additional responsibilities, such as working to support their households, interrupting their education and hindering their personal development. The United Nations reported in 2023 that over 700 million people are living in extreme poverty, struggling to meet basic needs such as health, education, water, and sanitation. One in five children is living in extreme poverty (UN, 2023).

Adding to this global situation, climate change represents another major contemporary challenge. The accelerated destruction of the environment, rising global temperatures, and extreme weather events – such as floods, tsunamis, droughts, hurricanes, tornadoes, wildfires, snowstorms, earthquakes, and volcanic eruptions – particularly affect adolescents. These phenomena generate intense feelings of helplessness, vulnerability, and profound psychic pain due to the loss of their places of belonging, their land, their culture, their daily lives, and, in many cases, loved ones.

The traumatic nature of these events lies in their suddenness – they strike without warning, without asking permission, and without discrimination, affecting infants, children, adolescents, adults, and families alike. Currently, approximately 1 billion children (50% of the global child population) are severely threatened by the effects of climate change, which increasingly endangers their health through heightened exposure to deadly diseases, such as dengue, malaria, and tuberculosis, as well as compromising their education and security (UNICEF, 2021).

Simultaneously, adolescents bear witness to how global warming is altering ecosystems, displacing entire communities, and jeopardizing vital resources such as water and food. Beyond the practical implications, these changes affect their perception of the future, generating feelings of risk, frustration, and powerlessness. Adolescents are acutely aware that the decisions adults make today will determine their quality of life in the coming decades. Movements led by figures like Greta Thunberg highlight how adolescents are taking an active role in combating the environmental crisis. However, this awareness also brings a significant emotional burden.

The Pandemic/Post-Pandemic Era and Virtuality: A New Horizon of Disconnection

Throughout history, human beings have had to face five major pandemics:

1 *The Plague of Justinian*, which occurred during the Byzantine Empire under Emperor Justinian, taking 40% of the population of Constantinople and the emperor himself;
2 *The Black Death* in the fourteenth century (between 1346 and 1353) where the European population went from 80 million to 30 million people;

3   *Smallpox*, considered eradicated since 1977 but which had a dramatic expansion period in Europe during the eighteenth century;
4   *The Spanish flu*, which first appeared in March 1918, during the last months of World War I, caused 50,000,000 deaths;
5   *The HIV/AIDS* pandemic, with its first documented cases in 1981, and which is estimated to have caused about 25 million deaths (Pané, 2022; Socci, 2023).

At the end of 2019, a new mutation of a coronavirus started in China and it spread all around the world in early 2020. More than 7.1 million deaths have been attributed to it (WHO, 2023a). The COVID-19 pandemic exacerbated many of the tensions already present in the contemporary world. While isolation was a challenge for all age groups, it proved especially traumatic for adolescents. The interruption of school activities, physical distance from friends and family, and uncertainty about the future created a fertile ground for increased anxiety, depression, and other psycho-emotional issues. Moreover, the pandemic accelerated the process of digitalization, shifting many social, educational, and work interactions to the virtual realm.

Although technology and social media had been integral to adolescents' lives long before the pandemic, they now became essential for daily existence. Video platforms, online chats, and social media provided a means of maintaining some form of connection, but this virtual environment also introduced new forms of alienation. The superficiality of online interactions, pressure to construct an idealized self-image, and constant exposure to misinformation, fakes, or biased content contributed to a growing sense of disconnection.

In Latin America, the post-pandemic period revealed significant social and economic repercussions. Eighty percent of parents reported job insecurity, and 32% faced financial concerns, resulting in high levels of anxiety and depression in 46% of surveyed individuals. This situation had a direct impact on the development of infants, children, adolescents, and their families. Furthermore, the interruption of education during the pandemic left lasting negative effects. According to UNESCO (April 2020), approximately 20 million primary school children in Latin America and the Caribbean were affected by school closures. One in three school-aged children could not access remote education, and there was an estimated global loss of two-thirds of a school year due to COVID-19 (UNESCO, 2020).

The unequal access to technological and educational resources widened existing gaps. While some adolescents had access to virtual classes and academic support, others were entirely disconnected, deepening inequities and jeopardizing their futures. The Pan American Health Organization (PAHO) (OPS, 2021) referred to this as one of the most severe educational crises in the region.

Technology: Blessing or Curse?

Technology is undoubtedly one of the most transformative forces in modern human life, but its impact on adolescents is often ambiguous. On the one hand, it enables access to information, connection with people worldwide, and the development of new skills. On the other hand, excessive use of technology can lead to a disconnection from reality and a decline in concentration abilities. Overexposure to screens, addiction to video games, online gambling, and social media, as well as the constant need for online validation, are some of the negative side effects of technology in adolescents' lives.

A particularly concerning phenomenon is early exposure to violent or explicit pornographic content via the internet. Adolescents often encounter images, videos, and messages promoting violence, drug use, hate, or sexuality entirely detached from love and tenderness. While many platforms attempt to regulate such content, the speed at which it spreads and the difficulty of establishing effective barriers make it challenging to protect young people from these risks.

Excessive use of electronic devices is also linked to health issues such as eye strain, sleep disorders, and postural problems. Adolescents, already undergoing significant physical and psychological changes, are particularly vulnerable to these consequences. In response to such concerns, Australia passed a law late last year prohibiting the use of social media for individuals under 16 years of age (Australian Human Rights Commission, 2024).

Drugs: A Temporary Refuge, a Deadly Trap

Finally, drug use remains one of the most serious problems for adolescents in the contemporary world. Whether alcohol, marijuana, synthetic drugs, or cocaine, psychoactive substances offer a world of sensations that produce a hypercathexis of the perceptual pole temporarily and a withdrawal from reality. We know well that it is the result of the full action of the death drive, but simultaneously they are dangerous traps in the service of cultivating the death drive, which jeopardizes the adolescent subjectivation process.

In 2018, the PAHO and WHO already reported that, worldwide, 26.5% (155 million adolescents) of all young people aged 15 to 19 were alcohol drinkers, with prevalence rates of 43.8% in the European region, 38.2% in the Americas, and 37.9% in the Western Pacific region, and between 50% and 70% began drinking before the age of 15. Cannabis consumption in the European Union reflects the global trend of higher consumption among adolescents and young adults (aged 15 to 24), with an estimated prevalence of 19.2% (UNODC, 2021, 2022).

The use of synthetic opioids has also increased, particularly due to fentanyl consumption, the leading cause of death today in the United States. The monthly median of overdose deaths among people aged 10 to 19 (adolescents) increased by 109% from 2019 to 2021. Its presence was recorded in 84% of fatal overdose cases among adolescents, which prompted a state of health emergency. Drug consumption in Latin America of MDA ("love pill"), MDMA ("ecstasy"), and pills mixed with different substances ("pasti") has also significantly increased (UNODC, 2021), especially in environments closely related, though not exclusively, to nightlife and the rise of electronic music concerts where, as a patient says: "Yo estoy en la mía y cada uno está en la suya!" ("I'm in my world, and everyone is in their own world!"). That is to say, in an isolated world, of narcissistic withdrawal and overestimation of the perceptual poles through substances, lights, and music.

Depression and Suicide: A Crisis No Longer So Silent

Finally, depression, self-injury (for example, cutting, lacerating, burning, etc.), and suicide are other major current problems that appear threateningly and quite frequently in our offices. According to UNICEF, in 2021 anxiety and depression globally accounted for 40% of the diagnosed mental health disorders that year. It is estimated that every year 45,800 adolescents die by suicide, which is more than one person every 11 minutes (UNICEF, 2021). According to the WHO, in 2019 suicide was the second leading cause of death in the 15 to 19 age group (WHO, 2023b). In the United States, a study conducted among adolescents aged 12 to 17 reported an increase in suicide attempts of 22.3% during the summer of 2020, and 39.1% during the winter of 2021 compared to the same periods in 2019 (Yard et al., 2021). In Europe and in Latin America and the Caribbean in 2019, anxiety and depression represented 55% and 50%, respectively, of disorders among adolescents aged 10 to 19; and depression and suicide were the second and third leading causes of death, respectively, among those aged 15 to 19. That is to say, the feeling of emptiness, hopelessness, and inability to envision a promising future are recurrent factors in many young people facing depression. Furthermore, the social stigma still surrounding mental health makes it difficult for many to seek help, and too many adolescents suffer in silence.

In summary, we cannot fail to see how all these threatening situations posed by the social and cultural phenomena of today's reality continue to be crucial in the becoming of the subject. Each of them leaves many children and adolescents exposed to a state of extreme threat, a state that closely resembles that initial state of helplessness and defenselessness that every human being experiences and must go through, and which we know is intimately linked to the potential to become a subject. In adolescence as in infancy, states of extreme helplessness highlight the need for the existence of another who, in the best case, satisfies the subject's needs for care, and through whom the possibility of living will be supported.

Psychic Reality and the Adolescent Process

One of the ways to understand the adolescent process suggests the need to understand the concept of psychic reality developed by Sigmund Freud. This does not refer to reality as it is perceived by the senses, but rather refers to the subjective experience that each individual has with external reality, with the world. This subjective experience will be mediated by their drives, desires, fantasies, traumas, and defense mechanisms. Repression and denial may intensify during the adolescent process, and new defenses such as asceticism and intellectualization may appear. In certain situations, more primitive defense mechanisms such as dissociation, projection and projective identifications, and splitting may be activated, especially today in response to the complex circumstances we have outlined regarding the reality of the modern world that the adolescent must face. Therefore, psychic reality is not a faithful reflection of external reality, but rather a dynamic and plastic construction.

On the other hand, adolescence is one of the most complex stages of human development, marked by profound bodily, psychic, emotional, and social changes. Numerous doubts, curiosities, and conflicting contradictions can reach varying degrees of coexistence with each other. Adolescence is characterized by undertaking one of the most difficult tasks of being human, which is exogamy, and the acquisition of true autonomy in which the adolescent seeks their own identity. The objects that represented security and love in childhood are no longer the same, and the adolescent must search for new ways to satisfy these feelings of security and love. At the same time, it is also a stage in which turbulence, uncertainties, and hesitations find room. These will be softened through the libidinization of social bonds with peer groups and through the development of ideals through major interests such as sports, music, arts, painting, writing, politics, environmental care, and social interests, among many others. It is a stage where creativity, love, and hope confront hatred, aversion, aggression, and violent feelings.

It is not only a stage of important bodily transformations but also a reconfiguration of psychic reality. The Pleasure Principle, which during childhood was linked to the immediate satisfaction of desires, comes into conflict with the Reality Principle. This urges the adolescent to adjust their drive demands to the requirements posed by the external world, a process that will serve the advancement of the subjectivation process.

In this regard, I think it is important to consider the place of the ego as one of the psychic instances that occupies a central place in this process of separation-individuation and subjectivation. If we consider the metaphor that Sigmund Freud used on many occasions to describe the ego, likened to "the clown who claims to have himself accomplished all the difficult feats of the circus" (Jones, 1955: 13); we are enabled to ask the following question: what more difficult feats could be presented to the ego than those that occurs during the dynamic, challenging, and turbulent moment of adolescence?

During adolescence, there are numerous domains of psychic work taking place. There is the attempt to consolidate the synthesis processes: enduring the onslaught of drives and the normative pubertal regression; the elaboration of castration and mourning for bisexuality; the elaboration and the dissolution of the Oedipus complex. The psychoanalyst of the adolescent attempts to support the adolescent during the crisis generated by the identificatory and disidentificatory processes, while simultaneously tolerating the onslaught of a superego that we know, at times, is extremely rigid and increases its sadistic, intimidating, and severe aspect through the imposition of demands that become unattainable. Another important task to be carried out, which is attempted toward the end of adolescence, is the passage from the omnipotent Ideal Ego, the heir to primary narcissism, to a mature Ego Ideal, consistent, with aspirations, goals, and objectives that are in service of the becoming of that adolescent subject (Hanly, 1983).

For these reasons, the psychic conflicts between desire, reality, fantasy, and external reality in the adolescent are often manifested through fears related to the emergence of disorientation, helplessness, abandonment, overwhelm, and emptiness. These feelings will be greater the more failed the initial processes of separation-individuation (Mahler, 1984) have been, as well as "the second individuation". Following certain theoretical lines of evolution, like the one proposed by Peter Blos, the latter occurs during adolescence along with the development of an important process of rehistoricization and the consolidation of the Ego Ideal, which will guide the process of second individuation through the discovery of ideals that anchor it in the world of reality and culture (Blos, 1971, 2004).

It is important, in light of the contributions of modern science, to mention that these metapsychological elaborations are not merely observational theoretical developments. They find strong support in the developments of modern neuroscience. Firstly, we now know that a quantitative or qualitative overload on the psychic apparatus impacts the biological matrix, notably interfering with the processes of development and remodeling of neural networks and maps. Using Sigmund Freud's words, in a subjective representational network, a matrix for encoding the psyche (Ansermet & Magistretti, 2020). Secondly, the maximum expression of the processes of neuronal plastic remodeling occurs during the first years of life and adolescence. Thirdly, an early and "attuned" attachment between mother and baby (Stern, 2005) helps to develop, through epigenetic mechanisms, the synthesis of certain proteins that participate in emotional regulation processes throughout life (Kuehner et al., 2019). During adolescence, this phenomenon of neuroplastic remodeling could perhaps be understood as a somatic correlate of what Peter Blos proposed as a process of rehistoricization, elaboration, and re-elaboration of the psychic network in order to carry out the second process of individuation. Why is this contribution important? Because this interdisciplinary vision makes it clear that it is the subjective experience that frees the human being from genetic determinism.

We can then ask ourselves how these lines of fire that the external reality presents today have been inscribed, are being inscribed, and will be inscribed in the adolescent's subjective representational map? And what cultural ideals are they encountering today?

External Reality versus Psychic Reality? A Complex and Inseparable Network that Adolescents Face Today

Some of Freud's writings, such as "The Future of an Illusion" or "Civilization and Its Discontents," have positioned psychoanalysis as actively observing and analyzing social and cultural phenomena (Freud, 1927, 1930). This theoretical intertextuality allowed Sigmund Freud to understand the subject as a social subject, where subjective constitution occurs with the "other." Concepts of the unconscious, drive, and sublimation as in relation to social and cultural phenomena have enhanced psychoanalytic clinical work:

> The contrast between individual psychology and social or group psychology, which at first glance may seem to be full of significance, loses a great deal of its sharpness when it is examined more closely. It is true that individual psychology is concerned with the individual man and explores the paths by which he seeks to find satisfaction for his instinctual impulses; but only rarely and under certain exceptional conditions is individual psychology in a position to disregard the relations of this individual to others. In the individual's mental life someone else is invariably involved, as a model, as an object, as a helper, as an opponent; and so from the very first individual psychology, in this extended but entirely justifiable sense of the words, is at the same time social psychology as well.
>
> (Freud, 1921 p. 69)

Some post-Freudian authors, such as the Greek–French psychoanalyst, philosopher, and sociologist Cornelius Castoriadis, tell us that the individual and the social are co-emergent:

> they are in turn irreducible to each other and truly inseparable. Socialization is not a mere addition of external elements to an unaltered psychic core; its effects are inextricably intertwined with the psyche that exists in effective reality. This makes the ignorance of contemporary psychoanalysts regarding the social dimension of human existence incomprehensible.
>
> (Castoriadis, 1997; my translation)

Therefore, as psychoanalysts of children and adolescents, we cannot remain indifferent to the numerous and constant challenges that phenomena of the current world confront us with, many of which constitute real threats to the becoming of the human subject. This perspective, sustained in my current clinical practice with children and adolescents, presents the difficulty of thinking about the process of subjectivation in adolescence in terms of the individual issues of the patient's internal world as well as the powerful ideals that prevail in the social fabric. Therefore, I understand a clear theoretical-clinical need for a psychoanalysis of infants, children, and adolescents that includes a balanced and adequate consideration of the intrasubjective, the intersubjective, and the transsubjective. The intersubjective is those experiences that present as the encounter and interaction with an "other." This domain refers to the existence, or lack thereof, of a space "between" and "with" the other, otherness, and language. The intrasubjective is the representations of the ego in relation to itself, its body, drives, desire, fantasy, and object relations. The transsubjective is that which contains representations of the real external world, in its social and physical dimensions.

Between Chaos and Alienation: Its Impact on the Adolescent Subjectivation Process

Chaos, in terms of Physics, refers to the phenomenon in which a deterministic system exhibits an apparently unpredictable behavior due to its extreme sensitivity to initial conditions. This behavior, known as the "Butterfly effect" introduced by Edward Norton Lorenz, shows how small variations at the beginning of a system can lead to large variations in its evolution in the long term (Lorenz, 1963, 1972). It is the complexity inherent in the system that makes its outcomes unpredictable. This concept proposed by the Theory of Chaos could be transposed to the psychoanalytic theoretical field if we understand that the development of a system of high complexity, such as the psychic apparatus and the process of subjectivization, are sensitive to the action of various aspects of the intra, inter, and trans-subjective experience. For example, the defenseless creature has anchorage based on a series of highly symbolic instances inscribed in culture: parental desire, the subject's nomination, the mother's gaze, the significance of the body, breastfeeding, the experiences of need satisfaction, among many others, that will intertwine with the emergence of desire. All these early experiences have the potential to develop a symbolic-symbolizing matrix, and will be strongly at play during the subjectivization process in adolescence, a moment in which the system is under the onslaught of drives accompanied by the explosion of significant hormonal and bodily changes. Therefore, the system's sensitivity in the initial stages of psychic apparatus structuring will imply that small variations during these processes can generate unpredictable consequences with disproportionate effects over time, which can then be experienced chaotically by the adolescent. Due to the inability to integrate these early experiences coherently into the conscious Ego, the adolescent will be more vulnerable to the impact of the various phenomena, images, and experiences posed by the current reality (wars, natural disasters, violence, immediacy, success, omnipotence, social media, virtuality, etc.), and consequently will potentially become more trapped and immersed in them. This entrapment could generate a disproportionate and chaotic effect with a clear interference in their subjectivization process. The adolescent thus trapped in such chaotic dynamics, influenced and encouraged by the media and social networks, can become alienated from a reality of ideals that serve what one has or should be (Ideal Ego). Such alienation can reinforce a pathological narcissism and interfere with their capacity to consolidate what "one aspires to" or "desires to be" (Ego Ideal), which would serve the creation of a coherent and stable identity (Erikson, 1968; Winnicott, 1965).

The psychoanalyst Haydée Faimberg described the concept of "alienating identifications" as those identifications in which the subject unconsciously submits to the identification from the stories of the "Other," heir to that parental narcissism that does not concern them, but to which they remain totally captive (Baranes, 1996; Faimberg, 1985, 1992). Therefore, could we talk about alienating ideals? Based on my clinical experience, I suggest the existence of "alienating ideals," as those ideals that come from that "Other," which could be embodied in a current narcissistic culture and internalized through unconscious identificatory processes. How do these alienating ideals act? They do so through the deployment of their mechanisms of intrusion and appropriation, which will force the adolescent into the establishment of these types of identifications.

In this context, the adolescent can be "identified with" and "being identified by" (Marucco, 1978) the images projected onto them by society, which do not necessarily correspond to their deep subjectivity. In this bidirectionality of the identificatory process with these representations of an empty or superficial image that does not reflect their internal experience, the adolescent may experience a paradoxical sensation. They feel filled with these alienating ideals that are becoming psychic flesh, and a sensation of "fullness" is generated in the subjectivation process, which will make the adolescent fully battle any instance of reality that threatens or risks these alienating ideals. However, even though they may have achieved certain social success, the lack of a solid internal foundation to sustain their identity, at the same time leaves them "excessively empty," with an existential void or a strong identity crisis. Strong incapacities are observed in them to deal with life, isolating themselves and sometimes experiencing feelings of meaninglessness in their life. On some occasions, this may lead to strong feelings of emptiness with the emergence of a severe adolescent "breakdown" which we observe frequently in many of the clinical situations we face in our daily work: panic attacks, sleep difficulties, hopelessness, depression, substance use, violence, self-harm, suicidal thoughts and/or actions, sometimes culminating in the adolescent's death (Laufer M & Eglé Laufer M, 1984; Laufer, M (1995)).

This cultural narcissism of the contemporary era promotes the image of a "perfect self," an Ideal Ego at the service of "should be" – an image of an adolescent without flaws, successful, attractive and always ascending. This narcissistic view, strongly disseminated through social media, fosters self-exhibition and constant validation through the gaze of the other. The "like" or virtual approval becomes a substitute for real validation and, more importantly, internal validation. In this framework, adolescents find themselves trapped, alienated in a constant search for a successful and desirable "self." The identification with these alienating ideals, as we have said, promotes identification with an Ideal Ego that reinforces pathological narcissism, favoring the establishment of a "false self" that responds to the expectations and demands of that Ideal Ego rather than the true internal needs and aspirational desires of the subject, which are linked to the formation of a mature and coherent Ego Ideal. The person, in an attempt to meet the narcissistic demands of the Ideal Ego, heir to that parental and cultural narcissism with which they remain totally captive, loses their authenticity and their ability to develop a solid and coherent identity.

The cultural Super-ego that is developing, influenced by the media and social networks, sets almost impossible demands and acts as an internal tyrant imposing an idealized image of the self (Ideal Ego) reinforcing pathological narcissism. The adolescent will constantly feel insufficient because they cannot reach the demanded perfection. This situation begins to create a tug between the false self and the true self, and between the life drive (*Eros*) and the death drive (*Thanatos*). Following the theory of drives, this progressive drive disentangling will allow the release of the death drive with its disobjectalizing function as a manifestation of its destructiveness. We clinically observe this phenomenon through the development of profound existential anxiety, criticism, persecution, inferiority, shame, and disgust feelings. In some cases, it may lead to extreme situations with a tendency toward disintegration, emptiness, and the dissolution of the self. These kinds of adolescent breakdowns, observed in many severe clinical situations, can include violence, self-harm, substance abuse, severe depression, and suicide (Green, 1991). All these clinical situations position us working directly with the line of fire, represented by the release of the death drive. For this reason, it is important that the analyst have a great "drive commitment" working with these patients, through transference, with the intention of co-creating an object-analyst (M'Uzan, 1994) that offers a "transformative function" (Roussillon, 1991). This will allow the analyst to recover the consequent objectalizing function of *Eros*, repair the failures in the processes of subjectivation and symbolization, and facilitate the creation of a safe "analytical working field" for the emergence of the unsymbolized (Baranger & Baranger, 1961).

Conclusions

This chapter aims to highlight how pandemic and post-pandemic adolescents face unique challenges that affect their process of subjectivation. The accelerated transformation of the world, marked by global crises, wars, climate change, technological advances, the omnipresence of social media, and the rapid assimilation of AI in younger generations, poses unprecedented complexities that, exacerbated by a perception of time dominated by immediacy, success-driven attitudes, and perfectionism, result in the increased incidence of extremely severe clinical situations with which adolescent psychoanalysts are faced today in the office.

The consideration of social and cultural phenomena, as today's cultural ideals, in our clinical practice today is of utmost importance. We need to take the necessary precautions not to become blinded by them as happens to some. If we do, we could postpone what lies at the core of our psychoanalytic identity: the unconscious and the psychic reality of the subject. Recognizing the complexity of the psychic reality in adolescents as a dynamic construction influenced by drives, desires, traumas, and current social phenomena, will provide better tools to face the alienation proposed by some cultural and narcissistic ideals presented by today's world. It is essential to recognize their effects specifically in the adolescent subjectivation process as it will serve the construction of their own identity. Due to the implications these phenomena have for individuals with a fragile and vulnerable psychic structure and reality, they can lead to the development of extremely severe and chaotic clinical situations: severe depression, substance abuse, self-harming, suicidal ideation and acts.

Working with these severely affected patients requires the analyst's fundamental ethical commitment, as well as a thorough understanding of technical specificities, which are beyond the scope of this chapter. Also, an interdisciplinary and integrative framework could be essential to comprehensively address the difficulties these adolescents face. It is important that psychoanalysts today respect the specific aspects of our own epistemological framework, within a collaborative and interdisciplinary network with educators, social workers, pediatricians, and other institutions that can provide more holistic support.

Psychoanalysis has an ethical responsibility to analyze how contemporary culture impacts psychic reality, and how the interaction of both realities places many infants, children, and adolescents today "in the line of fire." This viewpoint leads us to reinforce the idea of a preventive-active-interactive child and adolescent psychoanalysis, deploying all its power and resources in the service of the most vulnerable group in society: infants, children, and adolescents, who are nothing more and nothing less than our future!

References

ACNUR (2024). 1.000 días de guerra a gran escala en Ucrania: la Alta Comisionada Adjunta de ACNUR pide solidaridad con las víctimas inocentes. www.acnur.org/noticias/notas-de-prensa/1-000-dias-de-guerra-gran-escala-en-ucrania-la-alta-comisionada-adjunta-de-acnur-pide-solidaridad-con-las-victimas-inocentes.

Ansermet, F., & Magistretti, P. (2020). Neurosciences and psychoanalysis: a contemporary challenge. *IFP Bulletin*, December.

Augé, M (2009). Sobre modernidad. Del mundo de hoy al mundo de mañana. https://a sodea.files.wordpress.com/2009/09/auge-marc-sobremodernidad.pdf.

Augé, M (2014). *Où est passé l'avenir*. Panamá.

Australian Human Rights Commission (2024). Proposed social media ban for under-16s in Australia. https://humanrights.gov.au/about/news/proposed-social-media-ba n-under-16s-australia.

Baranes, JJ (1996). *Desmentida, identificaciones alienantes, tiempo de la generación.* Figuras y modalidades.

Baranger, M & Baranger, W (1961). La situación analítica como campo dinámico. *Rev. Urug. Psicoan.*, IV(1): 3–54.

Blos, P (1971). *Psicoanálisis de la adolescencia*. Editorial Joaquín.

Blos, P (2004). *La transición adolescente*. Amorrortu Editores.

Castoriadis, C (1997). El imaginario social instituyente. *Zona Erógena*, 35(4).

Erikson, EH (1968). *Identity youth and crisis*. WW Norton & Company.

Faimberg, H (1985). El telescopaje de las generaciones: la genealogía de ciertas identificaciones. *Revista de Psicoanálisis*, 42(5).

Faimberg, H (1992). A la escucha del telescopaje de generaciones: pertinencia psicoanalítica del concepto. *Revista de Psicoanálisis de Madrid*, 15.

Freud, S (1921). Group psychology and the analysis of the ego. In J Strachey (Ed.), *Beyond the pleasure principle, group psychology and other works*, vol. XVIII. Hogarth Press.

Freud, S (1927). El porvenir de una ilusión. In JL Etcheverry (Ed.), *Obras completas*, vol. XXI. Amorrortu Editores.

Freud, S (1930). El malestar en la cultura. In JL Etcheverry (Ed.), *Obras completas*, vol. XXI. Amorrortu Editores.

Green, A (1991). Pulsión de muerte, narcisismo negativo, función desobjetalizante. In *La pulsión de muerte*. Amorrortu editores.

Hanly, C (1983). *Ideal del yo y yo ideal. Revista de Psicoanálisis*, 40(1), 191–203.

Jones, E (1955). *The Life and Work of Sigmund Freud, Vol II: 1901–1919*. New York. Basic Books.

Kuehner JN, Bruggeman EC, Wen, Z & Yao, B (2019). Epigenetic regulations in neuropsychiatric disorders. *Front. Genet.* 10: 268. https://doi.org/10.3389/fgene.2019.00268.

Laufer, M (1995). *The suicidal adolescent*. Routledge.

Laufer, M., & Eglé Laufer, M. (1984). *Adolescence and developmental breakdown*: A psychoanalytic view. Yale University Press.

Lorenz, EN (1963). Flujo no periódico determinista. *Revista de ciencias atmosféricas*, 20(2): 130–141. https://doi.org/10.1175/1520-0469(1963)020<0130:DNF>2.0.CO;2.

Lorenz, EN (1972). Predictability: Does the flap of a butterfly's wings in Brazil set off a tornado in Texas. http://gymportalen.dk/sites/lru.dk/files/lru/132_kap6_lorenz_a rtikel_the_butterfly_effect.pdf.

Mahler M (1984). *Estudio 2: Separación. Individuación*. Editorial PAIDOS.

Marucco, N (1978). La identidad de Edipo: acerca de la escisión del Yo, de la compulsión a la repetición y de la pulsión de muerte. *Revista de Psicoanálisis*, 35(5), 853–900.

M'Uzan, M de (1994). *La bouche de l'inconscient: Essais sur l'interprétation*. Gallimard.

OPS (2021). Los niños, niñas y adolescentes están profundamente afectados por la pandemia de COVID-19, afirma la directora de la OPS. www.paho.org/es/noticias/

15-9-2021-ninos-ninas-adolescentes-estan-profundamente-afectados-por-pandemia-covid-19.

Pané, G. H. (2022). Grandes pandemias de la historia. https://historia.nationalgeograp hic.com.es/a/grandes-pandemias-historia_15178/7.

Roussillon, R (1991). *Paradojas y situaciones fronterizas del psicoanálisis.* Amorrortu editores, Buenos Aires.

Socci, AE (2023). *Avatares de la situation analítica: el encuadre ayer y hoy.* En: Transferencia, Contratransferencia, Encuadre. Perspectivas actuales. Colección Pilares del Psicoanálisis en la Contemporaneidad. Fernando M. Gómez (editor). APA Editorial, Buenos Aires. pp. 381–404.

Stern, DN (2005). *El mundo interpersonal del infante.* 1a ed. 4a reimp.- Buenos Aires: Paidós. 2005.

UN (2023). *Programa de las Naciones Unidas para el desarrollo.* Informe Anual 2023. Liberar el potencial y acabar con la pobreza. https://annualreport.undp.org/assets/Annual-Report-2023-Spanish.pdf)

UNESCO (2020). Oficina Regional de Educación para América Latina y el Caribe (OREALC/UNESCO Santiago), en abril de 2020. https://unesdoc.unesco.org/ark:/48223/pf0000374615?posInSet=28&queryId=3db8ab34-a41c-43d3-9713-e385c2275c27.

UNHCR (2024a). Sudan crises explained. www.unrefugees.org/news/sudan-crisis-expla ined/.

UNHCR (2024b). Global Trends: Forced displacement in 2023. www.unhcr.org/sites/default/files/2024-06/global-trends-report-2023.pdf.

UNICEF (2021). One billion children at "extremely high risk" of the impacts of the climate crisis – UNICEF. www.unicef.org/press-releases/one-billion-children-extrem ely-high-risk-impacts-climate-crisis-unicef.

UNICEF (2021). Estado mundial de la infancia 2021. www.unicef.org/argentina/m edia/12046/file/sowc2021.pdf.

UNICEF (2020). Latin and the Caribbean: The impact of Covid-19 on the mental health of adolescents and youth. www.unicef.org/lac/en/impact-covid-19-mental-hea lth-adolescents-and-youth.

UNODC (2021). Drogas sintéticas y nuevas sustancias psicoactivas en América Latina y el Caribe 2021 (septiembre de 2021). www.unodc.org/documents/scientific/21-02921_ LAC_drug_assessment_S_ebook.pdf.

UNODC (2022). *UNODC, Informe mundial sobre las drogas 2022.* United Nations.

WHO (2023a). Deaths. https://data.who.int/dashboards/covid19/deaths.

WHO (2023b). Suicide. www.who.int/news-room/fact-sheets/detail/suicide.

Winnicott, D (1965). *Los procesos de maduración y el ambiente facilitador: Estudios para una teoría del desarrollo emocional.* Editorial Paidos.

Yard E (2021). Emergency department visits for suspected suicide attempts among persons aged 12–25 years before and during the COVID-19 pandemic – United States, January 2019–May 2021. *MMWR Morb Mortal Wkly Rep*, 70: 888–894. doi:10.15585/mmwr.mm7024e1.

Chapter 13

SOS Brasil

Alicia Beatriz Dorado de Lisondo

Brazil's Socio-Political Situation: Context

> [I]t is possible to foresee that at some time or other the conscience of society will awaken and remind it that the poor man should have just as much right to assistance for his mind as he now has to the life-saving help offered by surgery; and that neuroses threaten public health no less than tuberculosis, and can be left as little as the latter to the impotent care of individuals of the community. When this happens, there will be institutions... to which analytically trained clinicians will be appointed, so that ... for children for which there is no choice but between running wild and neurosis, may be capable, by analysis, of resistance and efficient work.
>
> (Freud, 1918)

Brazil, being an underdeveloped country, has high levels of poverty, wealth inequality, and social and economic vulnerability. More than fifty percent of the population is considered poor, with twenty percent living in extreme poverty. Forty eight percent of Brazilians lack sewage utilities, and thirty five million people do not have treated water (Lazslo, 2024). Twenty nine percent of the Brazilian population between the ages of fifteen and sixty four is considered functionally illiterate, unable to interpret texts and use them in everyday life.

It is important to highlight the catastrophic three hundred years of slavery in Brazil, a stain still alive in the form of structural racism. Indigenous, black and brown-skinned people are marginalized by a perverse whiteness pact. According to 2022 demographic census, a majority of Brazilians, 55.5%, identify themselves as black or brown-skinned (Belo, 2021; Bento, 2022; Filho, 2021; Filho & Migueli 2022, Filho & Paim, 2023; Kon et al., 2017; Mondrzak, 2021).

The impact of structural racism in Brazil marks black people with biases severely rooted in our culture. They are systematically objectified and deprived of access to basic goods and services, perpetuating an evil and perverse circle of marginalization. Moreover, a degenerative pact denies the existence of such racism, which makes it difficult to be perceived and

DOI: 10.4324/9781003520757-15

confronted. It is such an unequal country in which there is great confusion. Some hold the view that Brazil is a meritocracy and feel that those who have wealth and power have earned it. They thereby can ignore structural inequalities and basic constitutional rights for all.

In order to occupy constitutionally demarcated indigenous lands for agribusiness exploitation and illegal gold mining, a genocide of such people takes place and proliferates due to water, soil and air contamination caused by mercury, used to separate gold from sediment. Illegal mining is responsible for deforestation in the Amazon, within indigenous territories and forestry reserves. It is also responsible for the silting up of rivers which leads to the loss of terrestrial and river biodiversity. It is also intertwined with sexual violence, rape of children and adolescents, organized crime, grooming of young indigenous people, and murders. Illegal mining deters communities from fishing, hunting and gathering or even health services. A frightening percentage of the population is excluded and discriminated against, without access to decent work, good education, health, housing, which perpetuates a sense of ontological inferiority transmitted transgenerationally (Kaës, 1991, 2007; Jesus, 2019).

The law does not make sense in Brazil's ancestral formation and, at the risk of continuing not to do so ad aeternum, it does not establish and stabilize the symbolic. It is an ambivalent and arbitrary rule insidiously oscillating between violence and rhetoric.

The historical invisibility of this poor population to authoritarian governments is responsible for the number of murders, femicides and violence in all forms of anti-social behavior breaking the social pact. A silent and unacknowledged war is responsible for the rise of violence reflecting Brazil's social perversion and cruelty (Verdi, 2024). Such bloody theatre is played between different well-organized groups of state violence with the participation and complicity of corrupt members of government. As in every war, many innocent people die as victims of planned savagery.

Such state sanctioned violence propels some of this population into highly organized gangs. The gangs offer an identity that has visibility, to fill the desperate gap of nothingness. The film City of God (Meirelles, 2002) is a faithful indictment of such horror. This sobering drama portrays life in Cidade de Deus, a Rio de Janeiro's favela, known as one of the city's most dangerous areas.

Many fanatical political ideologies seek to deny the state's responsibility to guarantee an environment favorable to living with dignity and to enable the mental development of its people. The social, cultural, political, economic, educational, anthropological and historical dimensions are key components of the individual's internal world. The internal and external world are inseparable, and it is necessary to understand the close correlation between mental health issues and sociopolitical problems (Borlle, 2024). Piera Aulagnier (1976), in conceptualizing the narcissistic contract, gives sociocultural discourse a structuring function in the child's psyche. For her, it is necessary to examine the singularity of each patient's mental configuration, both conscious and unconscious, crossed by culture at each historical moment.

In Brazil, COVID exacerbated an already catastrophic situation. It revealed the structural misery of poor people and the system in which they are imbedded. COVID was met by a government already well versed in refusing reality and promoting megalomania and omnipotence. It denied the danger of the virus and promoted false medicine as something miraculous. The government declarations during COVID illustrated disdain for science, appealed to religion, and spread of perversion. It led to a disaster in the fight against the virus; a disaster resulting in 713,000 deaths revealing the genocidal government's "face of absolute evil" (Gampel, 2024).

The catastrophe of COVID unfolded in a social fabric that was already profoundly impaired. It offered no stable emotional background and was already soaked in violence and deceit. Social violence erupted suddenly and massively, destroying families and the entire environment. Sol, the teenager featured below, exemplifies this rupture in existential continuity. The subject cannot represent its own fragmentation under these circumstances.

Working outside the walls of the consulting room became necessary for many Brazilian psychoanalysts. This level of human misery demanded it. To remain in our private offices would have been complicit blindness (Viñar, 2002, p. 163). These are situations where the subject cannot be present in its own fragmentation.

A Psychoanalytic Epistemological Foundation for an Emergency Project

SOS Brasil was a psychoanalytic response to this catastrophe. This project is an attempt to offer psychoanalytic work to open up the possibility of giving voice to unbearable pain. It was created as a call for expression of a psychoanalytic social responsibility. As the introductory quote illustrates, this call began with Freud.

Babies, children, and adolescents, those who fulfil parental functions, and health, education and the judiciary workers needed the psychic oxygen of psychoanalytic listening and containment to deal with pain, terror and help-lessness. Our commitment is to prevent more traumas from decimating these human beings. Far beyond the possibility of giving words to horror and opening up the possibilities of symbolization, the analytical encounter can be an unprecedented experience of *at-one-moment* (Bion, 1970) almost mythical, to allow us to immerse ourselves in pain and hopelessness (Gampel, 2024). In this dive, it is also possible to realize the beauty of life pulsating. Contemporary psychoanalysis recognizes the analyst's potential to offer an unprecedented experience, capable of opening new horizons and enabling new representations to take root in the psyche.

Our art-science, which has been around for more than a century and has a long history of clinical experience, has been rethought over the years, and today it has a refined and sophisticated toolbox. It contains a variety of effective devices from different schools of analytical thought. We, as psychoanalysts in the twenty first century, cannot limit our work to the private practice location. We should move beyond the walls of our consulting rooms. This change confronts us more directly with places where we feel impotent and forces us to work with that psychic reality. Under the safety of our workplaces, it is easier to rely on supposed knowledge. Moving out into the social and into our communities invites a deeper reckoning with the unknown. As Bion reminds us, this should be what we face towards as psychoanalysts.

Patients who are socially vulnerable know a great deal about psychological suffering and bleeding wounds in the soul. Until recently, they did not know about psychoanalysis and its transforming power. A teenager promises his analyst at the end of his agreed sessions: "I don't want to go to another analyst. Wait for me, wait for me. I'll study, work and treat myself with you again, with you. Wait for me!"

Project SOS Brasil Free, Emergency, Psychoanalytic Counselling

In January 2021, due to corruption, inefficiency and disregard for human suffering, Manaus, Brazil had no oxygen. Babies in the prenatal ICU were separated from their mothers or carers and transported to other states to receive oxygen, unfortunately, many of them would probably died on the journey. Latin American Baby Observation Association Esther Bick Method (ALOBB) and International Association for the Development of Baby Watching According to Bick (AIDOBB) intervened with international authorities to stop such an atrocity.

The despair of a young mother who watched her baby die from lack of oxygen, after her attempt to offer him mouth-to-mouth breathing, strongly shook this author who invited other colleagues, and this project was created. What is it that we as psychoanalysts could offer? How could we transform indignation into thoughtful professional action?

Aware of the project's importance to offer psychoanalytic care far beyond the pandemic, today we treat patients who suffer social vulnerability due to political and social violence. We seek to aid vulnerable populations in building their subjectivity in social emergency situations. The existential complaints are continuous, urgent, multifactorial.

Although the project started in response to COVID, we now treat populations affected by tragedies caused by climatic events such as the floods in Rio Grande do Sul, Bahia, Petrópolis or by vast woodfires around the country. We see adolescents who attempt suicide, patients suffering sexual and/or psychological abuse, and sexual violence victims. We see children and adolescents who have been raped, impacted by organized crime, and misdiagnosed with mental illness. Migration, sexual harassment, violence and drug addiction are some of the issues which motivate the research for this project.

In our research (Lazslo, 2023), we realized that COVID-19 and its consequences were not always the conscious and/or unconscious complaints of the patients in the SOS Brasil Project. The virus had brought up ontological, archaic dimensions of catastrophic helplessness (Lisondo, 2012) caused by the pain of the social violence and inequities in underprivileged population.

In 2020, the pandemic was an opportunity to allow what was submerged, sometimes silenced, to emerge: inhuman life that had been naturalized. As a metaphor for the pain of the social, Gampel (2022) appeals to the radioactive contamination of the psychic apparatus and the equipment of the social. It acts unpredictably and without control over its implantation and its effects.

Since the creation of the project, one hundred and twenty two psychoanalysts have treated seven hundred and sixty patients: children, adolescents, parents and caregivers. Our core project management has two axes, one of promotion and one of ateliers or workshops. All psychoanalysts participating in SOS sign up for one of their ateliers. There are a total of twelve ateliers split into five work areas: babies and families (aged 0–3), children (aged 3–11), adolescents (aged 11–21), adults (caregivers), institutions and the Body Group. The latter is a multi-disciplinary team of professionals from many areas: pediatricians, psychiatrists, speech therapists, physiotherapists, lactation consultants, osteopaths, social workers and occupational therapists, to meet the diverse demands child and adolescence care require. Each of these ateliers has a coordinator and a variable number of participating psychoanalysts. These working groups allow for clinical discussion and help the analyst develop their potential for creative, aesthetic and poetic thinking. The groups promote clinical thinking in the face of the strong anxieties that working on the frontiers can arouse. They aim to be containers that allow psychoanalysts to take creative care of their analytic function, to get in touch with the patient's inner world in the analytic field, and to avoid human epistemological sins.

As emergency care, we propose a small number of sessions, from three to eight. We seek psychoanalytic comprehension – emotional contact that can stitch a bit the holes caused by traumas. We offer emergency care to stop the bleeding of these severe psychic wounds. Whenever we notice the patient needs longer psychoanalytic care, either due to the severity of the situation or because of the possibilities for growth, the group's social worker looks for the local available resources in the patient's community.

In some cases, SOS has allowed the patient and/or their family to realize the seriousness and complexity of their suffering. These patients may then proceed with psychoanalytic treatment. When the patient is an adult, they can join the Solidarity Group, coordinated by a psychoanalyst who coordinates the Body Group. This group is made up of professionals who are not part of IPA, but who are well trained candidates who cannot yet take part in clinical seminars or begin official supervisions because of requirements of their institutes. In order to improve clinical work, the solidarity group is supervised by the IPA Committee on Psychoanalytic Assistance in Crises and Emergencies (PACE).

There are scientific meetings, held once a month, in which senior analysts discuss the presentations about patients from the different axes. A research project was also carried out with participating analysts about their insights of SOS Brasil. The following are themes that we have observed in each axis:

- In *Axis I* (babies and the family) (aged 0–3) we find maternal depression. The baby and the mother often lack paternal support. Sometimes an unwanted baby presents disturbances in emotional development due to conflicts within the primary relationship between mother and baby.
- In *Axis II* (children) (aged 3–11) we find many children with disorders on the autism spectrum, wrongly diagnosed as mentally ill or intellectually disabled. Also, we see psychosomatic disorders, school complaints, and trauma due to the inability to mourn parents' and caregivers' deaths.
- In *Axis III* (adolescents) (aged 11–18) we find suicide attempts, self-mutilation, sexual violence, drug addiction, apathy, withdrawing from school, and antisocial behavior.
- *Axis IV* (caregivers): Health, education, law professionals, parents and caregivers seek SOS when feeling helpless, disoriented and without psychological resources to deal with traumatic situations, severe depression, anxiety and social pain.
- *Axis V* (institutions): In Brazil, the COVID-19 pandemic left more than 200,000 registered orphans. Psychoanalysts offer psychic assistance to the professionals of the institutions responsible for these orphans. This assistance aims to enable these professionals to access their internal resources and symbolize their experiences. Hospitals, daycare centers, schools, orphanages and youth detention centers need to be privileged spaces towards facilitating and caring for mental functions, in order to prevent, as much as possible, the mental deterioration of developing human beings.

We have been trying to promote Project SOS through many interviews in social media, on Instagram, as well as presenting it in "Diário de la Calle de Fepal" and in "Conscientizar." As part of this effort, a book was released in 2023. This community project was awarded a first-place prize at the IPA Congress in Cartagena for its effective and invaluable contribution to the health area.

As the history of the psychoanalytic movement shows us, the perception and deepening of the professional's own life in the world at any given moment changes their being and their conceptual framework. Let us remember the impact and importance of the First and Second World Wars on the work of Freud and Bion. Psychoanalysis defines an ethical position that has emancipatory potential (Gherovici, 2021). Max Eitington, who made possible the establishment of the first free clinic in 1920, the Poliklinik in Berlin, observed Freud's idea of treatment being made available regardless of social class as "half a prophecy and half a challenge" (Danto, 2005, p. 3).

In view of the requirements for professionals to be able to carry out their duties in the community, it is important that the training of all analysts includes courses on the specificities of this work in the institutes' seminars and many community assignments for candidates. In this way, new generations will be able to commit themselves to the ethical duty of caring for people in need and/or traumatized by social pain. Ideally, institutional cultures will engage with scientifically discriminating the differences between psychoanalysis outside and inside the walls of the consulting room. The extension of the framework, in the extended clinic, enables theoretical-clinical, meta-psychological, epistemological and methodological contributions to psychoanalysis in general (Wald, 2021).

Clinical Illustration

S. was 14 years old during the tragedy in southern Brazil in May 2024. Her mother, a nurse, asks for help. The family had lost everything in Canoas. Her father is 48, a truck driver. Her only sister, 18, is studying for a medical degree. S. is a pretty, friendly, brown-skinned, sad adolescent. The family is temporarily living in an aunt's apartment in Porto Algre.

She tells me she has lost everything. Her mother doesn't want to go back to Canoas. Her parents are now unemployed, and roads are flooded. With regret, she complains having to look for another school in Porto Alegre. *Nothing is working.*

I tell her that, even with so much loss, so much pain, she is thinking about projects, another school. Connection is cut off. We reconnect. She says: "Difficulties with the internet."

I note the difficulties and the effort to connect with a new reality.

She shows me her small room full of suitcases, clothes, and packages. She also points her cell phone at the window. I admire the landscape with its green treetops contrasting with the grey, threatening sky.

I interpret that the room full of so many precious things and framed by the green of such trees. She can also dream – imagine windows, exits with other green possibilities, even in the midst of so many uncertainties and threats. She points her cell phone at the street. Floating in the brown, muddy water are debris caused by the disaster.

S. recognizes that people are compassionate and want to help. I show her that she is not alone. Now we can work together. She tells me she is having trouble sleeping because she has just watched the disaster on her mobile.

I tell her that she is curious, but that she wishes she was not aware of the disaster. I comment that the night can be dangerous.

A: You had to leave the house at night …
S: Luna, my cat, and Lucky, my dog, they are here with me. The others are elsewhere.

I emphasize the importance of being with animals, their warmth. She agrees and tells me her cat is like a son to her.

In the second interview, she tells me she's playing a house-building game. Her parents are looking for a house to rent using her father's trust fund. She dreams of a room of her own. She researches the cost of furniture, paint and decorations. She likes the yellow light but not the white light in her current place.

I emphasize that she does not like this cramped, crowded place, where tragedy is well lit. But she can dream, and she can imagine her newly decorated room.

S: My parents want me to be understanding, not to cry, not to be sad and mute or locked in the bathroom. In the end, we're alive and together. When I'm sad, I talk to no one. I hide in the bathroom to cry.

I claim her right to be sad, to cry, to be angry, and to not want or be able to talk. I tell her that now she can share her feelings with me, just by being together.

Excited, she tells me that she has lost her clothes, her make-up, special brushes for her wavy hair, hair clips, 10 different colored glosses.

S: My mother gets furious and shouts, "It is no big deal to be like this, for material things. I can't stand it anymore when you cry because of lip gloss."
A: I affirm your right to mourn these losses, which are not just material. There are stories in each of them, to be able to take care of your body that changes so much, your appearance, your hair … You can recognize what you have, your life, your family's … But you can be very sad.
S: It's difficult to communicate with my parents. Feels like a lump in my throat … I like to draw, write poems, dance. I used to write in a diary that got lost.

I encourage her to continue these important activities. She asks if she can share them with me. I agree. I reveal my interest in getting to know her better. Her work can be a portrait of her being. I feel I can broaden and strengthen the link between the analytical field and her dialogue with herself. She sends me many of her poems. She tells me that she had already asked her mother for therapy before the flood.

s: I felt fat, ugly, bad. I didn't eat. I was 33 kg. I'm sorry, I was wrong. Between 36 and 38. 41 when I got taller. I used to get dizzy doing exercises on YouTube to lose weight. I'd write it down and then tear it up and throw it away.

I tell her she can now keep what she writes and share it with me, feed her soul. She talks nostalgically about the joy of not having school. Her demands in terms of school performance are huge. S. is trapped within the context of her parents' ideals.

s: If I'm not a good student, my parents won't be proud of me.

Over the weekend, S. shares what she has written in her diary. Her sister broke her foot climbing the stairs to the fifth floor with a bottle of water. There was no electricity or water in her building.

s: Nothing looks like mine, the clothes I wear are donated, strange, used, I even got a T-shirt with Bolsonaro's picture on it! I lost everything in the fucking rain and I can't complain!

By the third session, the family has moved to a rented three-bedroom apartment. She is happy and shows me her room and the view from the window. She dreams about the decorations. Doubts and uncertainties about the new and unfamiliar school environment bother her. She complains her mother will not let her go out with her friends until she is 18.

s: 4G wasn't enough to keep me connected. We still don't have internet here.
A: Maybe you can connect differently with your family, with me. Change is possible. You're in a new room with new plans and dreams.

She apologizes for the connection difficulties.

A: Don't apologize! You're not to blame for such difficulties in our communication!

In the fourth session, she warns me and arrives late because they are connecting the Internet. She asks me if I have received her diary and poems. I had already replied to her via WhatsApp. I emphasize her interest in connecting with me and with her.

s: I'm very nervous. We're going to our house in Canoas. I don't know what to expect.

With an exciting narrative, she describes the house and we go through the floors.

S: It's very scary just thinking about what people are going to find. It's very hard to have to start all over again. School, friends, closet.

There's a knock on the door. Her dad wants to get a tool. She says, "I told you I'd be in therapy!" The father greets me with curiosity.

S. misses the next two sessions. One, she forgets the time, the other, she gets confused. She apologizes a lot in her WhatsApp messages. I ask myself: *Did my announcement of the sessions we still had, a prelude to the farewell, clarify her unconscious hatred, expressed in her absences?*

When we met, she told me her parents were alone in their destroyed house. They protected her by going in alone. She says, "Everything was rotten, destroyed, dirty."

S. tells an aesthetic story about the lost house and her forced exile. My parents will never return to Canoas.

I talk about their right to hate what happened to them and to hate our future separation. She tells me of her concern about not having a school. I reflect her anxiety about insecurity and uncertainty, but there is a plan.

S: My sister has removed the plaster. Life is better in this new apartment. My parents are calmer. Some roads are now open. My father has gone back to work.

I appreciate her strength and her family's ability to open up roads and get back to life.

S: Today I'm going to cook. Mom is going to teach me.

She talks about her favorite dishes and her joy at not having to receive donations. She laughs a lot. When I ask, she apologizes. She says she remembered a video she sent me. I tell her that she can laugh and cry without having to apologize.

After the session, I watch the video. A caption says: "Teacher: Your child is doing very well, he's a sweetheart." The same child appears in the house, blowing everything up with onomatopoeic sounds.

In the next session, I show her that she can have conflicting states within her: a well-behaved teenager and another who explodes with rage.

She associates this with an event in which her mother criticized her because she was distracted while feeding the dogs. She threw the glass jar of food on the floor. *Confusion. Glass, food, my mother shouting.*

S. finds a school and asks me to postpone a session. I'm worried about this request, but I'm happy to do it.

S: I started school today. It's awful. My mother is going to have knee surgery. She fell.

(What happens in this family with so many accidents?)

S: I'll put anything on the test. If I get a grade one, I'm winning. The teachers aren't even there. They didn't even realize I was a new student. I can't go to school on my own. I don't know the way; this condo is a maze.

I understand she is asking me to extend her sessions so she feels supported by something new. Her teachers told her that she is going to repeat the year. She says she has made two friends. She will miss school next week to help her mother after her surgery. She proudly tells me what she cooked with her sister. I point out she can learn and be happy with new achievements.

S: I used to Google subjects to study, listen to podcasts. My parents are very strict and against me dating. I won't be able to become friends with those colleagues. One of them has a boyfriend. My parents don't want me to have one.

(S. speaks very quietly, afraid of being overheard by them.)
I emphasize that what she feels and thinks is important. I also reflect how scared she might be about these dating issues and that we can talk. She replies that her friend's parents are open-minded. Her mother is a friend, and she talks to her daughter. She adds that it's the opposite at home. Dating is taboo. She decides to write because of the difficulties she has talking about it.

S: My parents are super against it. It's forbidden to them, they're very religious. My sister can't have a boyfriend. Imagine what it's like with me. Now I don't want to talk about it.

She plays with her beautiful hair. She curls it.
I point out that it is a difficult and complicated subject that makes her anxious and scared. There is a struggle inside her between wanting to be beautiful and the prohibitions.
In the last session, she is five minutes late. She tells me she was at the market. She's sucking, with great pleasure, on a sweet the shape of a stick. (I associate her mouth-vagina with this phallic signifier as a sweet encounter.)
She tells me that she went to school alone and that she got an eight on her math test. She says it was a good day. She'll have exams all week: "God help me!"
I point out she has been through many difficult challenges: changes, sadness, losses, but she also has new opportunities. She has meetings and goodbyes, like ours.
She says that today is her mother's birthday and that they are going to buy a cake to celebrate. I tell her that in life there is sadness, but also light, joy, celebrations and new friends.

s: I like funk music. I told mom with great fear. She told me she's going to buy me short funk clothes when this is all over. I was very happy to be understood. I even cried with emotion when we met.

I understand her joy at realizing that change is possible.

A: Your mother can be by your side to understand your teenage issues.

For Sol, this tragedy is a disruptive situation. Her psychic resources, in particular her symbolic capacity, allow her to keep this event from being traumatic (Benyakar, 2016). The consequences of the flood increased the psychic disorder of her adolescence, but she can differentiate between beauty and the damage caused by the catastrophic floods (Cardenal, 2021).

Art is a privileged way to symbolize, through poetry, the horror of her tragedy. Her writing, drawings, and poetry allows her a certain elaboration. Figurative interpretations allow S. to integrate emotions with representations in a shared emotional experience. She can also "build houses" as she feels contained in an internal and preserved space for herself and her objects.

S. almost completely regains the subjective continuity that is characteristic of the developmental moment she is going through, which even allows her to speak of the forbidden. Her subjectivity does not freeze, and life can continue. This is a general objective in this project of interventions.

The disruptive involves not only the power to disarm, dissolve or destroy, but it also enables an opening and an opportunity for the psyche (Torres, 2024). S. discovers new landscapes in the midst of catastrophe. As an analyst, I tried to help S. metabolize her emotions and allow her to live out her hatred, indignation, fear and love. Understanding S. as a subject implies not reducing her life to this tragedy. Without depreciating the current situation, how can I help open up existential paths? Over the course of this brief intervention, we traverse her anguish over potential love affairs; super-egoic threats over what is forbidden by her internal parents; concerns about her changing body; her identity framework; the right to experience a spectrum of emotions in a full way; and the change in her relationship with her mother.

She expresses her hatred for the first time at our announced farewell. She can nourish herself with her mother in less conflictive encounters, learning from her. She can allow herself to experience a range of emotions that were once almost forbidden. The gateway to a deeper analytical process is open!

Acknowledgement

I would like to thank the IPA, PACE, and Harvey Schwartz for their support.

References

Aulagnier, P. *La violencia de la interpretación*. Buenos Aires: Amorrortu Eds. 1976.

Belo, F. *Racismo, mensagem e desmentido: Diálogos entre Kilomba, Ferenczi e Laplanche*. São Paulo: Zagodoni. 2021.

Benyakar, M. *Lo disruptivo y el traumático: abordajes posibles frente a situaciones de crisis individuales y colectivas*. San Luis: Universitaria de la Universidad Nacional de San Luis (UNSL). 2016.

Bento, C. *O pacto da branquitude*. São Paulo: Companhia das letras. 2022.

Bion, W. R. *Attention and Interpretation: A Study of the Functions of Attention in Psycho-Analysis*. London: Tavistock Publications. 1970.

Borlle, J. M. *El sujeto y sue ex-sistencia. Contribuciones a la práctica psicoanalítica em dispositivos de salud mental*. Buenos Aires: Editora Letra Viva Libros. 2024.

Cardenal, M.Special Time: Working with street children. *Psychoanalytic Dialogues*, 31(4). 2021.

Danto, E. *Freud's free clinics: Psychoanalysis and social justice 1918–1938*. Columbia: Columbia University Press. 2005.

Filho, I. A. P. *Racismo: Por uma psicanálise implicada*. Porto Alegre: Artes e Ecos. 2021.

Filho, I. A. P. & Migueli, J. C. V. *Racismo e as trincheiras narcísicas implantadas pela branquitude [Palestra]*. Campinas: Sociedade brasileira de psicanálise de Campinas. 2022.

Filho, I. A. P. & Paim, A. M. *Racismo e psicanálise: A Saída da Grande Noite*. Porto Alegre: Artes e Ecos. 2023.

Franco, G. *El dolor de los márgenes (psicoanálisis y realidad traumática)*. Buenos Aires: Revista Latinoamericana de Psicoanálisis. 2015.

Freud, S. *Lines of Progress in Psychoanalytic Therapy*. London: Hogarth Press. 1918.

Gampel, Y. *A Dor Social*. São Paulo: Livro Anual de Psicanálise. 2022.

Gampel, Y.Entrevista a Yolanda Gampel. *Revista Controversias de APdeBA en Psicoanalisis de Niños y Adolescentes*, 34, 5–24. 2024.

Gherovici, P.A Psychoanalysis for the People. *Aperturas Psicoanalíticas*, 68, 1–17. 2021.

Jesus, C. M. *Quarto de despejo: diário de uma favelada*. São Paulo: Ática. 2019.

Kaës, R. *Rupturas catastróficas y trabajo de la memmoria: notas para uma investigación*. Buenos Aires: Centro Editor de América Latina. 1991.

Kaës, R. *Linking, Alliances, and Shared Space, Groups and the Psychoanalyst*. New York: Routledge. 2007.

Kon, N., Silva, M. L., & Abdul, C. *O racismo e o negro no Brasil: questões para a psicanálise*. São Paulo: Perspectiva. 2017.

Lazslo, A.The SOS Brazil Project: Coping with Covid-19 and its Social Impacts. *Journal of Psychoanalytic Studies*, 12(1), 45–62. 2023.

Lazslo, A.Os sentidos do trabalho do SOS Brasil e do Grupo de Estudos e Produção. In *Anais do Encontro de Psicanálise e Saúde Mental*. São Paulo: Associação Brasileira de Psicanálise. 2024.

Lisondo, A. B. D.O desamparo catastrófico ante a privação das funções parentais. Na adoção, a esperança ao encontrar o objeto transformador. *Revista de Psicanálise da Sociedade Psicanalítica de Porto Alegre (SPPA)*, 19(2), 367–393. 2012.

Meirelles, F. *Cidade de Deus*. Directed by F. Meirelles and K. Lund. São Paulo: O2 Filmes. 2002.

Mondrzak, V. S.Preconceito: razão e não razão. *Psychoanalysis Today*, 14. 2021.

Torres, J.Reunião Cientifica do grupo SOS Brasil com o Dr Moty Benjakar. August 4, 2024.

Verdi, M. *Crueldade social e compaixão. O trabalho psicanalítico como forma de desenvolvimento de nossa capacidade compassiva.* Rio de Janeiro: Fepal. 2024.

Viñar, M. *Psicoanalizar hoy.* Montevideo: Trilce. 2002.

Wald, A. *Notas sobre vulnerabilidad y desamparo en la infância.* Cátedra de Psico-pedagogía Clínica. 2021.

Chapter 14

On Psychic Envelopes and Spaces for Young Children during the Pandemic

Christine Anzieu-Premmereur

In 2025, we are 5 years into the most serious global health crisis of our lives. During the height of the pandemic, adults were experiencing fear of death and chaos in their professional and social lives, while mourning the loss of their friends and parents. Irrational decisions, emotional explosions and violence seemed to be the discharge of that excess of stress. What did babies and children experience?

The persistence of the trauma and confusion from the pandemic gave rise to what the French analyst André Green (1993) would call a disconnection between Eros and Thanatos, in which the death drive becomes stronger than the life drive. This imbalance led to the activation of primitive defenses. We observed externalization, projection, and denial. When psychic suffering could not be worked through, it was displaced into the other. Emotional withdrawal and avoidance of contact in the state of high alert are other characteristics of these situations.

During the pandemic, when confronted to the fear of contagion and death, many families were at their own limits. Dealing with those risks, some parents experienced primitive agonies which made them unable to contain their children. They communicated disorganization and lack of figurability to their children. Many children had distress about their body no longer being a reliable part of themselves, and I observed an epidemic of hypochondriac anxieties.

The sense of the virus as intruding into the body through oral and respiratory was associated with persecutive anxieties or with strong feelings of disgust. This led to fantasies of penetration into the oral and respiratory system associated with infantile sexuality. Some children who maintained a good level of autoerotic fantasies were very adept at drawing and playing with this kind of persecutory penetration. Drawing their fantasies about the virus attacking the body allowed some young children to play with fantasies of sadistic intrusion into their skin or body, as well as with their own aggression.

Babies and Toddlers

Toddlers were extra clingy, putting a great deal of pressure onto their parents. The more they had been put aside by exhausted parents who were confined at

DOI: 10.4324/9781003520757-16

home, the more they needed comfort. A child looks for physical contact when feeling disorganized, while the mother might be yearning for space. To be physically close is developmentally appropriate for a young child. The function of physical contact for babies is to establish a secure attachment and formulate a sense of self. Physical contact is extremely soothing. When the young child is in close connection with a mirroring figure, body contact offers not only connection, but also delineation that facilitates a sense of owning one's body and feelings.

It seems that babies did not mind looking at their masked mothers, but they did mind if she was not responding to their face, like in the still face experiment. Tronick and Snidman (2021) compared the baby's reactions when the mother interrupts interaction with those when the mother wears a mask. The masked mothers express in many ways their attunement and communication with their infant using their eyes, their voice, and holding. Babies reacted quite well to their masked parents and felt them smiling with their eyes.

As both Bowlby and Winnicott pointed out in their own ways, it's not separation that causes a baby to become depressed, it's the loss of hope. When an object that has disappeared, such as in the case of rapid weaning, an interaction with an inanimate mother, or an illness or surgery, all the links with the object can break, leaving the body and psyche destitute. The infant is atonic, withdrawn, slowed down, and rapidly psychosomatically disorganized. They may display the pathological defenses described by Selma Fraiberg (1982): avoidance of eye contact, behavioral paralysis or freezing, transformation of affects when instead of manifesting distress or anxiety. An abused baby manifests noisy joy and disorganized excitement; and self-aggression.

It was pleasurable for some children to share increased time with their parents when locked in their apartment, but this situation risked a sort of libidinal stagnation with increased hostility and unsatisfied demands from unavailable parents.

A 6-year-old girl was having intense tantrums and screaming that shocked her parents. She seemed to me to be quite a precocious girl, but she was extremely shocked that her busy, at home parents were unable to take care of her little brother, a toddler always in need for rapprochement. Full of rage, the girl tried to upend the life of her mother. A few sessions on a screen helped for a change. She reported having nightmares where the virus was attacking her parents. I talked about her anger, and she drew a frightening representation of a virus enjoying making her family helpless.

This amazing girl developed a new interest for spaces during her sessions on Face Time. She described how the different rooms in the apartment did not allow for privacy or intimacy. While I played at being squeezed in a tiny suffocating room, she discovered many secret spaces where she could spend time alone – reading, dreaming, regressing, and being creative. She labelled our common secret "the bubble," the shared space we had together in the virtual world. We created an envelope that gave the sense of being wrapped in the mind of one another, like a skin.

Families lived for several months locked down in fear, with no intermediary regulation such as school, caregivers, other children. The emotional capacity in adults in charge of children can have a mediating function in the way children cope with disruptions and traumatic events. Teachers, grandparents, caregivers are second attachment figures who fulfill vital support in daily life, and they are also motivators of novelty and renewed libidinal energy.

Social catastrophe produces a narcissistic injury that breaks the balance between strangeness befriended and total otherness. The latter is a source of persecutory anxiety, and it results in a lack of solidarity and a hatred for anything foreign. Isolation activates helplessness and vulnerability, provoking feelings of futility, distrust, and aggression. In the *Apres Coup* of this traumatic situation, the virus as an invisible enemy was a representation displaced into others named as bad, in a primitive splitting. This led to cognitive distortion and inadequate styles of coping when many were grieving. Poor libidinal ties with others caused a loss of empathy. Creating new spaces, even on a screen for session, was a way to try to respond to the longing for new ways of meeting and being understood.

Skin and Psychic Envelopes

This section addresses some fundamentals in early development through the metaphor of a psychic envelope. Such envelopes wrap one in a maternal space, a physical as well as mental skin.

During the pandemic, in adults and children and especially with adolescents, we observed issues of containment, stimulation and regression due to lack of touching and fear of being touched. Some were more on the regressive side and were looking for a metaphoric maternal space through passively clinging to each other, eating, drinking, smoking – all oral satisfaction that give a short relief. Infantile sexuality was replayed in parents and children, as an internal resource to be shared. Some tried to contain and integrate the frustration with masochistic cathexis into their own body, as a way of trying to control and master the lack of pleasure.

Working on a screen was a good tool for playfulness with the imagination deployed to deal with the lack of a real body. In order to maintain liveliness, in erotic as in aggressive capacities, even in a time of grief, analysts employed metaphors, with words maintaining a sense of touch.

The libidinal recharging that comes from this interaction with the mother and her body plays an important role in early development with consequences later on. When there is limited satisfaction in a specific body area, autoeroticism, a self-induced sensorial activity with no link towards an object, is used to provide satisfaction without the presence of an object. This autoeroticism can be either masturbation with no representation, or it can be inhabited with representations of libidinally invested internal objects. Any deficiency in the autoerotic capability will lead to self-soothing activities associated with the need for strict mastery over a concrete object. Since this does not create any internal object of satisfaction, this must be repeated compulsively. Some examples of this are intense pacifier sucking repeated, over and over and associated with insomnia in infancy; endless swaying; or pulling of hair. All of these can be attempts to compensate for the emptiness of the mother's absent body.

A combination of stimulation, containment, attunement, mirror and reciprocal pleasures depends upon the mother's unconscious infantile sexuality, and on her narcissistic and oedipal cathexis into that baby. She arouses, filters, and offers new experiences, adjusted to the baby's capacity and development. She shares the experience, repeats it with different modalities, and plays with rhythm and intensity.

At the beginning of life, the body envelope is still fragile, and maintaining the sensations and the experience of being alive through movements and through the action of the own body is essential. Restoring bodily dialogue is urgent. It is the economic aspect of maternal psychic functioning that is communicated to the child in the dance of preverbal games. Prosody, the rhythm of the utterance as well as of the porterage, the intensity of the gaze and of the voice are essential signals to which the baby responds with its own register of primitive drive.

A 4-year-old child, who I have seen as a floppy infant in the arms of a borderline distressed mother who couldn't provide her baby with firm and secure holding, told me in a psychotherapy session how distressed he was when being forced to take a bath. He felt that his body was melting like powdered sugar. In order to hold himself when facing those primitive agonies, he was compulsively drawing animals with multiple skins. He was attempting to feel like he was being wrapped in a solid container.

The patient, who cannot express with words the neglects or failures of the early environment will signify the defect and its consequences through his body and its sensations. It is a mistake to understand this solicitation as sexual. The therapist does not have to give the patient what the mother did not give and does not touch the body. It is by an active, verbal response that the analyst transforms the body signals from the patient into symbolic communication. This work allows the patient to communicate not only through his immediate sensations and body experiences, but also to think and talk about them. When a mother cannot echo the baby's experiences, no meaning appears. The capacity for experiencing oneself as a feeling being is connected to the mirror the mother offers by feeling in herself what the child lives. The analyst holds the patient with her attention, preoccupation and active interventions. By echoing the patient's archaic experiences, the analyst gives a circular relation where the perceptions and sensations are shared, contained, and expressed through words.

Transitional Space and its Potential Collapse

This section describes the baby and young child's experience of transitional phenomena and the developmental implications when these phenomena collapse. During the pandemic, while some children were able to maintain transitional space, for others this space collapsed, often leading to fetishistic activities.

For Gaddini (1970), the precursors of transitional objects – the sensations of keeping something inside the mouth – seem to maintain the self-integrated when the baby is separated from the object. The sensations of vertigo, understood as the trace of an internal object trying to keep the self-cohesion, is eventually used in play when the body is shaken like a rattle. Later in adolescence, this is played out through spinning and thrill rides. The pleasure shows here the integration of an internal space with bodies in an external space. A very anxious hypochondriac patient, who started to speak about his representations of death and disorganization, calmed himself by playing with his hands in his hair, unaware of this new connection with the internal object.

There are different kind of spaces in analytic therapy. There is a narrow space where the analyst is stuck and the patient feels him/her as intrusive, while maintaining a need for symbiosis. There is also more comfortable space when both creatively share imaginative free-floating associations.

In analytic interventions with parents and infants, the triangulation associated with the presence of the therapist in front of the dyad modifies the libidinal investment of the baby for a "non-mother" character who has an essential anti-depressive impact by modifying the constellation of anxieties and defenses.

The great models of play are organized around absence – *fort–da*, hide and seek are games of controlled loss. The creation of pleasures in all registers and the re-sexualization of life always show an improvement in a baby. This is why it is essential to intervene quickly and early when we see worrying signs in babies. The consequences of primary collapses are multiple throughout life, from early childhood disorders, such as developmental delays or attention difficulties, to serious obstacles to access to oedipal conflict, to damage to adolescence, and traumatic repetitions in adulthood.

For the child analyst, transitional object is only the visible part of a more general transformation and process that structure the psyche. Winnicott observed the children's behavior and discovered the creation of a new space between mother and child. The transitional object manifests the existence of space between internal and external, object and self. The object represents more than the environmental mother than the erotic object. This is important for the analytic technique, since therapists work at developing that space within the frame.

There is also a transition between the first state of undifferentiation to the internalization of the maternal object. The transitional object is at first an object of addiction like the mother was. Then the transitional object ushers in another stage, when this new space in between is created and owned by the child. Then, the child will move again to another stage, leaving the transitional object that is not needed anymore to experience the existence of oneself in a state of aliveness. The transitional space has been created and will last.

Transitional space is in daydreaming and the fluidity of associations that link unconscious fantasies to present representations. It is the source of the richness of mental life, protecting the body from discharge of loaded emotions and affects. It is the source and the pleasure in arts and culture.

In transitional space, the maternal function is introjected and is displaced into the object. This process diminishes neediness and aids in the development of the capacity for symbolic equivalences. This is the source of the dreaming capacity. An object that is used in an addictive way plays a different role, since it cannot replace the mother's presence and is used as a denial of the lack of her presence and the need for her when she is absent.

This precious object in the new intermediate space shows the beginning of a new capacity to generate states of being (Ogden 2001). This capacity allows the child not only to be able to live in the absence of the mother; it also enables him to discover the feeling of being oneself.

This nascent capacity comes from the illusion of "creating the breast," in an "invisible oneness" of mother and child. This creates a personal space, containing emotions and allowing for total relaxation.

I again compare it with addictive behavior that tries to recreate the mother as an object for the needs, not as an environment. The infant stays fully dependent upon the actual mother as an object.

In *Playing and Reality*, Winnicott (1991) made us aware of a neutral space between the internal reality, as described by Melanie Klein, and the external reality. This third area, labelled as the area of Illusion, is vital for maintaining a sense of being. Transitional phenomena constitute a space for overcoming the split between internal and external. This is a space in which there is no question about me and not me, and playful activity can be discovered.

When there is no compliance towards the environment, a new field of experience opens. The analytic setting is the mean for the actualization of this area of illusion. Transference offers this illusion that will eventually help for a better adjustment to reality. Winnicott's psychoanalysis is about providing in the analytic set up a facilitating environment that permits the emergence of the patient's creative potential.

Freud (1941) wrote that "psyche is extended; knows nothing about it." The perception of space is a product of the extension of the psyche. From the common Skin with the mother to the Skin Ego (Anzieu 2016) separated from the maternal skin, a continuous space of shared sensation leads to symbolic equivalences. This movement is important for the capacity for auto erotic activities instead of self-soothing or fetishist ones which are sources of compulsive repetition.

Here the drive activity interferes with a space in which the maternal object is absent, or benefits from the creativity of symbolic equivalence of her presence.

This was about the sense of being, of feeling alive in a continuity. Between moments of integration and periods of non-organization, infants experience more and more that they can be reassembled and unified around bodily activities, inside the maternal preoccupation. Then the pleasure principle can be maintained without helplessness.

The Psychoanalyst's Fatigue

Narcissistic patients are wounded in their infantile omnipotence. The move to thinking about the other through the transference can be difficult. The global prevalence of depression and anxiety in children and adolescents during COVID has doubled. The prevalence of anxiety has increased 25% compared with pre-pandemic estimates, and the most aggravated population is adolescents.

The analytic couple is in a space where the patient can feel their Being, while coexisting with the analyst. This is what André Green (1993) has called the desire for the One is about erasing traces of desire for the other. An Ego can become a unity in the awareness of a spatial separation that created a delay for getting satisfaction. The reality principle and the fundamental taboos interfere with the omnipotent wishes. In Green's view of the space, there is an active part from the subject which is the negative hallucination for the mother. This primitive process gives form to an internal space while the mother is present and negated. A representation of the mother can be built. In her absence, her representation is maintained. Autoeroticism starts from there, as opposed to all the addictive systems.

Psychoanalysts paid a high price in maintaining their much needed work in front of a screen, or with the telephone, exhausted, with patients very motivated to do analytic work., but in serious need for being securely held by their analyst. As a French analyst wrote (Gustin 2021), paraphrasing Freud: "And the shadow of the confinement fell on the Ego," stressing how close we were to the issues of mourning and melancholia. For almost a year, we repeated the same sense of loss that makes some fragile patients at risk for disorganization.

I had analytic sessions with a young, often dissociated, male patient obsessed with death, who started to write novels instead of using drugs, soon after I interpreted his wounded pride when acting out at abusing alcohol and stimulants. This feeling of being injured made him remember dreams full of horror as if he were a burn victim, with his painful skin falling into pieces, a representation of his wounded lonely self. He was then able to associate with bodily experiences and felt alive again. He moved to a new capability for sublimation, writing novels about his moral values.

How disturbing is the anguished countertransference of the analyst facing near-death sensations and explosions of psychotic destructiveness of fragile patients in the never-ending pandemic? The unreachableness of vulnerable patients in front a screen is close to a sense of ego dissociation facing autistic withdrawal.

I had that same pandemic fatigue with children who were avoiding even eye contact and tried to play videogames while faking to be in the session with me. My movements of decathexis of those patients, immediately followed by feelings of anger and guilt, are in a dance of back and forth away from them followed by a big effort to make them alive again.

Psychoanalysts became more aware of body-mind dissociation during the pandemic. This is a complex clinical figure, related to a variety of splitting and issues with the self and the psychosomatic organization. Dissociation is a complex defensive system, when the psyche abstracted from the body and lost its source of liveliness, the drives. The quality of embodiment of the psyche, the awareness of the sensorial life can change through time and during an analytic encounter,(Lombardi 2020).

Peter Goldberg (1995) emphasizes altered states of consciousness that can mask the body-mind dissociation. These are states of distraction, self-stimulation or hypnoid states that can be passing and are different from structured pathological forms of retreat. Each requires a different approach from the analyst, who must extend his attention to the frame and the analytic space.

The reconnection of mind and soma can unfortunately be a coercive connection that is not an integration. During the multiple rich connection between drives, emotions, and thinking, speech is fluid, but, when pseudo integration maintains a rigid control over psychosomatic experiences, this can result in a "pervasive inauthenticity in self-experience," (Goldberg 2020). Goldberg associates this with Winnicott's "segregation of two psychical qualities," that are not so much psyche and soma, but two split spheres of psychical experiences.

(The sense of frozen time leaves our analysands hypersensitive to being "touched" by physical contact in case of meeting in person, and to emotional mirror without containment when on a screen.)

The brutality and magnitude of the COVID pandemic has upset political agendas as well as individual lives. The virus has forced us to cancel, postpone, and adopt an unpleasant position: waiting. Our profession recognizes the ferocity of the imperatives of absolutism, and we try to argue in favor of compromise formations.

It's Time to Promote Mental Health Priorities

Our current situation is a crisis of resources, values, and priorities. We observe now a loss in playfulness and a new need for fetishistic objects. Porn, abuse of screens in loneliness, and anxieties about origins and identity create an insecurity that adults have trouble dealing with. Eroticism has been devalued, and sexuality is more action than relation with feelings.

Yet, young people will bring a lot of creativity and innovation in society through technology and artificial intelligence, if adults provide them with tools to think clearly and ethically It is also necessary in our increasingly virtual world that psychoanalysts keep their acute sense of emotional attunement and connection with the unconscious fantasy world.

It's time to promote mental health from the beginning of life, to support parents and infants while they build a new brain and a young Ego that can think and not only manage anxieties. When relatedness is disappearing, psychoanalysts are good at using their own awareness of their affects to maintain communication and pleasure at being together. Playfulness is at risk; psychoanalysts are the ones who know about the essence of libidinal creativity.

The uncertainty and all the limitations on our bodies became a reality with the pandemic. This has an emotional impact on the analyst, as on analysands.

Outside the analytic setting, reading the patients and their nonverbal language of emotions is a now very difficult exercise. We have observed quite a few patients adjusting well and enjoying staying at home, concentrated on themselves, but also there has also been an increase of destructiveness in more vulnerable patients.

Self-harming in adolescents and suicidal ideation in children now look like another epidemic. The effects on mental health are determined by a cascade of risk factors linked to prolonged confinement as well as depletion of adaptive capacities.

Understanding these issues without giving in to our aspirations for happiness is not an easy task. For comfort, let us remember the words of Freud in *Beyond the Pleasure Principle*: "Besides, a poet can console us for the slowness with which our scientific knowledge progresses: What we cannot achieve by flying, it must be reached by limping. Limping, says the Bible, is not a sin" (Freud 1920).

References

Anzieu, D. 2016. *The skin ego*. Trans. N. Segal. Karnac Books.

Fraiberg, S. 1982. Pathological defenses in infancy. *Psychoanal Q.*, 51: 612–635.

Freud, S. 1920. Beyond the pleasure principle. In J. Strachey (Ed. and Trans.), *The standard edition of the complete psychological works of Sigmund Freud*, vol. 18. Hogarth Press.

Freud, S. 1941. Findings, ideas, problems. In J. Strachey (Ed. and Trans.), *The standard edition of the complete psychological works of Sigmund Freud*, vol. 23. Hogarth Press.

Gaddini, R. 1970. Transitional objects and the process of individuation: a study in three different social groups. *Journal of the American Academy of Child Psychiatry*.

Goldberg, P. 1995. "Successful" dissociation, pseudovitality, and inauthentic use of the senses. *Psychoanal. Dial.*, 5(3): 493–510.

Goldberg, P. 2003. Life narcissism death narcissism. *Canadian Journal of Psychoanalysis*, 11: 246–248.

Goldberg, P. 2020. Body–mind dissociation, altered states and altered worlds. *Journal of American Psychoanalytic Association*, 68: 769–806.

Green, A. 1993 *The work of the negative*. Free Association Books.

Gustin, P. 2021. Des bebes dans une coque de noix ... Naitre en temps d'epidemie. *Le carnet psy*, 240(1): 26–30.

Lombardi, R. 2020. Will coronavirus and social distancing shift psychoanalytic focus towards our bodies? Reflections on remote analysis, pandemic panic and space-time limits. *JAPA*, 68: 55–462.

Ogden, T. H. 2001. Reading Winnicott. *The Psychoanalytic Quarterly*, 70(2): 299–323.

Tronick, E., & Snidman, N. 2021. Children's reaction to mothers wearing or not wearing a mask during face-to-face interactions. https://papers.ssrn.com/sol3/papers.cfm?abstract_id=3899140.

Winnicott, D. W. 1991. *Playing and reality*. Psychology Press.

Part III

Climate Change

Kristin Fiorella

Introduction

In the novel *Bewilderment*, by Richard Powers, a nine-year-old boy agonizes over the lack of adult responsiveness to our climate crisis. Diagnosed with Asperger syndrome and ADHD, and unable to comprehend this indifference, he spirals into rageful and severely depressed states. His father, an astrobiologist who feels desperate to help him, enrolls the child in an experimental neuro-feedback program in which the child communicates with states of mind previously recorded from his late mother. During the period of calm while he is undergoing this therapy, the boy watches videos documenting the destruction of various animal species. Later, after his treatment is interrupted and he unravels again, this kind of witnessing drives him to beat his head against a wall until a welt appears. Yet, during this period of equanimity, he asks his dad:

> *Do people know this?*
> "I think so. Mostly."
> *And they don't fix it because...?*
> The standard answer – economics – was insane. I'd missed something essential in school. I was still missing something. I stroked the crown of his head ... "I don't know what to say, Robin. I wish I knew."
> He reached up blindly to clasp my hand. *It's okay, Dad. It's not your fault.*
> I was pretty sure he was wrong.
> *We're just an experiment, right? And you always say, an experiment with a negative result isn't a failed experiment ... Don't worry, Dad. We might not figure it out. But Earth will.*

When I read this scene, it reminds me of an adult patient telling me about a conversation with her young son. One night, while she rubbed his back to help him fall asleep, he asked why people who know the planet is dying don't do more to save it. "If the government knows we're all going to die," he said, "will they tell us?" My patient told me that, a half hour after this

DOI: 10.4324/9781003520757-17

conversation, she crawled back into her own bed and had a panic attack imagining the world her son will inhabit as an adult.

Our world is so damaged already. Oceans acidify; forests burn; rivers run thick with pollution; plant and animal species vanish at unprecedented rates. If we are fortunate enough not yet to experience extreme weather events, is it possible not to be aware, at least dimly, of the millions who have already been displaced, impoverished, or killed by fires and floods, tornados and hurricanes? In human history, there has never been a global threat like this.

As the chapters in Part III reflect, many children and adolescents are acutely aware that we must live with our planet in a radically different way if we are to continue to inhabit it. To many of them, our inaction is incomprehensible. In Chapter 15, by Caroline Hickman, a child describes contemplating suicide in the event that nothing begins to change in our world's response to the climate crisis. "That is how it feels," the child says, "To live among those who don't make sense."

In the past, as in the present, certain humans were deemed expendable -usually people of color and those living in poverty. Kathryn Yusoff in *A Billion Black Anthropocenes or None* writes, "The Anthropocene might seem to offer a dystopic future that laments the end of the world, but imperialism and ongoing (settler) colonialisms have been ending worlds for as long as they have been in existence" (Yusoff, 2018). After centuries of such exploitation and violence, the proverbial chickens are coming home to roost, and they are not sparing those historically in the dominating positions. Although it still seems that the affluent will have some protection, we have entered a period when the dispensable lives are potentially all future lives. Unsurprisingly, for children and adolescents, this evokes intense rage, terror and despair. Our ruthless disregard for future generations and for the more-than-human world *is* stunning and disorienting. There is no adequate reply to Greta Thunberg's cogent embodiment of the fury of future generations, "How dare you?"

Understanding why we are not reacting with anything near the scale required is complex and multiply determined. Certainly, the list of entangled factors – annihilation anxieties, shame, guilt, narcissism, capitalism, geopolitical power, colonial history – is vast. Among myriad contributions, many writers, ecological theorists and philosophers are pointing out that our concept of "nature" has made Earth inaccessible for modern people. We have mischaracterized it and gone about destroying it with a staggering disregard.

Several chapters in this section take up the ontological cut between the more-than-human and the human, which is itself deeply rooted in colonialism. In this nature/ culture paradigm, which is intimately enfolded with other well-worn binaries, nature is mute, governed by ahistorical laws, and without agency. Culture is the domain with agency, thought, representation, and history. The deadly implications of this ontological rupture are increasingly visible in the climate crisis. Reworking this cut could not be more urgent.

Amitav Ghosh in *The Nutmeg's Curse* traces the physical and conceptual violence that was involved in this creation of nature,[1] during the European conquest of the Americas and the trans-Atlantic trade of enslaved Africans (Ghosh 2021). Prior to this subjugation and repopulating of the Americas, the great majority of Europeans, like Indigenous peoples in the Americas, believed themselves to be embedded in a living world. They did not have a modern ideological construction of nature as something that they were outside of. Dividing the complex phenomenon of being embedded in the world into the balkanized enclave of human and nature drags with it qualities that reinforce and produce systems of power.

Many philosophers and anthropologists posit that nonhuman entities, including forests, rivers, and glaciers have agencies. They represent and communicate in signs, although not linguistic ones (Kohn 2013; Cruikshank 2005). In psychoanalysis, as in other Western epistemologies, humans have erroneously been conceptualized as the only being capable of representation. Thinking of the more-than-human world as the inert background container for the human who represents only reifies the Cartesian cut between intelligibility and materiality that still haunts psychoanalysis. If we can acknowledge modes of representation that are not linguistic, as psychoanalytic theorists have begun to do with the body,[2] how does that shift and decenter the human and make the familiar world more strange? Could this reconfiguring invite a more loving relationship to the alterity of the more-than-human world? Both Ryan Lamothe and Kristin Fiorella's chapters take up some implications of this ontological rift for children and adolescents.

Overview of Chapters

Part III begins with Caroline Hickman's research and clinical material exploring how eco-anxiety can be felt and thought about by children and young people (Chapter 15). The chapter then discusses the importance of understanding the range of climate anxiety found in practice ranging from mild to medium to significant, severe, and critical climate distress. Frequently eco-anxiety is perceived as a singular emotional and cognitive experience, and all too often seen through adult eyes. Hickman's scale, drawn from clinical practice, attempts to differentiate between the different experience children and young people can have by drawing on international clinical and research evidence. The chapter looks at the individual, relational, collective and planetary trauma as experienced by children and young people.

In Chapter 16, Kristin Fiorella explores the material and discursive intraplays between gender, Right-wing gender discourse, and the climate crisis. Some children and adolescents wrestle with being born into a world with powerful signifiers, soaked with violence and incapable of responsiveness to our unfolding ecological disaster. In the midst of a world that is radically unstable and breaking down, these signifiers seem embedded and structuring, but not immutable. Gendered embodiment is a domain in which, for some, the dismay, despair, and reconfigured possibilities of our current world gathers. Fiorella uses clinical examples to suggest that the disruptive elements constitutive of embodiment, as well as its potential porosity with the more than human world, make gender a possible site of resistance and ethical response-ability in the context of the climate crisis. Child and adolescent patients wrestling with gender sometimes become sensitized to an embodiment that does not unify the self and that is in touch with an alterity that is inside/outside. At the same time, as our usual coordinates are lost or we know that they will be, the immersive panic that ensues also creates attempts to reinforce a threatened heteropatriarchal world, as is seen in far right ideologies. Attacks on trans children frequently center on defending "the natural" and protecting these children from "mutilation" and the "contagion" of gender ideology. The chapter explores the way in which these claims project destructiveness while perversely distorting several terrors entangled with the climate crisis – about "nature," violence, and permeability.

Chapter 17 looks at the intimate connection between climate anxiety and climate grief. Panu Pihkala applies interdisciplinary research and especially theories of grief and bereavement to climate anxiety in order to illuminate encounters between children and adults in relation to climate change and other environmental crises. Pihkala points out that intangible losses can be involved may be difficult to notice, and that grief over these losses is often disenfranchised grief. Climate change also produces nonfinite loss, which is difficult to live with. The literature on grief research can help discern these forms of loss and guide constructive reaction. Pihkala further applies to climate grief the framework of maturational loss. While even normal developmental changes evoke sadness, climate change can intensify this, because "climate maturity" brings many things that are difficult to bear. At the same time, climate anxiety raises the possibility of adversarial or post-traumatic growth in ways that likewise demand further attention.

In Chapter 18, Ryan LaMothe draws on Levinas and Heidegger to explore philosophical notions of existential homelessness. He then argues that the homelessness that Heidegger and Levinas describe are symptoms of the ontological rift between the human and the more-than-human. This rift is produced by Western semiotic apparatuses that construct human beings as superior and in possession of power over the more-than-human world, while constructing other species and the Earth as inferior and valued primarily for their instrumental value to humans. This ontological cut produces ways of dwelling in the world that accompany a sense of not being at home in the world. He explores what it might mean for the psychosocial development of babies, children and adolescents to reimagine this rift and to move to forms of dwelling characterized by reciprocity and adaptation with other species.

In the final chapter, Christine Franckx explores the impact of our environmental catastrophe on adolescent development (Chapter 19). For the first time, there is a real possibility of a future breakdown in which life is untenable. Will the parental and societal identificatory objects survive adolescent attacks? Can the former generation model a transformational process? The anger of adolescents reveals the violence they feel subjected to in discovering both global warming and the incapacity of their parents' and grandparents' generations to react appropriately. The adolescent facing this situation is confronted with a reality without an external object to challenge, with a temporality that is unreliable and therefore threatening, and with internal parental objects that cannot provide necessary containment.

The iconic adolescent in this context, Greta Thunberg, addressed world leaders and pointed out that this catastrophe was created by former generations. She also calls out their inability or unwillingness to take joint action and limit the damage. This is the world upside down – instead of parents reprimanding adolescents for irresponsible behavior, adolescents now remind adults of the need to contain their insatiable infantile drives of greed and to face reality.

Franckx describes clinical vignettes with adolescents struggling with eco anxiety. She recommends a tension between acknowledging our collective anxiety and suffering while also maintaining the transferential dimension in treatment. In this way, the analyst working with adolescents may offer a reliable identification figure, strengthen ego functions, and help reestablish connections with good internal objects.

Notes

1 As I noted in the Forward, many other scholars have explored the link between colonial history, modern ideas of "nature," and the climate crisis. See, for example, Orange (2017), Latour (2017, 2022), Mbembe (2019), and Gentile (2021, 2024).
2 For example, Elise (2017), Markman (2020), Goldberg (2012), Fiorella (2023).

References

Cruikshank, Julie. 2005. *Do Glaciers Listen? Local Knowledge, Colonial Encounters, and Social Imagination.* Vancouver, Toronto: UBC Press.
Elise, Dianne. 2017. Moving from within the Maternal: The Choreography of Analytic Eroticism. *Journal of the American Psychoanalytic Association* 65: 33–60.
Fiorella, Kristin. 2023. Thinking in a Marrow Bone: Embodiment in Vajrayana Buddhism and Psychoanalysis. *Journal of the American Psychoanalytic Association* 71: 277–309.
Gentile, Katie. 2021. Kittens in the Clinical Space: Expanding Subjectivity through Dense Temporalities of Interspecies Transcorporeal Becoming. *Psychoanalytic Dialogues* 31: 160–165.
Gentile, Katie. 2024. Destroying the "Human" for the Survival of the World: A Proposal for a Temporally Dense Psychoanalytic Subject. *Psychoanalytic Dialogues.* 34: 533–541.

Ghosh, Amitav. 2021. *The Nutmeg's Curse*. Chicago, IL: The University of Chicago Press.

Goldberg, Peter. 2012. Active Perception and the Search for Sensory Symbiosis. *Journal of the American Psychoanalytic Association* 60: 791–812.

Kohn, Eduardo. 2013. *How Forests Think: Toward an Anthropology beyond the Human*. Berkeley, CA: University of California Press.

Latour, Bruno. 2017. *Facing Gaia*. Cambridge: Polity Press.

Latour, Bruno. 2022. *How To Inhabit the Earth*. Cambridge: Polity Press.

Mbembe, Achille. 2019. *Necropolitics*. Durham, NC: Duke University Press.

Markman, Henry. 2020. Embodied attunement and participation. *Journal of the American Psychoanalytic Association* 68: 807–834.

Orange, Donna. 2017. *Climate Crisis, Psychoanalysis, and Radical Ethics*. New York: Routledge.

Powers, Richard. 2022. *Bewilderment*. New York: W. W. Norton and Company.

Yusoff, Kathryn. 2018. *A Billion Black Anthropocenes or None*. Minneapolis, MN: University of Minnesota Press.

Chapter 15

Eco-anxiety in Children and Young People

A Rational Response, Irreconcilable Despair, or Both?

Caroline Hickman

Introduction

There is a growing understanding of the impact on mental health – the distress, confusion, and anxiety that follows increased awareness of the climate and bio-diversity crisis (Ogunbode et al. 2022; Pihkala 2020) with concern increasingly centered on how this is affecting children and young people (Hickman & Hoggett 2019; Hickman 2020; Hoggett 2019a, 2019b; Lawton 2019; Aspey et al. 2023).

Recent Intergovernmental Panel on Climate Change (IPCC) reports and other research have highlighted the mental health impact and relational aspect of this distress with children and young people suffering severe emotional and mental upset because of climate change, knowing about it, witnessing, directly experiencing its impact, and fearing for their increasingly uncertain futures (UNICEF 2021). They are suffering from prolonged psychological and physical stress. Relationally they have symptoms of depression, grief, and anxiety, feeling betrayed, abandoned, and dismissed by the people in power and governments who they expected to protect them (Hickman et al. 2021; UNICEF 2021). Research has examined the impact of direct exposure to traumatic events (wildfires, floods, or extreme heat) as well as indirect adverse experiences observed by witnessing the harm being caused to others, such as news reports showing animals and people fleeing wildfires or listening to stories told by survivors of traumatic events (Lawrance et al. 2021; Obradovich et al. 2018).

In 2022 an IPCC report first addressed the mental health impact of climate change, whereas prior to this the emphasis was on physical health impacts (IPCC 2022). The recognition of post-traumatic stress disorder, trauma, depression, and anxiety validated what many have understood and felt for some time: that the effect of the climate and biodiversity crises on mental health is significant, and widespread, and affecting children the most. At the report's launch, António Guterres, the Secretary-General of the United Nations, responded as follows:

DOI: 10.4324/9781003520757-18

Today's IPCC report is an atlas of human suffering and a damning indictment of failed climate leadership... delay means death ... I know people everywhere are anxious and angry; I am, too ... every voice can make a difference ... now is the time to turn rage into action.

(Guterres 2022)

I have been working with children and young people in clinical psychotherapy practice alongside international research developing an understanding of eco-anxiety for more than ten years. This chapter will present some key findings and clinical approaches developed from this work. However, given that I am confident in arguing that eco-anxiety is an emergent mental health crisis, I would also offer the caveat that despite immersing myself in the study of this for ten years, I am only starting to understand it, because it changes continuously; it is linked to changing environmental, social, and cultural factors, and it cannot be wholly predicted. We are in the most changeable of changeable times, and we need to be fluid and flexible, and have humility and curiosity to keep learning about what it means to be fully alive and awake in the world today.

Two Important Frames for This Chapter

Before examining the nature of eco-anxiety in more detail there are a couple of points I want to note as important frames for the following discussion. Firstly, we need to walk a careful line when thinking about eco-anxiety in relation to children and young people, so we do not individualize and therefore pathologize what is in effect an emotionally mentally healthy response to environmental problems in the world with a social and relational root system (Marks & Hickman 2023). This constitutes a clinical dilemma in practice because the climate emergency is not going to recede or be fixed in the near future. It follows, therefore, that eco-anxiety will also continue to increase in response, and any attempt to fix it or remove it in children and young people risks invalidating their fears rather than supporting them. We do of course also need to respond to the distress that is being felt and support children and young people to navigate their distress while validating their emotionally healthy and congruent response. Meaning can be found in the dilemma and therapeutic responses.

The second point to note is how I have noticed requests for support from young people with eco-anxiety have started to change in practice. This is something I started to see increasingly around the time of COP27 in Glasgow in 2021 but have been hearing more consistently in recent months. Young people initially were asking for help with their eco-anxiety, but now they often start to question how to live in a world that evidently does not care about the future of children and young people. Their anxiety can move from anxiety about climate change to anxiety about the people in charge or in powerful positions in the world. They ask, "Am I crazy to feel this way?" Within a few days in July 2023, Antonio Guterres stated that we have entered the world of global boiling, not global warming. Rishi Sunak, the UK prime minister at the time, stated that he is on the side of car drivers rather than clean air projects in cities that are needed to give children a healthy environment, and announced that he is going to "max out" oil and gas reserves in the North Sea, against all the advice from his own scientific advisors. It is little wonder that children are left confused about who to trust, wondering how to process and make sense of these contradictory arguments where there seems to be no priority given to their needs for a safe future. Increasingly I hear climate-aware children and young people say things like "Tell me how to stay alive and sane in a world that clearly just doesn't care about us having a future" and "Well, if he is for motorists then he may as well say he is against children breathing clean air," and "I feel like I'm just going mad now, there is no way that I can feel OK about this, all the therapy in the world is not going to make me feel OK about feeling so hated by the people running this country." More and more I have seen children and young people who are knowledge-able and aware of the climate crisis fall into increasingly unresolvable despair as they perceive that their futures are being abandoned and betrayed. The increase of this expression of anguish has been marked. And it's hard to know how to respond to this because these are irreconcilable positions.

I feel it is important to note these points, because while we do need to develop a therapeutic approach to supporting individuals and groups of chil-dren and young people with their despair, this needs to be done in a context that includes recognition of a contemporary social, political, cultural, and planetary lens through which we need to look, in order to fully understand how this feels, and not individualize what is a collective hurt.

Eco-anxiety in Children and Young People

Research tells us about the impact of the climate and biodiversity crises on children and young people worldwide as it examines how these crises affect children emotionally, cognitively, in their daily lives, and their trust and faith in government action (Hickman et al. 2021). This research told us that 67% said climate change makes me feel sad and afraid, while 62% felt anxious. While we might be able to reassure ourselves that children in Europe are currently somewhat protected from the impact on daily living with only 26% in the UK, 31% in Finland, 35% in France identifying that it impacts on going to school, spending time with friends or in nature and studying, as opposed to the children in the Philippines and India (74%), not that we are reassured by this global injustice. But it is when we look at the figures show-ing cognitive impact that we can see how children globally may have more in common with each other than difference. It is when we look at how they see their futures, how they navigate the world today, what hope they have for the future that we see how powerfully children are being affected even in countries that are not facing the worst impact of climate change, yet 75% worldwide and 73% in the UK think the future is frightening, and over half think that humanity is doomed (56% worldwide and 51% in the UK).

It is therefore crucial when thinking about climate distress with children and young people that we take a global or planetary perspective, encompassing and respecting cultural differences and sensitivity as well as thinking about the immediacy or distance to the threats being faced. It is in the similarities that we see a powerful message about how this is affecting all children and young people, and for me this means that an intergenerational frame or child-centered climate crisis lens is essential if we are to understand what this can mean for all children.

While research has largely framed eco-anxiety as an emotionally healthy and congruent response to environmental crises there remain contrasting and competing theories surrounding its nature and treatment, with eco-anxiety variously framed as pretraumatic or posttraumatic stress, collective trauma, intergenerational trauma, and Adverse Childhood Experiences. While there is a general recognition from professional bodies that eco-anxiety should not be diagnosed or pathologized (RCPsych 2021) there are also attempts to categorize and differentiate between "normal" and "abnormal" forms of eco-distress. To be certain, what could be argued is that eco-anxiety is an emergent mental health problem that we are learning about amid the unfolding climate and bio-diversity crisis itself.

Counseling and psychotherapy have been increasingly addressing these issues with development of "climate aware therapy" models in clinical practice (Davenport 2017) and through groups such as the Climate Psychology Alliance. Climate-aware therapeutic models would argue that rather than diagnosing and pathologizing eco-anxiety we should be supporting people in making sense of their distress, finding meaning in their changing world, developing community and collective approaches for support, challenging climate denial, and embracing the range of emotions found through this including grief, despair, depression, radical hope, and empathy. A depth psychology approach (Hoggett 2019a, 2019b ; Hollis 1996; Lertzman 2015; Randall 2019; Weintrobe 2013) would argue that it is this very descent – depression, experience of grief and loss – that gives meaning to the experience of waking up to the climate crisis.

Looking under the surface requires us to be curious, to use deep listening (Hoggett 2019a, 2019b), show respect to all forms of emotional expression (Hickman 2020; Pihkala 2020; Weintrobe 2021), show humility to feelings (Hollis 1996) and develop climate-aware psychological models that can help us to navigate these unprecedented challenges, both internal and external in the world today.

Based on clinical case studies and research, climate anxiety can be defined and categorized on a scale ranging from mild to critical. While much of this is generalized from clinical practice and can be applied to any age group, I have also included a specific note to each category reflecting the way in which I see each category apply to children and young people specifically. I first published this scale in 2020 and immediately started to see it as inadequate or incomplete because it ended with the category of Significant. I realized that a further category of Critical was needed as I started to see more suicidal and self-harming behavior appear in practice in relation to climate distress. The scale represented here includes my developed clinical understanding from the last three years.

Eco-anxiety Scale

Mild Impact on Daily Functioning

Some feelings of upset, weekly or less often, sometimes after watching news of climate problems in the world. Little interest in reading news articles about climate change or watching documentaries or engaging in deeper discussions. Often will only discuss climate change in a limited way and, while interested, will not want to dwell on the problems too much "let's not spoil the party or be negative."

Can still spend time in nature without any focus on the losses or impact of climate change on nature.

Can think that flying is not ideal but may be unwilling to face the disruption of avoiding flying or making significant changes to lifestyle. Largely hold others responsible for "fixing" the problems.

Thoughts

Strong beliefs that other people (such as scientists, business, and governments) hold answers to the climate emergency – the belief which provides reassurance and relief from distress, "It's in their hands." Generally, thinks it is just warming (temperature rise, ice melting, and flooding), doesn't see the broader systemic nature of climate change, and looks for evidence that things are not that bad. Strong attachment to "positive" news. Little disruption in cognition/thinking.

Feelings

Feelings of distress include initial anxiety and some upset, but not developing into panic or stronger dysregulated feelings. The feelings are not overwhelming and recede quickly. Feelings of distress can be transient and respond to reassurance from others who do not have to hold authoritative positions such as scientists or politicians. Feelings of anxiety can be reduced by a focus on relatively small individual and local actions such as food choices (eating less meat, reducing dairy consumption) and recycling. There is a focus on optimism and hopeful solutions and avoiding painful feelings such as depression or despair. Can show evidence of reacting strongly to other's fears, upset briefly but then making jokes, silencing the anxious person, or changing the subject quickly. Psychological equilibrium is achieved by dismissing the fears and avoiding evidence. Can become cross and irritated if someone persists in talking about their climate anxiety leading to pejorative dismissal, such as "don't be such a tree hugger."

Defenses

Defenses of disavowal are often present and can border on denial (such as minimizing the scientific evidence and feelings of anxiety and emergency). Strong evidence of managing fears through reassurance by others (leading to strong rationalizations such as: "It will all be OK, humans have faced crises before and survived, we survived a world war, we can survive this"). These defenses could be considered to be naive. Fears may be largely unconscious.

Children and Young People

When experiencing mild eco-anxiety children and young people will depend fully on the adults around them such as parents, carers, and teachers to guide them in how to regulate their emotional responses. They will very much take their lead from authority figures and be reassured that there will be solutions to these crises that they do not need to be worried about. Their trust will be largely intact.

Medium Impact on Daily Functioning

Slightly increased impact with a willingness to make some changes in lifestyle. More recognition of personal responsibility to make some changes but with some bargaining "I'll do this if everyone else does the same." The beginnings of doubt in the solutions of "others" to the emergency but still retains fundamental beliefs that technological solutions will be found before it gets too bad. Making some lifestyle changes such as reduction in flying/meat eating/dairy consumption, but still choosing to mostly maintain life as before with minimum significant changes. Can be reassured by discussion with others. Some limited knowledge about facts and figures, but not obsessed or preoccupied. Happy to "leave the science to the scientists." Likely to be flexitarian or vegetarian.

Thoughts

Some disruption in cognition/thinking, but not preoccupied by the emergency.

Feelings

Stronger feelings of anxiety, sadness, some guilt, and shame more likely to appear, particularly in relation to own family. Feeling upset more frequently (weekly).

Defenses

Psychological defenses are less efficient in minimizing distress. For example, some intrusive thoughts may break through defenses, especially after watching news or listening to mixed messages about solutions to climate change. Rationalized beliefs can still be strong, such as "If things were that bad then authority figures would take action as an urgent matter – wouldn't they?" Can be defensive if challenged, such as redirect challenge to displace anxiety "Well what about China? We are doing enough here" or "The cost-of-living crisis has to be taken seriously, we need to be reasonable and pragmatic." Splitting the economy from taking action on climate change. More conscious awareness that this is a complex problem, but defenses largely intact.

Children and Young People

In this category children and young people will still largely have trust in adult figures in their lives but will likely have started to initiate more conversations at home, to challenge their families' narratives about the crisis, to question if they are doing enough. They are likely to be seeking out information for themselves online or from the news and become increasingly aware that their parents or carers may not have all the answers.

Significant Impact on Daily Functioning

Daily upset and feelings of distress – minimal defenses against guilt, grief, and fear. Some fears of social collapse can be seen alongside fears about climate change. Often, they have made many practical changes in their world such as making conscious choice to be vegetarian or vegan. Increasingly little faith in "others" finding or acting on solutions. Willing to end relationships with people who are in denial about the climate emergency including friends at school or within the family. Partnerships and marriages can be threatened, especially if one partner is in denial or minimizes the emergency. Conflicts about the practical action needed to tackle it are seen in families and close relationships. Significant impact on lifestyle. Tendency to try to reduce own carbon footprint and then feel it is never enough, leading to further and further reductions, none of which lead to satisfaction "I can never feel as if I can do enough." Often turns to nature-based work/ leisure for emotional support such as joining wildlife trusts, rewilding projects, growing vegetables. Often start to imagine different futures such as moving to rural location, joining an eco-community, running an insect farm. Support school strikes, marches, or other forms of direct action, with some willingness to break the law/be arrested, or if not personally willing, wants to support others in doing so.

Thoughts

Increase in signs of cognitive/thinking changes such as guilt and shame in relation to children and grandchildren (their own and/or others), their futures being damaged, and anxiety about children's futures. Hesitancy to have children of their own in the future because of climate change. Much stronger feelings that they should "do something." Some intrusive thoughts and dreams, but infrequent in comparison with more severe category. Some breakdown in ability to trust information and others.

Feelings

Anxiety is much harder to mitigate by reassurance, for example reassurance will have to be repeated frequently and trusted authority figures will be sought out. But this is a very changeable category when people can really struggle with who to trust and believe. Frequently feel insecure and anxious, but group actions can reduce anxiety to a more manageable level. Much more anger with feelings of being let down by powerful others. In children and young people some significant anger about the choices made in the past by adults destroying children's futures.

Defenses

Defenses against distress are much more fragile. Intolerant of climate deniers. Still holds some beliefs in government and democratic political solutions, but with some increasingly cynical and doubting views.

Children and Young People

Upset that can seriously interfere with their daily lives including fear about human extinction. Struggle with believing the messages of reassurance from adults vs their own research and knowledge gained often from external sources. Much more willing to challenge the values and beliefs of their family and school.

Severe Impact on Daily Functioning

Sleep affected and preoccupation with the climate emergency leading to a struggle to enjoy any aspect of life because of fears for the future. May affect daily functioning – be unable to go to school or to work some days. Can be driven by a need to gain knowledge of climate change facts and figures and spend hours reading information online.

Thoughts

Severe cognitive/thinking changes such as intrusive thoughts. People can manage daily life by compartmentalizing their feelings and thoughts, but they return frequently and can be intrusive (force their way into consciousness even when the person really doesn't want them to do so for example during conversations, meetings, or in a class at school). Strongly held beliefs (certainty) that the climate emergency will lead to social collapse and social breakdown.

No belief or trust in "others" (authority figures/governments) to act sufficiently swiftly or decisively to take steps to adapt or mitigate against climate change – this is supported by evidence as they fail to act or fail to take what is believed to be sufficient action to protect people. Supported as well by considerable knowledge about the crisis (climate, political awareness, literature, scientific).

Knowledge of the climate crisis as a systemic global injustice with a preoccupation with "fairness" and "unfairness," and injustice. Extensive awareness of the impact of climate change both nationally and internationally. Thoughts about being unable to have children of their own either now or in the future as the world will be inhospitable or uninhabitable.

Feelings

Anticipation of extinction of human species leading to terror rather than anxiety.

Can be felt as though people who don't understand or empathize, or care are being abusive and cruel. Can be intergenerational tensions such as children who are angry with their parents' generation for failing to act on climate change.

Unable to manage or contain strong emotional responses all the time – these will come and go – (such as crying a lot, impulsive anger at people who don't seem to care about the climate crisis).

Feelings of guilt and shame and heightened distress when thinking about their own potential future children, or the impact on their current younger siblings having to grow up in such a dangerous world. Sometimes there is less personal concern, and the anxiety is directed toward more global anxiety, empathy, guilt, and shame.

Defenses

No defenses against the terror felt, or the defenses are easily overwhelmed.

Dreams about climate change occur. Some fleeting thoughts about suicide as a future option if things get worse, but no immediate plans to act.

Children and Young People

Can see considerable anger and feelings of disbelief, betrayal, and abandonment in relation to parental figures which will include psychotherapists if they try to offer reassurance about climate change or tell the child not to worry so much. This can lead to the child avoiding discussion with the parent or therapist. They can hide their distress and only present what they consider to be allowable or rational fears. The child will maintain the close relationship with the parent or therapist at the expense of their feelings.

Critical Impact on Daily Functioning

Daily functioning can be affected significantly. People can have an obsessive need to get information about climate change such as checking the IPCC report repeatedly (may be seen behaviorally occurring every hour) or checking weather reports repeatedly throughout the evening in anticipation of the following days weather looking for evidence of worsening climate figures related to heat, rainfall, air quality, temperature. Intrusive thoughts about human survival – such as looking at a group of people on a train or in classroom at school and wondering which ones will survive and which will die first from heat or starvation in the group.

Thoughts

There may be thoughts of suicide in children and young people as well as adults. Some discussion about mass suicide as a reasonable and rational option, as the only way to get away from the endless fear and irreconcilable terror that is felt. Children can worry about having to kill their pets or set them free to fend for themselves and catch their own food to save them from death by starvation as food supplies run dry and they are no longer able to provide food for them.

Distress and crying at the thought of never being able to have children of their own in the future as they will be unable to protect their children from harm.

Feelings

Feelings of anxiety can frequently turn into panic. Extreme guilt and grief about the impact of climate change on all parts of the world "nowhere is safe" with very high levels of guilt felt toward populations more vulnerable to the emergency. There can be a powerful feeling of alienation from parents or carers if they do not understand or cannot tolerate the level of distress felt by the child at this point.

Defenses

Very few defenses in this stage against terror and overwhelm, both cognitively and emotionally. Real risk of suicide.

Children and Young People

In this most critical category children and young people may feel extreme alienation from peer groups and family as they can be left feeling that no one understands their distress. Can be faced with fears that they are going mad, or everyone else in the world is going mad. Extreme insecurity and distrust. No tolerance of rationalized defenses, these are seen as further abandonment. Compounded distress on multiple levels. Suicide or extreme climate activism is one of the only regulating emotional states available.

Dreams

Sally Weintrobe said that "we need a dream and a nightmare to understand what is wrong and to imagine a better world" (Weintrobe 2021, p. 5). Children are increasingly relating dreams or nightmares that suggest they are fearful unconsciously and unable to find respite from reasonable rational reassurances. Children are telling me about dreams or nightmares of having to kill pets, younger children, or elderly parents as food sources run out and social breakdown threatens social order, or of being killed themselves by parents or older siblings. Dreams that their parents have collected drugs so the family could commit suicide together. Dreams about fuel shortages leading to them being unable to travel to get to loved ones such as parents, partners, family, or to safety more generally, for example breaking down in unknown places and being threatened by strangers who live there. Dreams or nightmares of this nature about climate change are very stressful as they frequently cause the child or young person to wake up full of fear and there will be a period of time when this is believed and felt to be real before they are fully awake and realize it was a dream and can start to feel calmer. The loss of sleep and dysregulation from nightmares is particularly damaging as the child is unconscious when asleep and so is unable to have any control over these experiences. Children and young people tell me that this feels like torture, there is no respite from fearful emotion and distressing thoughts, even when asleep.

Moral Injury

Given the arguments above about the relational nature of the distress felt by children and young people regarding the climate crisis we need to have a way to understand and respond to their distress that offers containment without pathologizing. We need to understand it through their eyes. The very people who are supposed to be protecting children and young people are often the people who are causing their distress, often unwittingly. But worse, they are simultaneously telling children to trust them, that they have their best interests at heart, and that their fears are disproportionate. Weintrobe (2021) identifies this as a moral injury, a climate and ecological trauma (p. 241). A violation of "what's right."

In Practice

Empathize, try to see things through children and young people's eyes. While this might seem obvious and everyday practice for therapists, the climate crisis means that we are facing our own anxieties about survival alongside those of our clients. This makes the therapeutic alliance absolutely crucial as a container for the fears of both the therapist and client.

Therapists are skilled in working in the liminal spaces between fantasy, dream, imaginal, and reality. The climate crisis requires us to tolerate multiple uncertainties over a period of time. We need support to absorb and assimilate our own fears alongside those of our clients.

Facing the realities of the climate crisis requires us to acknowledge feelings of grief and loss at what is already lost as well as hold the hope that much can be done to mitigate or adapt to these crises. It requires us to adopt a both/and approach and be careful of splitting into an either/or stance.

I would argue that all adults need to say "sorry" to children and young people for the crises we are facing. And while not taking on disproportionate guilt and responsibility (we are not Exxon Mobil), all adults do have a culpability for the position we are in and have known about for many years.

We need to support children and young people in developing emotional biodiversity (all feelings are relevant here), emotional intelligence, internal as well as external sustainable activism.

We cannot remove or replace feelings of eco-anxiety from children and young people, as that would constitute a further betrayal. What we can do is to bear their feelings of frustration and despair, so at least they are not alone with these feelings, and that may reduce the sense of betrayal.

Finally, I am suggesting a psychoeducational approach, and reframing of eco-anxiety. Children and young people feel this often acutely as a measure of the care they feel for the world, themselves, and the animals and children around the world. Instead of feeling anxious about their eco-anxiety, they could feel proud that they care.

Hoggett (2019b) beautifully summarized what it means to be alive in the world today when he reminds us that we are "witnessing" and "living with" a tragedy without precedent that is unfolding in front of our eyes. As therapists we are human beings experiencing the same climate breakdown as our clients. We are both witnessing and experiencing it, for ourselves and our families, and we need, alongside this, to stretch our hearts, minds, and imagination widely enough to understand it from children's points of view, even when they do not fully fit with our own.

References

Aspey, L., C. Jackson, & D. Parker (Eds.). 2023. *Holding the hope: Reviving psychological and spiritual agency in the face of climate change.* London: PCCS Books.

Davenport, L. 2017. *Emotional resiliency in the era of climate change: A clinician's guide.* London: JKP.

Guterres, A. 2022. UN press release – IPCC report. April 4.

Hickman, C. 2020. We need to (find a way to) talk about … eco-anxiety. *Journal of Social Work Practice* 34(4): 411–424. doi:10.1080/02650533.2020.1844166.

Hickman, C., & P. Hoggett, Eds. 2019. *Climate psychology: On indifference to disaster.* London: Palgrave Macmillan.

Hickman, C., E. Marks, P. Pihkala, S. Clayton, R. E. Lewandowski, E. E. Mayall, B. Wray, C. Mellor, & L. van Susteren. 2021. Climate anxiety in children and young people and their beliefs about government responses to climate change: A global survey. *The Lancet Planetary Health* 5(12): e863–e873. www.thelancet.com/action/showPdf?pii=S2542-5196%2821%2900278-3.

Hoggett, P. 2019a. *Getting real*, vol. 29. London: New Associations.

Hoggett, P. 2019b. *Climate psychology: On indifference to disaster.* Basingstoke: Palgrave Macmillan.

Hollis, J. 1996. *Swamplands of the soul new life in dismal places.* Toronto, Canada: Inner City Books.

IPCC. 2022. *IPCC climate change 2022. Sixth assessment report – impacts, adaptation and vulnerability.* Geneva: IPCC.

Lawrance, D. E., Thompson, R. H., Fontana, G., & Jennings, N. 2021. The impact of climate change on mental health and emotional wellbeing: Current evidence and implications for policy and practice. www.imperial.ac.uk/grantham/publications/all-publications/the-impact-of-climate-change-on-mental-health-and-emotional-wellbeing-current-evidence-and-implications-for-policy-and-practice.php.

Lawton, G. 2019. I have eco-anxiety but that's normal. *New Scientist* 244(3251): 22. www.newscientist.com/article/mg24432512-900-if-we-label-eco-anxiety-as-an-illness-climate-denialists-have-won.

Lertzman, R. 2015. *Environmental melancholia: Psychoanalytic dimensions of engagement.* London: Routledge.

Marks, E., & C. Hickman. 2023. Eco-distress is not a pathology, but it still hurts. *Nature Mental Health* 1(6): 379–380. doi:10.1038/s44220-023-00075-3.

Obradovich, N., R. Migliorini, M. P. Paulus, & I. Rahwan. 2018. Empirical evidence of mental health risks posed by climate change. *Proceedings of the National Academy of Sciences* 115(43): 10, 953–910, 958. www.pnas.org/cgi/doi/10.1073/pnas.1801528115.

Ogunbode, C., 2022. Climate anxiety, wellbeing and pro-environmental action: Correlates of negative emotional responses to climate change in 32 countries. *Journal of Environmental Psychology* 84: 101887. doi:10.1016/j.jenvp.2022.101887.

Pihkala, P. 2020. Anxiety and the ecological crisis: An analysis of eco-anxiety and climate anxiety. *Sustainability* 12(19): 7836. doi:10.3390/su12197836.

Randall, R. 2019. Climate anxiety or climate distress? Coping with the pain of the climate emergency. https://rorandall.org/2019/10/19/climate-anxiety-or-climate-distress-coping-with-the-pain-of-the-climate-emergency.

RCPsych. 2021. Information for parents and carers. www.rcpsych.ac.uk/mental-health/parents-and-young-people/information-for-parents-and-carers/eco-distress-for-parents?searchTerms=climate%20anxiety.

UNICEF. 2021. *The climate crisis is a child rights crisis: Introducing the children's climate risk index*. New York: United Nations Children's Fund.

Weintrobe, S., Ed. 2013. *Engaging with climate change: Psychoanalytic and interdisciplinary perspectives*. London: Routledge.

Weintrobe, S. 2021. *Psychological roots of the climate crisis. Neoliberal exceptionalism and the culture of uncare*. London: Bloomsbury.

Chapter 16

The Climate Crisis and the "Unnatural" Body

Onto-epistemological Possibilities of and Threats to the Genders of Children and Adolescents

Kristin Fiorella

Entangled Existence

The ecological crisis has made it frighteningly clear that existence is entangled. We are entangled with the more-than-human,[1] with the past and the future, and with the complex effects of the disavowal of our entanglement. As theorized by philosopher and quantum physicist Karen Barad, beyond separate things being intertwined, entanglement suggests that things are not ontologically separable in any final sense. Differentiations between phenomena are produced and stabilize over time, but these differentiations arise out of a fundamentally entangled state of being.[2] Barad's theories about entanglement, which I will describe in more detail shortly, highlight that the more-than-human world is *in* us, just as we are in it. Her theories offer ways of understanding that move beyond the ontology of isolated individuals enclosed in discrete bodies and separated from a passive, external world.

Many of the norms of modern life, sedimented with histories of violence, domination and acquisition, have supported a conceptual baggage that blinds us to our embeddedness in the world. Far from reckoning with our entanglement with the environment, for the past several centuries, Western history and knowledge making practices[3] have produced and described the division and hierarchizing of complex phenomena with an inert, mute Nature positioned as an outside. The more-than-human world has been theorized as the other to culture and thinking, including within psychoanalysis, and it has been viewed as something we could subdue and extract from for our interests. The disastrous implications of this characterization are erupting. The postcolonial writer Amitav Ghosh suggests that future generations, should they exist, might find it remarkable that for a period of about three centuries, humans believed the more-than-human world was an entity, without agency, that could be dominated. He writes, "Quite possibly, then, this era, which prides itself on its self awareness, will come to be known as the time of the Great Derangement" (Ghosh 2017).

DOI: 10.4324/9781003520757-19

Our ecological emergency is pressing up against our usual frames of reference, and they are beginning to tremor. What happens to norms that govern reality when it becomes increasingly clear that they threaten the very possibility of future life? For many children and adolescents, what is as, or perhaps more, troubling than the crisis itself is the collective incapacity of adults to care enough to take action (Hickman 2020, 2022). Caroline Hickman writes that many young people are left in a borderland (Hickman 2022). They are existentially connected to adults and the current structures of a world that are refusing transformation in the name of protecting and accelerating a way of living that could destroy their future.

This incapacity to respond raises questions about how we have organized life, how we have accorded value, and the onto-epistemological[4] categories with which we have built worlds. The narcissistic state of human exceptionalism is so stultifying that it cannot recognize its own limit, even as we are living its disastrous implications. In my own work, I regularly hear from children and adolescents who feel our familiar categories and ways of thinking got us into this disaster, and they are not helping with anything near an adequate response. Some children and adolescents are seeking, as transgender studies scholar Susan Stryker writes, to "change cultural imaginaries" and find "speculative visions of livable futures in which bodies and difference function within new economies of meaning" (Stryker & McCarthy 2024). Rather than the view, prominent in our current heteropatriarchal world, that alterity is a threat to be mastered, new economies of meaning for a more livable future might invite a more radical, even loving, relationship to alterity.

Given the intensifying reality of the climate crisis, we are all in transition. "Deep down," writes Bruno Latour (2022), "we all know that we have changed worlds." Whether we are consciously registering this or not, we are in a caesura between a colonial capitalist patriarchal world and whatever may come next, which will always be entangled with what came before. This would suggest a kind of global catastrophic change that would have multiple and diverse manifestations. Some of these will certainly be, and already are, terrible, but it is possible that some could be generative. I am arguing in this chapter that gendered embodiment is a domain in which, for some, the dismay, despair, and reconfigured possibilities of our current world gather. I will use clinical examples to suggest that the disruptive elements constitutive of embodiment, as well as its potential porosity with the more-than-human world, make gender a possible site of resistance and ethical response-ability[5] in the context of a world that is radically unstable and ecologically breaking down.

Gender soaks up meanings from multiple domains, some of which are sedimented with inequalities and violence. These meanings are constraining – they are in us and become us, creating lasting propensities to perceive, think, and feel in certain ways. Yet, there is something in excess of these sedimented meanings. Particularly at moments of transition, in cesuras between forms or between worlds, gender can become a powerful medium for an intra-play between the density of the familiar and an alterity that is inside/outside.

Intra-action

In recent decades, recognition of the entanglement of human and more-than-human life[6] is prompting a reorientation within many academic disciplines (Morton 2010; Povinelli 2022). Although psychoanalysis has been engaging with entanglement with the social in recent decades, only a few psychoanalytic theorists have begun to theorize the implications of ecological entanglement.[7] Given the urgency of our global emergency, it is incumbent on psychoanalysis as a field to follow the lead of other academic disciplines and rethink the ethical and ecological consequences of our theories. I highlight some of these in the Implications section below.

For Barad, individuals and phenomena emerge through "entangled intra-relating." Barad uses "intra-action" to highlight ontological inseparability of phenomena, which she distinguishes from the usual "interaction," which relies on "a metaphysics of individualism" (Barad 2007). Interaction happens between or in the midst of differentiated subjects or phenomena that exist before their encounter. In intra-action, entities lack an independent, self-contained existence (Barad 2007).

Intra-action gives up on linear cause and effect and looks for how ways of knowing and being, conscious and unconscious, are affected by the social and more-than-human and have an effect on them. It is also sensitive to how ways of knowing and being are sedimented with history, inequality and power. These sedimented realities inhabit matter and bodies as well as discourse.

Intra-action is a generative lens for exploring the material and discursive intra-plays[8] between gender, the current right-wing discourse around gender, and the climate crisis. Many factors, anxieties, desires, and hatreds constellate around gender and the climate crisis. Gender enacts boundaries that are sedimented with histories, and it also becomes a field for reconfiguring possibilities. Attacks on trans children frequently center on defending "the natural" and protecting these children from "mutilation" and the "contagion" of gender ideology. These claims perversely distort several terrors entangled with the climate crisis – about "nature," violence, and permeability. The anti-trans discourse works to project destructiveness, as I will explore in an upcoming section on the far right. Trans children, rather than a history of extraction and violence, become the embodiment of moral perversion. Their parents are the ones who are guilty of endangering these children's future and a strong patriarchal state is needed to defend the future well-being of children.[9]

Matter and Meaning

For Barad, matter and meaning are co-emergent at every level – in the quantum, the individual, the social, and the natural. This entanglement of matter and meaning has profound implications for embodiment. Bodies emerge through and as a part of their entangled intra-relating with the social and material world to such an extent that bodies have different physiological capacities in different entanglements. Philosopher Vickie Kirby gives the example of bodies in the Hindu ritual festival of *thaipusam* to engage with this point. In this ritual, practitioners are pierced in their skin, cheeks, tongue and internal organs with metal spokes. In other contexts, this practice would cause bleeding, scarring, and possibly lethal injury. For these practitioners, none of these occur. "[W]hatever the … structural frame, or cultural text – call it what you will – through which this man's body is ciphered and located as 'being in the world,'" Kirby writes, "one can only presume that this information also informs the very matter of his body's material constitution" (Kirby 1997, p. 3).

This example of embodiment, among others,[10] demonstrates that the socio-symbolic environment is materially *in* our bodies, not merely represented in our minds. Stacy Alaimo, building on Barad's work, extends these material linkages between corporeality and the social to include ecological forces through her concept of transcorporeality. Transcorporeality is aiming to account for material and discursive intra-actions between human bodies, more-than-human bodies, and climatic bodies. The environment is materially in our bodies – in our lungs, in our microbiome, in toxins that create disease and chemical sensitivities. This intra-action between bodies and the world constitutes an integral part of the life of the body (Alaimo 2010). Transcorporeality challenges certain psychoanalytic notions that bodies are in a vertical and dependent relationship with minds which represent an external world.[11] In a transcorporeal account, embodied minds are *of* the world, not merely *in* it.

Transcorporeality is a useful concept to describe the ways in which patients register climate disturbance in and through their bodies, as I will describe in a clinical example. In San Francisco, where I live and practice, heat waves are hotter each year and we are regularly contending with smoke from burning fires. These rapidly evolving changes are entangled with histories of violence and exploitation of the planet and peoples. Although our bodies register this material and discursive transcorporeal communication whether we are conscious of it or not, embodiment can be more or less attuned to this communication, as I will elaborate shortly. Alaimo and Barad are highlighting the forces that flow between human bodies, nonhuman bodies, and ecological systems. In this catastrophic time, for many of us, it may not be possible to directly apprehend the magnitude of the destructiveness – past, present, and future – bound up with the climate crisis. In the midst of the unremitting problem of collective dissociation and the unassimilable dimensions of potential extinction, we might corporeally register something at the perimeter of this vast field of destructiveness that initially occludes conscious recognition.

Implications for Psychoanalysis

Although it is outside the scope of this chapter to delve more deeply into the potential ramifications of Barad's work for psychoanalysis, I will briefly note the following three points in this section. First, entanglement and intra-action add dimension and complexity to psychoanalytic theories of the social unconscious. Second, they also highlight that psychoanalysis's entanglement with colonial history enacts certain onto-epistemological boundaries. The climate crisis makes it clear, if it was not already, that these boundaries are destructive and urgently need to be reworked. Finally, Barad's theories on the entanglement of matter and meaning could expand psychoanalytic thinking about embodiment in the context of the climate crisis. Barad's work also has implications for gender and gendered embodiment, which I will expand on throughout the chapter.

Theories of the social unconscious address the social and collective dimensions of the unconscious. They take up various ways that culture and ideology shape the unconscious (see, for example, Dajani 2017; Guralnik & Simeon 2010; Layton 2006). Although they are valuable, they are not enough for theorizing in the face of the ecological crisis in a couple ways. First, they are centered on the human and replicate the foundational division between the human and the natural. In the face of planetary threat, we need ways of thinking that more radically locate us in and with the world and that theorize the ways that subjectivities have been shaped by the disavowal of entanglement with the more-than-human.

Western philosophies and practices have made an ontological cut between nature and the social/ culture-producing human. In this division, which is intimately enfolded with other well-worn binaries, nature is mute, governed by ahistorical laws, and without agency. The culture-social is the domain with agency, thought, representation, and history. Postcolonial scholar Amitav Ghosh traces the physical and conceptual violence that was involved in dividing the complex phenomenon of being embedded in the world into the balkanized enclaves of human and nature (Ghosh 2017, 2022). In a variety of ways, the creation of "nature" as the passive container for the human/social, during early colonial conquests, enabled a logic of horrific violence, acquisition and hierarchy that is entangled with our current ecological breakdown. Psychoanalysis needs to expand its thinking on the social to include the more-than-human world.[12] If we do not do so, we are producing and participating in ideologies of disconnection that deeply impair our capacity to respond to the climate crisis

Psychoanalysis is entangled with its own emergence during nineteenth-century European colonialism in ways that structure some of our metaphysics. Intra-action highlights the ways in which this history becomes sedimented into psychoanalytic understandings of subjectivity and configure what will be possible to imagine, what will be obscured, and what will be enacted. During

Freud's lifetime, evolutionary anthropologists created notions of "primitive cultures" of Indigenous and African peoples to justify their racist claims of superiority. Freud read and borrowed from some of these anthropologists[13] and these influenced elements of his representation of mature development (Brickman 2003). One of these anthropologists, Edward Tylor, characterized Indigenous peoples' embeddedness in webs of kinship with the more-than-human world as primitive animism. Freud borrowed this characterization and linked it with the earliest infantile stages of development. For Freud, these animistic beliefs were narcissistic because they were beliefs in one's own wishes. This primitive was always in us, but, in healthy development, there would be a movement away from this embeddedness with the more-than-human. The subject who remains embedded in a world that is alive with its own intelligence is the primitive Other to the mature, culture – producing, differentiated Freudian subject.

From where we are now, it is somewhat disorienting to reckon with this embeddedness being read as always already narcissistic and differentiation from this as inscribed in healthy development. How would our metaphysics and onto-logical boundaries be reconfigured if we absorbed that we are embedded in a more-than-human world that has its own agencies?[14] While there are certainly pathological states of fusion and adhesive identifications that psychoanalysts have rigorously described,[15] psychoanalysis has undertheorized the human's sensory embeddedness in the world.[16] Psychoanalytic entanglement with colonial history deserves more attention than I can give it here, but I am briefly outlining it because it is important to the points I will be making about embodiment's role in ecological thinking in this chapter. When individuals do not have experiential access to being rooted in a more-than-human world, dissociative and narcissistic processes reign. It is ethically necessary that psychoanalysis rethink its participation with and production of onto-epistemological cuts that obscure ecological thinking and that are ensnared with colonial epistemologies.

Embodiment is dynamic, contingent, and co-emergent with meaning-making practices. I have argued elsewhere, drawing on views of embodiment in Vajrayana Buddhism, that, if we have a sedimented, conceptual view of our bodies as bounded at our skin and separate from others and the more-than-human, this conceptual stance influences our embodiment and perceptual boundaries (Fiorella 2023). It lends itself to a psychophysical posture that is based, to varying degrees, on self-aggrandizement and self-survival. Once we are in subtly disembodied state, perceptual shifts galvanize a tendency to objectify reality and set ourselves radically apart, which iteratively encourages disembodiment. The self that considers itself bounded and separate still has to contend with its porousness to the world, but such a self is more likely to dissociate or narcissistically seal off from this embeddedness. From within the meaning-making practices of Barad, Buddhism and Indigenous peoples, my body is emphatically not *mine* and is potentially open to its own alterity as well as that of the more-than-human world.

Like embodiment more generally, for Barad, gender configures the world with particular meanings and boundaries. Our ecological crisis is exposing that those ideological categories that were enfolded into the "natural," including race and patriarchal gender binaries, were part of the conceptual repertoire that erased the otherness of the world (Ghosh 2017; Barad 2015). Gender becomes one domain of an ongoing (re)configuring of the world in the context of the climate crisis, as I will show in two clinical examples below. It also becomes a contested terrain for attempts to maintain the hetero-patriarchal colonial capitalist intra-actions.

World of the Worldless People

Andrea, an adolescent patient, came in on an unseasonably hot day in San Francisco and threw off her hoodie.[17] She had stripped down to a sports bra underneath, and I was able to see, for the first time, the cuts on her arms that I knew she connected to her despair over our ecological crisis. Within a few moments, she told me about "the neutered conversation about climate change we had at school today." I asked her what it was like, and she looked at me with exasperation:

> How are you supposed to tell kids that they have no future? Adults can't handle it so they talk about recycling and Earth Day. *"We're doing everything we can. Yes, we should have acted sooner but we feel* so *bad."* Face it, it's a disaster. And they don't even have the decency to take an honest look. Why should I want to hear anything about it from these stupid, selfish cowards?

It was not the first time we had talked about adult indifference to the climate crisis, and it was clear to both of us that, in the transference, I am one of the adults who can't take an honest look. I am also an adult who, like her, struggles with a dizzying sense of an incompatibility between the human and the world. When the emotional reality of the sheer scale of destruction of the climate crisis penetrates a more ordinary state of knowing about it, as it often does for me when Andrea speaks about it, it seems impossible to go on with life as usual.

As the urgency of our planetary emergency is becoming clear to children and adolescents, the task of wrestling with it is formidable. In one study of 10,000 children and young people (under 25) from ten countries (Australia, Brazil, Finland, France, India, Nigeria, Philippines, Portugal, the UK, and the US), 75% of the respondents said that they feel that the future is frightening and 83% said that they think people have failed to take care of the planet. 59% said they were very worried or extremely worried (Hickman 2020). In the face of a catastrophe of this scale, how is it that we have not collectively undergone a transformation in our relation to the world? Far from it. In May 2024, the UN released a report that the planet had experienced twelve consecutive months of record high global temperatures. The report described this as unprecedented and of urgent concern: "there can be no more playing for time," UN Secretary-General Antonio Guterres said, "we're out of time" (Paddison 2024). Months later, the US elected President Trump, who has made reversing progress on climate change to "Drill, baby, drill!" a central feature of his presidency. In his first two weeks, Trump withdrew us from the Paris Climate Agreement, moved to ramp up oil and gas production, and removed incentives for electric vehicles. Many of my child and adolescent patients cannot comprehend how this backward slide is happening.

Andrea describes feeling the climate disaster "in my bones." For her, intimations of where we are and what is coming are inescapable and can intrude at any moment. "How are we all walking around like zombies? I constantly feel a scream in my throat. I get in my Dad's car and turn on the news ... I'm always expecting that any moment it's all anyone will talk about." She is enraged by the casual indifference to poorer countries where people are dying and poised to die first. "They've done the least to cause this. We're just going to let them die, while we debate if we can bear to transition to electric cars? Are we *this* callous?" She registers this as transcorporeal communication – in hotter Septembers and in more destructive fire seasons when she ambivalently concedes to her mom's request that she wear a mask. She feels that she is living a lie, acting as if it matters what grade she gets in chemistry while her body registers both the violence and the suffocating indifference in the intensifying heat and fire in her world. "My Dad is getting central AC installed. In San Francisco? Does no one notice? Like let's just turn that into a home improvement project and think about improved resale? Are we actually this stupid?"

A few months into treatment, she told me about her cutting, which helped calm her panic attacks. Pointing to her covered arms, she said, "This feels more honest. At least I'm marking what's really here."

"On your skin – one place where you feel the violence of this."

"Yes, it's already here. This makes me feel that at least I'm not just completely asleep."

Andrea is sensitized to the climate crisis, as well as the violence and power relations that are entangled with it. She is also psychically and materially torn between power differentials in her family. Her mother, Kai, is Taiwanese and significantly younger than her white, wealthy father, Ryan. Her parents were briefly married, but they divorced when Kai was pregnant with Andrea. Kai struggled with drugs and alcohol when Andrea was a baby and eventually lost custody when she was two. Kai, who has an extensive trauma history of her own, eventually sustained sobriety and re-entered Andrea's life more consistently when she was eight.

When Andrea was a baby and toddler, Kai would leave her alone for hours while she was passed out in her bedroom with the door locked. An aunt reported this behavior, which led to her losing custody. Andrea is fiercely protective of her mother, and she struggles to tolerate her rage and despair with her mother's failures. She is acutely aware of the power differential between her parents and is enraged with her father for his insensitivity to this reality. She also feels a heavy guilt that she has significantly more wealth and opportunities than her mother.

I frequently thought of Andrea's internal objects as she would describe her responsiveness to our ecological disaster. There were so many clear genetic links between her response to the climate crisis and the injustice of her mother's fate and her own. When Andrea described her rage and terror at an oblivious world that she depends on to survive, I sometimes thought of her as a baby and toddler alone with her drunk mother. Initially, I tentatively made some of these links. Andrea did not disagree with me, but she registered these connections as somewhat beside the point. I privately felt that these links were too fraught and painful, and they might become thinkable in time.

As Andrea describes evidence of our unfolding crisis nearly everywhere she looks, I often feel my own rising panic and despair. Images – of tortured landscapes, migrant mothers and children, my own young child – would float in my mind. I came to realize that, while I was no longer interpreting the climate crisis's resonances with her internal objects, I was steadying my own anxieties by orienting towards understanding them. I started to experiment, when I could bear it, with relinquishing these reference points. I could feel myself sink into dread and overwhelm that has uncanny dimensions. We know the nature of the catastrophe that is already in our midst. We know it will hit all of us in one way or another and that it will continue to kill millions, if not a sizable portion of the human population. The scale of ignorance and destructiveness is staggering. At times, I felt at risk of a "pain – so utter – It swallows substance up" – where objects, self and world disintegrate (Dickinson 1976). When I can risk experiencing this with her, I feel on the cusp of a freefall into a kind of nowhere-ness (Durban 2017) that was foreclosed by seeking to understand.

Unsurprisingly, Andrea seems to register when I am able to more fully open to what it is to live in a world that is dying and to bear the incomprehensible dimensions and disintegrating anxieties of this. She eventually began to describe in her own words the ways in which her experiences in her family "make me more sensitive to what's happening. I'm glad they do. It's too easy to be asleep." Her sensitivity to the vertiginous quality of the crisis holds within it a fracture that sometimes opens to a sense of the Otherness of the world. She is beginning to feel that coming to terms with her own personal history could be in service of opening to this alterity.

For an array of reasons, she fears her experiences as a heterosexual, cis-gender person align her with systems of power that, for her, are destructive and linked to the climate crisis. She feels that heteropatriarchal gender norms are no longer congruent with the world and are one of many social realities that need to be transformed. She attempted to feel desire for women and also experimented with her gender. At this point, these attempts have not taken hold. While she is still struggling with this and anxious about the implications of her gender and sexuality, she is provisionally coming to inhabit a gendered body that is female and heterosexual. We have spoken about how painful it is to bear her entanglement with violence, as well as her sensitivity to it. None of us can solve this entanglement through the fantasy that what we are and what we put into the world can be fully regulated and determined by one's will. Her despair over the violence in her, in her family and in the world is valuable to her, as she does not want to lie to herself about the horrors around her. In her attempts to tenuously hold the relevance of life in the midst of violence and death, she has made an animated film that centers on multiple selves, all female, within one body. Initially these selves are at war with one another and then learn to hate and love one another while coexisting.

Composition – De/recomposition

Bruno Latour, who wrote about the climate crisis for decades, argues that as we are bombarded by horrible climate news each day, the "natural world" has lost its fragile capacity to unify us and to offer the illusion of norms. For him, it never had this capacity, but, prior to our planetary emergency, ideals could be located in the normative dimension of "nature." In the face of climate breakdown, many are now being driven mad rather than relinquish constructs that are so obviously insufficient. He writes that we need a more open concept than nature/ culture, with all its accompanying bifurcations, that would be sensitized to the Otherness of the world and its multiple existents (Latour 2017).

Nature, which he refers to as a cosmological figure, offered modern people the illusion of living in a pregiven world unified by a system of ahistorical laws. That "composition" failed. It is decomposing, and our new ecological reality will require new compositions. For Latour, these new compositions will involve multiple transformations, some of which are underway and some, he argues, have not yet been imagined. In my own practice, for some of my child and adolescent patients, including Andrea, the climate crisis has a radically "decomposing" effect in Latour's language. In psychoanalytic language, it is unbinding (Laplanche 1999).

As more children and adolescents are experimenting with a variety of gendered forms, could we recognize this as a means of re-composition? Susan Stryker writes:

> I increasingly understand trans-ing as an exploratory and experimental practice for surviving within, and pointing beyond, the biocentric world order constructed in the wake of European colonization and racial chattel

slavery… It needs to be a resource for transforming the Anthropos itself, the figure of Man that orchestrates our era, in whose name and for whose needs we now destroy the planet.

(Stryker & McCarthy 2024)

Gender is one domain where material and discursive possibilities can be reconfigured. Some efforts take hold at the level of the unconscious, as I will explore shortly, and, as with Andrea, some do not. At the same time, our planetary crisis is also prompting reactionary backlash. As our usual coordinates are lost or we know that they will be, the immersive panic that ensues creates attempts to reinforce a threatened heteropatriarchal world, as is seen in far right ideologies in the United States and elsewhere. Heteropatriarchy is doubling down with ferocious and formidable intensity. As more children and adolescents are coming out as trans or nonbinary, they are indeed a threat to a patriarchal ordering of the world, and are therefore subject to legislative and physical violence.

Anti-Gender Right

In 2023 in the United States, lawmakers in 37 states introduced 142 bills to restrict gender affirming healthcare for trans people. Four fifths of these restrict care for trans children under 18. In 2024 652 bills were introduced, more than in any other year. For five consecutive years there have been record breaking numbers of anti-trans bills each year. Donald Trump signed several executive orders targeting trans people in the first few weeks of his second term – insisting that federal documents have gender markers that align with sex at birth, directing agencies to prevent gender-affirming care for people under nineteen, and attempting to dictate to students, parents, and educators that they cannot affirm a trans child's gender.

Many of these bills claim to be protecting children from abuse and destruction. For example, in 2022, Governor Abbot of Texas signed a bill into law that accuses parents who seek gender affirming care for trans children of child abuse. In 2023, Governor Ron DeSantis of Florida signed a law that blocked gender affirming care for anyone under 18. He claimed to be protecting children from being "mutilated" (Beaty 2023). In his executive order blocking care for trans young people under nineteen (January 28, 2025), Trump claimed "medical professionals are maiming and sterilizing a growing number of impressionable children."

Conservatives in the United States who treat trans children and their parents as the embodiment of moral corruption are dismissive of climate change. In recent years, they have mostly stopped straightforwardly denying that climate change exists, with the notable exception of President Trump. As recently as August 26, 2024, Trump said in an interview about climate change, "You know, they have no idea what's going to happen. It's weather" (Ryan 2024). More often, the right treats the climate crisis as a phenomenon that has been sinisterly and hysterically exaggerated by the Left. For example, Ron DeSantis stridently asserts that he does not "believe in politicizing the weather" (Millman 2023). Senator Lindsay Graham of South Carolina does not deny climate change but asks not to be lectured about it (Friedman 2023). The conservative host Matt Walsh grounds his essentialist gender claims in "truth and science." He simultaneously claims that scientists "pull these numbers out of their asses" with regards to the climate crisis. He bizarrely asserts that 1.5 degrees warming limit does not make sense because you "can't set a limit for a thermostat of the Earth" (Iadarola 2024).

How is this reckless impenetrability to the threat of the climate crisis achieved? If anyone over the age of eight had a question about why 1.5 degrees is significant, googling it would answer that in a matter of minutes. What kind of psychic work is required to disavow something that is clearly happening in order to justify the way one is living?

Perversion is doing some of this work. I am not referring to sexual perversion, which is sometimes how the far right describes LGBTQ. The meaning of perversion I am using here refers to states of mind organized around a hatred of understanding and growth that enacts extremely rigid ways of knowing and relating (Stein 2005; Bonner 2006). In perverse relating, there is a pretense of communication, as is very much the case in Matt Walsh's documentary which I explore below. Words and actions are intended to manipulate and control both the other's and one's own state of mind to defend against intolerable levels of anxiety.

In the face of terrors related to the clear existential threat of the climate crisis, the far right is using a perverse mode to fetishize trans children. Their tactic has several features that often accompany perverse states – the nullification of the other, the twisting of meaning, and the excitement of moral sadism. In framing gender affirming care as mutilation and abuse, they externalize violence at a moment when the scale of destructiveness that is entangled with a colonial capitalist world is increasingly visible. They locate monstrosity outside in trans children and their caregivers, not within their escalation of a global threat to protect their way of life for another decade or so. They then excitedly insist on the need for a strong patriarchal state to restore order and reestablish a moral floor. They can then claim to be protecting children, while blatantly imperiling the viability of future life.

What is a Woman?: Anti-trans propaganda

In the beginning of Matt Walsh's documentary *What is a woman?* (Folk 2022), he identifies himself as a father and tells his audience that he is concerned about the messages around gender identity that young people are "bombarded with." He intends to investigate what a woman actually is so that he can help children make sense of gender. We do not hear from any children or adolescents in this film, which is fitting, as the film only pretends to be interested in them. The real intent of the film is to use claims about the child and the adolescent to make ontological, essentialist claims about hierarchal differentiation.

The film opens with a montage of two fraternal twins, around 10, at their birthday party. These are presumably Matt's children, and, as we hear Matt speak his introductory thoughts in a voiceover, the twins open presents. The boy is given a BB gun and later a football. The girl is given a tiara and a tea set. Both are elated by their gifts. Then they are both given fishing poles. The boy is delighted, and the girl seems at a loss as to how to hold this foreign, phallic object.

Then, we see Matt walking with his own fishing pole towards a lake. He describes how he likes to spend time in nature because "nature never lies." In this section, he is conveying that there is one true Nature. In a variety of comments and images, he collapses biological reality with patriarchal stereotypes of gender – girls like tiaras and are emotionally labile; boys like football and are simpler. Everything else is falsely dressing up what is patently given and real. After naturalizing the binary and linking this to patriarchal gender norms, he then pretends to engage in questions to find out what a woman is. He flies around the US to ask this question of gender affirming pediatricians, therapists, an obgyn, a professor of gender studies, and many trans women. Most of those he interviews pick up that his questions are manipulative pretense and that he is actually engaged in degrading them. He is signaling to his audience with his tone and editorial technique that his interviewees are idiots, and he is taking pleasure in doing so.

A number of those he interviews respond to his questions about biological reality. They emphasize that it exists and matters but that it is not determinative of gender. Walsh frames any complexity in an answer as intellectual nonsense. Pull down your pants, and we can settle this question.

Over the course of the documentary, he claims to be increasingly concerned about the contagion of transgender phenomena. Despite the fact that he has told his audience from the beginning where he stands, he pretends to experience a growing moral outrage at the inability of these people to answer a simple question – "What is a woman?" He interviews several therapists and a psychiatrist who share his fear over this gender business. Towards the end of the film, he makes a speech to a school board who are allowing trans children to use the bathrooms that correspond to their gender. He accuses them of allowing this contagion to spread.

Fear of contagion is a frequent part of anti-gender rhetoric and is part of what Judith Butler refers to as anti-gender's inflammatory syntax (Butler 2024). Non-normative genders become a disease that a child or adolescent could catch and must be protected from. Given that gender circulates as a complex phenomenon that is clearly social as well as intrapsychic, there is a sense in which contagion is part of how gender, normative as well as nonnormative, works. The most cursory glance will show that, in any given time, some gendered forms are intelligible that would not have been at an earlier time, and these will shift again in the future. The fear around children and adolescents exposing one another to non-normative genders, including trans, involves many anxieties. Most obviously, the far right is anxious that children and adolescents will be swept up in something corrupting. By grounding their gender claims in the "natural," they affirm that patriarchal norms are built into the fabric of the way things are. They are using this protest against gender complexity as an appeal to an imaginary past where things were safe and predictable, and they are claiming that a strong patriarchal state could take us back there. This appeals to the fears that many parents have that their children might inhabit a world unlike the one they have known. Despite the fantasy that we could return to a fictional predictable past where no one need feel disrupted, children will be in a different world. That transition is already under way, as the most recent UN report highlights, and, no matter what changes we are able to mobilize, it will intensify.

One could speculate which is the more intense annihilation anxiety for members of the far right – potential extinction of the species or the loss of certain structural power relations. Their think tanks and military leaders are aware of what is coming with the climate crisis and are planning for ways to protect power (Ghosh 2017). At the very least, there is likely an unthought known among many in the far right that aspects of the modern composition are proving so destructive they will either transition in some way or destroy most, if not all, of us. To retain power, they need to maintain the illusion that white supremacy and patriarchal gender binaries are built into the structure of reality. As the climate crisis robs "nature" of its fragile, illusory normative dimensions, the far right is anxious to affirm that "nature" is there – still a passive, static entity governed by regulatable laws that humans have abstracted and mastered. Trans children provide them a medium to stake this claim. This boundary between the human and the natural (re)organizes human exceptionalism, patriarchal norms, and a dominance over the "natural" world. Such a dominance is entangled with capitalist colonial power over the non-normative, the non-white, and the poor.

Within these anxieties of contagion is a glimmer of recognition that we are all entangled with multiple practices that others have a role in shaping and through which we are shaped. There is also an intra-action with the disavowed reality that our permeable bodies are bound to the Earth. As both the pandemic and the climate crisis make clear, we fundamentally cannot get out of our porous interdependence with the more-than-human world. It is clear that there are fantasies that wealth and power will be prophylactic against our interdependence and that the poor will be the ones who die. There may also be fantasies of Western superiority as a kind of immunity (Ghosh 2022). While it is true that poor nations, who have done the least historically to cause this, are the most vulnerable, it is already clear that wealthy nations will also be profoundly damaged.

Onto-epistemological Medium of Gender

In the past decade, many children and adolescents in my practice in San Francisco have been exploring a range of gendered possibilities. Unsurprisingly, for these patients, gendered embodiment reverberates with desires, anxieties, conflicts, and fantasies. As they have wrestled with gender, some of these patients have also been disrupted by how elusive, enigmatic, full of holes, and strangely coherent gender is. One adolescent patient described, "Gender is slippery. Period. What even is it? And it's very real. It's weird how those things go together." As more of my child and adolescent patients have been exploring gender, they have helped me become more sensitive to the aspects of being and knowing that gender potentially focuses and exposes.

Gender can provide a medium for experiencing the gap between claims of identity and the lack of an underlying unifying cohesion that identity suggests. It is a psychic and collective phenomenon that can gather an onto-epistemological weave that can become a psychoanalytic object. In exploring gender, a child or adolescent may come to know something about their gender. They might create or find a narrative that has more intelligibility to them and feels more livable. Yet, the knowing that gender exploration can invite is one that is potentially in touch with a becoming that will always disrupt what we think is stable.

As Buddhism and many Western philosophers have pointed out, knowing has a grasping, proprietorial quality that can have trouble sensing its own limit (Spivak 1987; Trungpa 1973; Nishitani 1987). Knowledge, even knowledge that is initially subversive to the ego, can be folded into the ego's structure, leading to what Merleau-Ponty (1969) called "thinking that basks in its own acquisitions." When an experience of being breaks an epistemological horizon, one might shake free momentarily from the self-capture of the ego and glimpse an otherness within that cannot be assimilated. Such experiences can loosen the prison of self-aggrandizement and self-protection that the ego typically involves. One then will usually (re)find a resting point in knowing but potentially with a growing sense of something open-ended, something that cannot be controlled or mastered. Child and adolescent patients wrestling with gender sometimes become sensitized to an embodiment that does not unify the self and that is in touch with an alterity within.

Mia

In a period of treatment during which she was anxious about changes in her body, Mia, a ten-year-old trans girl, began to speak of her sense that there would always "be a hole" in whatever she felt she knew about herself. She was afraid to tell me because she worried I would read this disclosure as evidence that she was not "really trans," despite the fact that she had been clear that she is a girl since she was two. I felt she was grappling with something more than her uncertainty of what she knew about herself, and I told her so.

She then had the following dream: "I was holding onto a Rubik's cube. It's wet and I'm trying to hold on. I'm above water and it feels like I'm going to fall." She draws this image from her dream. When she's drawing the Rubik's cube, she says: "And there's a piece under my hand that I can't see. So I can't solve it. I don't know what color is under my hand, and, if I let go to see, I wouldn't see. I'd fall."

I said, "A Rubik's cube is something that seems like it has an answer. Even if you can't get there, there's a solution. Your dream makes me think that this Rubik's cube isn't something that can be solved."

She agreed and replied, "There is always something that doesn't fit what you know." Over time, this lack of ground became important to her sense of an unfolding in herself. A remarkably sensitive and intelligent, even philosophical, child, she expressed wanting to know things about herself but "there is something that's not that." As she was approaching puberty and on the cusp of taking puberty blockers, her biological body was unsettling to her, and she felt anxiety and grief around this discrepancy between her biological body and her experience as a girl (Saketopoulou 2014). This discrepancy was also entangled with her sensitivity to the dynamic, elusive enigma of gender.

Stephen

Stephen, a late adolescent, was almost three years into their analysis before gender became a focus. They had struggled to live up to their father's expectations that they be more athletic and popular than they were. They had previously questioned their sexuality and had had some sexual experiences with adolescent boys and girls. They were "mostly heterosexual" but felt trapped. They said in their second year of analysis, "I'm looking at all these pretty girls. I'm cramped inside. There's no room to actually want them because I'm supposed to have them. For status." It became clear that they were strangled by an introject that was an amalgam of their father, other important caregivers, voices from sports teams, TV shows, and the cultural surround. This introject kept them in a narcissistic and melancholic encapsulation – they could not live up to its demands, and could not stop trying. Although they often wavered on this, they envied girls and women as they felt women were allowed more latitude for emotional expression.

They were able to productively explore this introject and envy in the transference, yet they often fell into despair about how intractable these dynamics were. This ravenously acquisitional object was never satisfied with whatever accolades they could feed it, and there was no room to desire. They said they were "begging inside myself ... I'm excited but I'm not turned on." Although they were increasingly aware of the contours of their entrapment, I also began to feel hopeless about their pernicious enclosure.

Their gender was saturated with sadomasochistic, rigid patterns – girls were on bottom, boys on top. It felt impossible and naïve to apprehend a world outside of this top/ bottom logic, and every experience was eventually organized around it. Stephen was failing gender's mandates, and so they were on bottom with the girls. Yet, girls, including me, were actually superior to boys and knew it, so we were secretly on top. Stephen envied me and wanted to become more like me, only to feel trapped and humiliated.

They were frequently swept up in a manic, envious attempt to garner status. They did not speak of the climate crisis during this period of treatment but I would sometimes associate to it. Despite the envious misery entangled with it, this logic of acquisition seemed to endlessly breed itself, as it so often does in various ways in our cultural surround. Their rigid gender binary circled around who has power and how, set against who does not have it, and this oriented them to other people as resources in their desperate quest. They would reify the naturalness of these patriarchal norms by pointing to the more-than-human world. "Look at primates. They know who the alpha is. It's just how things are."

Gradually, they begin to relate to how profoundly miserable they were. They had new memories about their idealized and hated father that left them with a growing despair about an intergenerational transfer of gendered failures. They began to mourn the ways in which they could not exist in their body in this world, much less feel pleasure. They began to feel that neither "male" nor "female" felt like a home to them.

As they began exploring a non-binary gender, something began to loosen. Their thoughts and associations began to move in more directions, and they were less riveted by accumulating people and experiences. Phenomena that we had been living and I had been interpreting for years finally began to open up. They still could be swept up in familiar envious states of mind, but they could begin to think about them rather than be strangled by them. About a month into this shift, they said of being non-binary, "it's a way of saying something is fluid about me." In après coup, they began to feel that they had always been non binary, although they also felt it was not something they could not have imagined on their own. Non binary was both "always there" behind their desperate struggle and an emergent solution that was dependent on exposure to this gendered form as a possibility. They found this paradox strange but pleasurable.

Even as they were experiencing an enhanced sense of aliveness and internal flexibility, they also ran into a kind of existential dread. They felt unmoored and anxious that their sense of self could unravel. "When you pull out the fabric, gaps open and you could fall through." There were certainly dimensions to this anxiety that involved their personal history and specific content. They had dreams and fantasies that they felt were monstrous, and this frightened them. They were also becoming increasingly sensitive to a disruptive dimension of being that they could not dominate or instrumentalize. They were experiencing an otherness and indeterminacy inside, that, while frightening, loosened a self-imprisonment.

They also began to experience a sensuous energetic communication with the more-than-human and within their own embodiment. They would describe the quality of light while walking down a city block; the sadness in someone's eyes who they talked to at a bus stop; brushing their young niece's hair and then her brushing theirs. They said, "I've never really felt my body before. I thought I had, but I had no idea." They described experiencing "tenderness," including with me. Neither of us could have imagined their finding that word prior to their gendered transition, and it was surprising to both of us how their rigid states of mind were opening. They wept as they reckoned with their experience of non-appropriative relations to others and felt grief for lost time and relationships. They felt exposed but could finally desire, and described dancing with an adolescent girl at a party: "I didn't *have* to dance. I could move. There was space and I actually wanted her." "Not just for status," I said. "Exactly," they replied.

They had begun college during this gendered transition, and, in the year following their becoming non binary, they gradually became consumed with the climate crisis. They were contemplating majoring in environmental science and spoke frequently of oceans acidifying and species lost, as well as the rampant consumerism that fueled this. While there were many factors at play in their evolving relationship with the climate crisis, I thought that their gender transition was a crucial element. Through this, they had finally worked their way out of a ravenous and claustrophobic logic into an embodied recognition that they are bound and exposed to the other.

During this period, they came in one afternoon telling me about a book by Donna Haraway they were reading for a class in which she describes how gendered and racial assumptions influence how primatologists read the behavior of primates. I waited, anticipating that they would bring up their numerous appeals to primates in the past. They did not, so I did. "I was a different person then." I agreed. I shared with them my sense that their gender transition had opened them to the climate crisis. They shrugged. "Chicken or egg. I was becoming freaked out about the world at the same time my gender was shifting. Even though I guess all we ever talked about was my gender. They kinda happened together."

As we reflected more on the time of their transition, an intra-action between their gender and the climate crisis came more into view. Their gender reconfigured their embodiment and their responsiveness to the world. Simultaneously, the climate crisis brought them face to face with practices of boundless extraction and consumption. The mode of knowing and being entangled with this felt nauseatingly familiar, which further encouraged their gender transition. During this time, there had been a circulation between knowing, sedimented with familiar constrictive desires for power, and a being that disrupted it. "Knowing" became more open to creative (re)articulation. Through their transition to a different gendered embodiment, they loosened rigid hierarchies to apprehend a world that held a tenderness and aliveness that were more precious than ownership. This was a world that they passionately wanted to protect from destructive forces they knew intimately.

Prototype of an Ethical Experience

For Barad, intra-action is an ethical matter, as much as it is an epistemological and ontological one (Barad 2007). We cannot escape our responsibility for the worlds we bring forth through material and discursive intra-actions. Knowing and being are not bounded practices; we are always shaping and being shaped by these enacted worlds. She turns to Levinas at the end of her brilliant book, *Meeting the Universe Halfway*, to draw out the ethical implications of her thinking.

In thinking with Levinas about embodiment, gender potentially becomes a site of ethical responsiveness and resistance to patriarchal domination and to the climate crisis. Barad draws from the feminist philosopher Ewa Ziarek's uses of Levinas's late work to describe "an ethos of becoming in the context of an ethics of alterity" (Ziarek 2001). Levinas proposes an irreducible responsibility to the Other that does not rest on a self-contained individual self. It is an asymmetrical relation where the Other places a demand on the subject that is pre-discursive. This responsiveness to the Other cannot be fundamentally known in thought because this entails pulling the Other into the domain of the self and, in that sense, is an act of domination. In Levinas's later work, this ethical relation to the other is predicated on a being in one's embodiment as it is bound to the Other, what he calls "being in one's skin" (Ziarek 2001).

For Levinas, sedimented history, including violence, inequality and power, are interwoven with the body's becoming. Yet, embodiment is not reducible to this. There is an element of embodiment, disruptive to the "I," that involves exposure to the Other. This alterity within embodiment is akin to "having the other in one's skin" (Barad 2015: 392). It is an alterity that is inside/ outside.

Levinas's sense of the alterity that is constitutive of embodiment offers a modality of relation that might allow us to live with or even love what is not identical to ourselves. It suggests a non-appropriative relation to the Other – in oneself, human others and the more-than-human world. Beyond coherence and mastery, there is the potential to open to an over-flowing world in which there is always more than can be grasped. If we are open to this world, it should reshape us. We could nurture the cracks in self cohesion where we find them, including within gender, not with a moralizing demand that everyone perceive them but with an insistence that those who do will live without the threat of violence. Now perhaps more than ever we need an emergent multiplicity of forms and modes of existence that are potentially less enfolded with our histories of dominance and boundless consumption (Latour 2017).

Conclusion

As we are suspended between worlds – the colonial capitalist world and the unknown to come – there is a fragile possibility of an ethics that is rooted in experiencing our ontological entanglement. In addition to consciously intended just actions, this ecological reality could encompass an ethics based on an alterity that is inside/outside and that cannot be instrumentalized. If our collective relationship to Otherness shifted, there might be rubrics by which to bring forth worlds that render mastery, coherence, and ownership impoverished, at least at times.

As it becomes ever more apparent that the patriarchal colonial capitalist ordering is more than capable of destroying us all, the challenges to this ordering will likely keep coming. For some children and adolescents, our current ways of being and knowing are not congruent with future life. As these challenges to the modern composition continue, the colonial capitalist world will fight to maintain itself and will endanger future generations and compositions. Levinas reminds us that we are all implicated in how this unfolds and what subjects are able to come into being.

Notes

1 "More-than-human" refers to everything in the world that is not human, including other species, the planet, all of the material world, and the technological. This language is problematic as it reproduces a binary split between the human and everything else, even as it attempts to gesture beyond this split by highlighting the world that the human is embedded within and dependent on (Morton 2010; Gentile 2021). For the sake of clarity in this chapter, which is already introducing several terms that may be unfamiliar to psychoanalysts, I am using this phrase rather than some of the more experimental language that attempts to subvert the binary (e.g, Morton 2010).
2 "Entanglement" has resonances with the Buddhist idea of co-dependent origination as well descriptions of interdependence in Indigenous epistemologies.
3 "Knowledge making practices" is Barad's term to underline that knowledge is made out of engagement with the world. This concept challenges the idea that knowledge *represents* a pre-given world.
4 Barad challenges the classical separation of epistemology and ontology. We know because we are of the world, not by standing apart from it. For Barad, we know *in* being.
5 Barad (2007) hyphenates "response-ability" to emphasize the capacity to respond or lack thereof.
6 Timothy Morton (2010) calls this "ecological thinking."
7 For example, Bodnar (2008, 2024), Clough (2023, 2024), Kassouf (2017), Orange (2017), and Weintrobe (2014). Katie Gentile has most radically theorized the implications of our planetary emergency for psychoanalytic subjectivity (Gentile 2021, 2024).
8 One way of visualizing intra-plays is the spreading of waves as in the patterns of waves in light or water. Intervening in such patterns creates new patterns. Barad tracks in detail the way these "diffraction patterns" operate materially and discursively, in the so-called natural world and among social phenomena.

9 The initial ideas for this chapter arose from listening to the far right's speech as well as listening to the gender explorations of my patients. As an onslaught of anti-trans legislation was coming out in 2022–2023, I began listening to some far-right podcasts to hear their descriptions of trans children and gender. I was struck by how often these right wing politicians and pundits would attack trans children and then shortly after move to something dismissive about climate change. I began to wonder about the associative logic between these two phenomena in the far right's imaginary.

10 Vajrayana Buddhism, with which I am familiar, has many examples of practices in which bodies, "perform biology differently," as Kirby puts it. Within that tradition, these examples underline that the ontology of the body is not given.

11 There are a number of psychoanalytic accounts that also challenge this rendering. See, for example, Winnicott (1949), Goldberg (2012), Markman (2020), and Elise (2017), among others.

12 Patricia Clough (2024) argues that psychoanalysis should expand the social to include both the ecological and the technological, "the entangled whole of humans and more-than-human."

13 E.g. in *Totem and Taboo* (Freud 1913) and *Moses and Monotheism* (Freud 1939).

14 Many philosophers and anthropologists posit that nonhuman entities, including forests, rivers, and glaciers, have agencies. They represent and communicate in signs, although not linguistic ones (Kohn 2013; Cruikshank 2005).

15 For example, Ogden (1989), Mitrani (1994).

16 Peter Goldberg's notion of the sensory commons is a notable exception (Goldberg 2012).

17 The title of this section is taken from *The Ends of the World* (Danowski & Viveiros de Castro 2017).

References

Alaimo, Stacy. 2010. *Bodily Natures*. Bloomington, IN: University of Indiana Press.

Barad, Karen. 2007. *Meeting the Universe Halfway*. Durham, NC: Duke University Press.

Barad, Karen. 2015. Transmaterialities: Trans/Matter/ Realities and Queer Political Imaginings. *GLQ: A Journal of Lesbian and Gay Studies* 21(2–3): 387–422.

Beaty, Thalia. 2023. Florida's Ban on Gender-Affirming Care for Minors also Limits Trans Access for Adults. *PBS*, June 4.

Bodnar, Susan. 2008. Wasted and Bombed: Clinical Enactments of a Changing Relationship to the Earth. *Psychoanalytic Dialogues* 18:484–512.

Bodnar, Susan. 2024. The Fierce Urgency of Now: The Case for Including the Environmental Field in Clinical Work. *Psychoanalytic Dialogues* 34–36: 795–811.

Bonner, Svetlana. 2006. A Servant's Bargain: Perversion as Survival. *IJP* 87: 1549–1567.

Brickman, Celia. 2003. *Aboriginal Populations in the Mind: Race and Primitivity in Psychoanalysis*. New York: Columbia University Press.

Butler, Judith. 2024. *Who's Afraid of Gender*. New York: Farrar, Straus, and Giroux.

Clough, Patricia. 2023. What is the Social? *Psychoanalytic Dialogues* 33: 140–156.

Clough, Patricia. 2024. Psychoanalysis as "Studio Practice." *Psychoanalytic Dialogues* 34: 190–199.

Cruikshank, Julie. 2005. *Do Glaciers Listen? Local Knowledge, Colonial Encounters, and Social Imagination*. Vancouver: UBC Press.

Dajani, Karim. 2017. The Ego's Habitus: An Examination of the Role Culture Plays in Structuring the Ego. *International Journal of Applied Psychoanalytic Studies* 14: 273–281.

Danowski, Deborah and Viveiros de Castro, Eduardo. 2017. *The Ends of the World.* Cambridge, MA: Polity.

Dickinson, Emily. 1976. *The Complete Poems of Emily Dickinson.* Boston, MA: Little, Brown and Company.

Durban, Joshua. 2017. Home, Homelessness and Nowhere-ness in Early Infancy. *Journal of Child Psychotherapy* 43: 175–191.

Elise, Dianne. 2017. Moving from within the Maternal: The Choreography of Analytic Eroticism. *Journal of the American Psychoanalytic Association* 65: 33–60.

Fiorella, Kristin. 2023. Thinking in a Marrow Bone: Embodiment in Vajrayana Buddhism and Psychoanalysis. *Journal of the American Psychoanalytic Association* 71: 277–309.

Folk, Justin (Director). 2022. *What Is a Woman?*The Daily Wire [distributor].

Friedman, Lisa. 2023. Delay as the New Denial: The Latest Republican Tactic to Block Climate Action. *The New York Times*, June 22.

Freud, S. 1913. Totem and Taboo. In J. Strachey (Ed. and Trans.), *The standard edition of the complete psychological works of Sigmund Freud*, vol. 13 (pp. 1–162). London: Hogarth Press.

Freud, S. 1939. Moses and Monotheism. In J. Strachey (Ed. and Trans.), *The standard edition of the complete psychological works of Sigmund Freud*, vol. 23 (pp. 7–173). London: Hogarth Press.

Gentile, Katie. 2021. Kittens in the Clinical Space: Expanding Subjectivity through Dense Temporalities of Interspecies Transcorporeal Becoming. *Psychoanalytic Dialogues* 31: 160–165.

Gentile, Katie. 2024. Destroying the "Human" for the Survival of the World: A Proposal for a Temporally Dense Psychoanalytic Subject. *Psychoanalytic Dialogues* 34: 533–541.

Ghosh, Amitav. 2017. *The Great Derangement.* Chicago, IL: The University of Chicago Press.

Ghosh, Amitav. 2022. *The Nutmeg's Curse.* Chicago, IL: The University of Chicago Press.

Goldberg, Peter. 2012. Active Perception and the Search for Sensory Symbiosis. *Journal of the American Psychoanalytic Association* 60: 791–812.

Grand, Sue. 2000. *The Reproduction of Evil.* New York: Routledge.

Guralnik, Orna and Simeon, Daphne. 2010. Depersonalization: Standing in the Spaces Between Recognition and Interpellation. *Psychoanalytic Dialogues* 20:400–416.

Hickman, Caroline. 2020. We Need to (Find a Way to) Talk about… Eco-Anxiety. *Journal of Social Work Practice* 34(4): 411–424.

Hickman, Caroline. 2022. *Saving the Other, Saving the Self. In Vakoch, D. and Mickey, S. (Eds.) Eco-Anxiety and Planetary Hope: Experiencing the Twin Disasters of COVID-19 and Climate Change.* Switzerland: Springer.

Iadarola, John. 2024. Matt Walsh PROVES How Clueless He Is With Dumbest Stunt Ever. *Damage Report*, January 12.

Kassouf, Susan. 2017. Psychoanalysis and Climate Change: Revisiting Searles' The Nonhuman Environment, Rediscovering Freud's Phylogenetic Fantasy, and Imagining a Future. *American Imago* 74: 141–171.

Kirby, Vickie. 1997. *Telling Flesh*. New York: Routledge.

Kohn, Eduardo. 2013. *How Forests Think: Toward an Anthropology beyond the Human*. Berkeley, CA: University of California Press.

Laplanche, Jean. 1999. *Essays on Otherness*. New York: Routledge.

Latour, Bruno. 2017. *Facing Gaia*. Cambridge: Polity Press.

Latour, Bruno. 2022. *How To Inhabit the Earth*. Cambridge: Polity Press.

Layton, Lynn. 2006. Racial Identities, Racial Enactments, and Normative Unconscious Processes. *Psychoanalytic Quarterly* 75: 237–269.

Markman, Henry. 2020. Embodied Attunement and Participation. *Journal of the American Psychoanalytic Association* 68: 807–834.

Merleau-Ponty, M. 1969. *The Essential Writings of Merleau-Ponty* (Ed. A. Fisher). New York: Harcourt.

Millman, Oliver. (2023). DeSantis Accused of "Catastrophic" Climate Approach after Campaign Launch. *The Guardian*, May 6.

Mitrani, J. 1994. On Adhesive Pseudo-Object Relations. *Contemporary Psychoanalysis* 30:348–367.

Morton, Timothy. 2010. *The Ecological Thought*. Cambridge, MA: Harvard University Press.

Nishitani, Keiji. 1987. *Religion and Nothingness*. Berkeley, CA: University of California Press.

Ogden, Thomas. 1989. *The Primitive Edge of Experience*. NJ: Aronson.

Orange, Donna. 2017. *Climate Crisis, Psychoanalysis, and Radical Ethics*. New York: Routledge.

Paddison, Laura. 2024. UN Chief Says World Is on "Highway to Climate Hell" as Planet Endures 12 Straight Months of Unprecedented Heat. *CNN*, June 5.

Povinelli, Elizabeth. 2022. *Between Gaia and Ground*. Durham, NC: Duke University Press.

Ryan, Shawn. 2024. President Donald J. Trump – Make America Great Again. *The Shawn Ryan Show*, August 26 [podcast]. https://podcasts.apple.com/us/podcast/127-president-donald-j-trump-make-america-great-again/id1492492083?i=1000666636330.

Saketopoulou, A. 2014. Mourning the Body as Bedrock: Developmental Considerations in Treating Transexual Patients Analytically. *Journal of the American Psychoanalytic Association* 62: 773–806.

Spivak, Gayatri. 1987. *In Other Worlds: Essays in Cultural Politics*. London: Routledge.

Stein, Ruth. 2005. Why Perversion?: "False Love" and the Perverse Pact. *IJP* 86:775–799.

Stryker, Susan and McCarthy, Dylan. (Eds.) 2024. *The Transgender Studies Reader Remix*. New York: Routledge.

Trungpa, Chogyam. 1973. *Cutting Through Spiritual Materialism*. Boston: Shambhala Publications.

Weintrobe, Sally (Ed.). 2014. *Engaging with Climate Change: Psychoanalytic and Interdisciplinary Perspectives*. New York: Routledge.

Winnicott, D.W. 1949. Mind and its Relation to the Psyche-soma. *British Journal of Medical Psychology* 27(4): 201–209.

Ziarek, Ewa. 2001. *An Ethics of Dissensus: Postmodernity, Feminism, and the Politics of Radical Democracy*. Stanford: Stanford University Press.

Chapter 17

Climate Anxiety, Maturational Loss, and Adversarial Growth

Panu Pihkala

Introduction

Climate anxiety is increasingly impacting people of all ages, but it is especially prominent in many children and youth (e.g. Hickman et al. 2021; Ogunbode et al. 2022; Sangervo, Jylhä, & Pihkala 2022; Galway & Field 2023). Broadly speaking, climate anxiety is related to many kinds of distress and many difficult emotions which are significantly related to anthropogenic climate change (Pihkala 2020a; Hickman 2020; Clayton 2020). Underneath climate anxiety is a dimension of being able to notice threats which are diffuse and include some uncertainty, and this kind of constructive worry or "practical anxiety" has been explored in relation to ecological issues (for an overview, see Ojala et al. 2021; for practical anxiety, see Kurth & Pihkala 2022). Climate anxiety as a concept has been partly politicized and it is possible to use various kinds of terminology (Wardell 2020), but what is essential is to realize the extent and depth of climate change-related distress.

Climate anxiety provides many challenges for therapists and other psychological professionals (e.g. Silva & Coburn 2022; Lewis, Haase, & Trope 2020). It is difficult for any human being to encounter the severity of current and predicted environmental damage (e.g. Weintrobe 2013, 2021; Dodds 2011). Children's fears, anxieties and sad feelings are painful to encounter for adults, and at the same time the adults need to wrestle with their own emotional reactions and mental states (Hickman 2019, 2020). However, this situation also means that adults have profound possibilities to support children and youth, if they find ways to engage with difficult emotions together. Furthermore, supporting children may at the same time also support the caring adults. Overall, what is needed is growing together in community.

In this chapter, I am focusing on certain aspects of the lived experience of climate change and other ecological problems, and the ways in which these aspects affect encounters between children and adults. Therapeutic encounters are a major part of this, but many dynamics which I discuss are relevant also for any climate conversations. I am especially focusing on feelings of loss and sadness, which can be here captured with the overarching terms climate grief

DOI: 10.4324/9781003520757-20

and ecological grief (Cunsolo & Ellis 2018; Pihkala 2020b). With the help of research in grief and bereavement studies, I explore various forms of climate-related loss and grief that especially children and young people may feel. I argue that these feelings are intimately connected with climate anxiety and that they can affect child-adult encounters in profound ways.

I also explore how the dynamics of growing up in the midst of the climate crisis can be intertwined with climate grief. To my knowledge, this chapter includes the first application of the framework of maturational loss (Walter & McCoyd 2016) into climate grief and lifespan development amidst climate change. The concept and framework of maturational loss draws attention to the fact that people often experience both positive and negative emotions when they grow up and mature. I point out that the climate crisis brings new severity to these kind of feelings of loss: growing up amidst climate change can evoke many intangible losses, especially if and when the adult world does not care enough about climate change. I argue that the framework of disenfranchised grief can help to understand some aspects of this, and I briefly discuss a historical comparison with the existential threat produced by the looming possibility of nuclear war amidst the Cold War as described by psychologist Robert Jay Lifton.

In the final section of the chapter, I discuss the challenges and possibilities inherent in child-adult encounters amidst climate emotions. There are also possibilities for adversarial growth or post-traumatic growth arising out of climate anxiety, in addition to the profound distress. Adults' developmental tasks also need attention, and parents and grandparents are examples of groups which need further social support. This also helps to provide better support for children. Sometimes children can be agents of this support, but care must be taken so that they do not end up as carriers of others' issues.

A caveat must be mentioned. Climate anxiety and grief are constantly evolving phenomena with a plurality of possible manifestations and dynamics, which makes research about them both extremely important and very challenging. Everything is changing rapidly and new research emerges constantly. I am drawing in this chapter from many years of focused study of the topic area (e.g. Pihkala 2017, 2018, 2020a, 2020c, 2020d, 2022a, 2022b) but the reader should acknowledge that more research is needed on the nuances of the topics covered here. (Some of my earlier studies about children and eco-anxiety are in Finnish and I try to explain in English the main results of those that I cite here; Pihkala, Sangervo, & Jylhä 2022.)

Earlier research is relatively scarce both about children's varieties of climate grief and about developmental dynamics amidst climate change. There are some important in-depth interview studies of children's climate emotions (esp. Hickman 2019, 2020, 2023), useful studies about adults' observations of children's climate emotions (e.g. Baker, Clayton, & Bragg 2021; Verlie et al. 2020), and general observations about the impacts of climate change on children's development (Vergunst & Berry 2021; Burke, Sanson, & Van Hoorn

2018). In the vast literature on environmental education, there are many observations of children's climate emotions (for an overview, see Pihkala 2020c), but much less in-depth explorations of grief dynamics or developmental tasks. Many studies in environmental education and environmental psychology do provide empirical data which can be analyzed in relation to theories of grief and loss. Education and psychology researcher Ojala has studied the general topic of eco-emotions for a long time (e.g. Ojala 2007, 2016), and she has briefly discussed the impacts of these studies for developmental psychology (Ojala 2022). Many psychoanalytic and psychosocial scholars have made important contributions for understanding related dynamics especially among adults (e.g. Randall 2009; Hoggett 2019; Lertzman 2015; Gillespie 2020), and the results of these inquiries are here applied in many ways to the topic at hand.

The results of this chapter can be helpful for both therapists and any adults who wish to explore deeper the emotions and dynamics related to climate anxiety. I am mentioning several psychodynamic concepts in the chapter, but the results also provide many opportunities for further in-depth explorations of related themes by various professionals. (I am not a psychotherapist myself, even though I have some therapeutic training and experience, and I partly draw from discussion groups and workshops about eco-anxiety which I have co-facilitated with various psychological professionals.)

I will start by exploring various forms of climate-related loss and grief, and then proceed to discussing the dynamics of growing up amidst the climate crisis.

Various Forms of Climate-Related Loss and Grief

Tangible and Intangible Losses

What do contemporary children and youth feel to be lost in relation to climate change? There is an increasing number of studies which touch on aspects of this question (e.g. Hickman et al. 2021; Diffey et al. 2022; Coppola & Pihkala 2023), but more attention would be needed to the plurality, depth, and complexity of these losses, and psychodynamic thinking can be one important tool for such exploration (e.g. Lertzman 2015; Nicholsen 2002; Randall 2009; Hickman 2020). Relevant research frameworks include ecological grief, eco-anxiety / climate anxiety, climate distress, and solastalgia (for overviews, see Pihkala 2020a, 2022b), but many more could be named. Practically, any good studies on the lived experience of climate change can add to our understanding of related issues, but studies on the seldom named and easily concealed aspects are very much needed.

The conceptualization of tangible and intangible loss in grief research (Harris 2020b) offers helpful tools for investigating people's climate losses. Tangible losses are those which can be rather easily noticed, at least to members of a same culture. These are often the visible aspects of losses, or otherwise noticeable with human senses. Intangible losses are those which can be totally invisible for other people, or at least not easy to first notice. These may be things that others don't know that have existed, or they may be aspects related to tangible losses which others do not realize.

An example of tangible climate change-related loss is the receding glacier nearby one's place of residence. However, with the glacier, many other things may be felt to be lost at the same time, and these intangible aspects may be different for various people. There may be cultural, psychological and symbolic significance which is tied with the glacier. The loss of the glacier may generate intangible loss of income, for example via loss of hunting opportunities or tourism opportunities. And furthermore, the loss of the glacier may resonate with more global losses and anxieties: a receding local glacier may be an important focal point of climate anxiety and worry related to the whole global climate (for glaciers, climate change and different significances, see Brugger et al. 2013; for resonance of local and global, see Pihkala 2022b).

Table 17.1 shows numerous kinds of intangible climate change-related losses which scholars Tschakert and colleagues (Tschakert et al. 2019) have charted from studies around the world.

A key point is that children and young people in various parts of the world may feel numerous kinds of both tangible and intangible losses in relation to

Table 17.1 Subjects of climate-related loss and grief.

Culture, lifestyle, traditions, and heritage
Physical health
Mental and emotional well-being
Human mobility
Indirect economic benefits and opportunities
Sense of place
Social fabric
Ecosystem services, biodiversity, and species
Productive land and habitat
Knowledge and ways of knowing
Human life [span]
Identity
Self-determination and influence
Order in the world
Dignity
Territory
Ability to solve problems collectively
Sovereignty

Source: Tschakert et al. (2019)

climate change, and that these aspects may be combined in profound ways (see also Goldman 2022, pp. 27–31). Adults, including therapists, must pay attention to these and especially for the possible intangible losses. The intangible aspects may be very serious: their range includes the felt loss of whole futures, lifepaths, and dreams, as will be discussed more below.

Nonfinite Loss and Ambiguous Loss

The climate losses that children and young people feel may be further complicated and made more intense by the nonfinite and sometimes ambiguous character of the losses. Again, concepts and frameworks from grief research may provide help.

Nonfinite loss as a concept and research framework has been applied for example to people experiencing life-changing disabilities in either themselves or others (Schultz & Harris 2011; Harris 2020a). These are losses which continue to evoke feelings of sadness: they remain present, even though a stronger process of grief may be experienced and processed at the time when the loss is generated or faced. Many characteristics of nonfinite loss are easily discernible in studies of ecological emotions, even while as a concept and framework nonfinite loss has not yet been much used there (see, however, Kevorkian 2020; Pihkala 2023). Table 17.2 shows some of these characteristics.

Nonfinite loss can have close relations to ambiguous loss, a term developed by grief scholar Pauline Boss (for an overview, see Boss 2020). Ambiguous loss is characterized by simultaneous absence and presence, leading to unclarity and uncertainty. A classic example is a soldier missing in action. Boss notes that there can also be physical presence but psychological, ambiguous absence, such as in cases of dementia: is the personality still there or not?

Table 17.2 Cardinal features of nonfinite loss according to Schultz and Harris (2011) and Harris (2020a).

"There is an ongoing uncertainty regarding what will happen next. Anxiety is often the primary undercurrent to the experience."
"There is often a sense of disconnection from the mainstream and what is generally viewed as 'normal' in human experience."
"The magnitude of the loss is frequently unrecognized or not acknowledged by others."
"There is an ongoing sense of helplessness and powerlessness associated with the loss."
"Nonfinite losses may be accompanied by shame, embarrassment, and self-doubting that further complicates existing relationships, thereby adding to the struggle with coping."
"There are typically no rituals that assist to validate or legitimize the loss, especially if the loss was symbolic or intangible."
"chronic despair and ongoing dread"

It is more difficult to grieve if one cannot be certain about the loss or if the loss fluctuates, and this has been observed to happen in relation to some ecological losses (Cunsolo & Ellis 2018). Since many losses are expected to happen in the future, the role of anticipation also becomes often intertwined with the threats and losses. Many things are already partly lost because of climate crisis; because there are processes of change happening all the time, people may find it difficult to estimate whether some things are already going to be totally lost or whether they may still be partly saved. This connects ecological grief with discourses about anticipatory grief and mourning (for anticipatory grief in general, see e.g. Worden 2018, pp. 204–208; for anticipatory climate grief and "pre-traumatic stress," see e.g. Babbott 2023).

As regards types of grief, nonfinite loss often generates what is called "chronic sorrow" (Ross 2020; Harris 2020a), a non-pathological grief which is different from what is commonly called chronic grief. Chronic sorrow is characterized by both persistence and fluctuations of intensity, among other attributes, and its descriptions have much in common with aspects of climate grief.

Disenfranchised Grief

Disenfranchised grief is a term developed by grief researcher Kenneth Doka, referring to griefs which are not allowed to gain space (for an overview, see Doka 2020). There may be various reasons for this kind of social behavior, but it is fundamentally related to some kind of difficulty in accepting the grief in question, and includes power dynamics. Sometimes the loss is not acknowledged as valid, and sometimes the griever is excluded from the status of those allowed to mourn. Disenfranchising may be either silently operating or more maliciously, a result of straight use of power (Attig 2004).

Grief researchers have observed that certain types of loss and grief often generate disenfranchised grief. These prominently include many those attributes which are often found in relation to ecological grief: intangible loss, nonfinite loss, ambiguous loss, anticipatory grief, and chronic sorrow (Harris 2020a). Others may not notice the losses or they may devalue the losses because they are not in touch with them themselves. Disenfranchised grief has been observed in relation to ecological grief (Cunsolo & Ellis 2018), but its nuanced dynamics deserve more attention. It is closely related to what has been discussed in relation to ecological grief with the help of philosopher Judith Butler's term "grievability" (Cunsolo Willox & Landman 2017; Barnett 2022).

Grief researchers discuss the possibility that people may self-disenfranchise their grief (Doka 2020), and there are examples of this in relation to ecological grief (Nicholsen 2002). In disenfranchised grief of any kind, it is important to note that it may be the intangible aspects which are disenfranchised, even if tangible aspects are recognized.

Results for Climate Encounters

The aforementioned aspects of loss and grief produce significant impacts for climate change experiences of people of various age. Both children and adults have to wrestle with difficult cognitive evaluations of types of loss, often amidst significant emotional disturbance. Questions which may be asked include: What aspects of this loss are total and what are ambiguous? For example: will the summers always be this hot and dry in the future, too? In other words, is this loss nonfinite and if so, how can we deal with it? Grieving has in general been found to be very difficult for contemporary people (e.g. Horwitz & Wakefield 2007; Levine 2017) and climate grief operates on a scale which brings more difficulty, due to its existential impact (Rehling 2022; Budziszewska & Jonsson 2021; Passmore, Lutz, & Howell 2022). The potential complex dynamics of grief and loss, which were discussed in the previous section, bring further difficulty.

Children and youth often report that they feel that adults do not understand or validate their climate-related losses, including significant intangible losses (e.g. Diffey et al. 2022; Jones & Davison 2021). From a psychodynamic perspective, it can be seen that these losses are often so severe and threatening that they easily generate defenses in adults, including therapists (Silva & Coburn 2022; Haseley 2019; Kassouf 2017). Bringing the various aspects of loss into daylight, with the help of concepts and frameworks from grief theory, is one important effort in developing more resources to encounter them (similarly Hickman 2023).

It is important to notice that people may not themselves recognize (a) ecological grief in general or (b) particular aspects of it, including intangible aspects (e.g. Barnett 2022; Weintrobe 2021, pp. 161–167, 237). In addition, they may or may not recognize how strongly disenfranchised grief affects them (see e.g. Kretz 2017). Safe exploration of various aspects of loss and related experiences is a major service that an adult, including a therapist, can provide for the child – or for another human being of any age, but there is a significant community responsibility to help children in this regard. More compassion and safe spaces would be needed for this engagement, which shows for example in the growing popularity of climate-related death cafes (see Weber 2020).

The negative impacts of an inability to mourn are a classic theme in psychodynamic thinking (Mitscherlich & Mitscherlich 1984), and this has been discussed also in relation to ecological losses (Jones, Rigby, & Williams 2020; see also Nicholsen 2002; Lertzman 2015). The ability to grieve collectively various climate losses may have a very important political aspect, too.

Growing Up amid Climate Change

Maturational Loss

Growing up has always included the potential for both enthusiasm and loss. In societies driven by an ideology of progress, grieving can be generally disvalued, and the loss aspects of maturation may be disenfranchised. Grief scholars Walter and McCoyd are among those who have drawn attention to the ambiguity of growing up and the potential simultaneous existence of loss and gain. They describe their framework of maturational loss: "Normal maturational changes are recognized not only as growth, but also as a special form of loss in which one is expected to delight in the growth and ignore the loss aspect of the change, a perspective we challenge" (Walter & McCoyd 2016, p. 1).

Climate change brings an additional aspect to maturational loss. For numerous people, growing up now causes more losses than before, because they have to deal with the intangible losses generated by the climate crisis. By accepting climate reality, there may result for example a felt loss of carefree youth, along with many other difficult intangible losses such as loss of earlier dreams and plans about the future. There may result also tangible losses, such as conflicted relationships and/or an inability to fully enjoy many carbon-intensive activities, thus resulting in an omission of some of them (e.g. ability to enjoy travel, for those who could afford it). "Climate change maturity" comes with a price, even when such maturity is desperately needed for the sake of Earth's ecosystems and human wellbeing.

When is climate change maturity reached, and what is the role of age in its attainment? Experienced climate psychotherapist Sally Gillespie (2022) discusses group methods and "maturing conversations" in relation to climate change, pointing out that many adults are in need of more maturity in this regard. There are always differences between children and youth, but many of them currently reach climate change maturity at a rather early age, even while many of the adults close to them have not yet managed to encounter climate reality and engaged in such maturing (e.g. Hickman 2020, 2023). This brings additional stress for the children and youth, both because of the weight of this knowledge and the complex social dynamics encountered with less enlightened adults. Hickman argues that it is exactly the inability of the adult world to engage constructively with climate change which produces such heavy psychological impact in children, and she finds similar dynamics in child abuse and moral injury.

Even in normal circumstances, children and youth may experience much longing towards their earlier conditions, and for people of any age, there is always a psychological need to grieve for the ending of bygone life eras (Walter & McCoyd 2016). One can only imagine how much the climate crisis can strengthen and complicate this kind of longing and sadness, especially in societies where both normal maturational loss and climate grief are predominantly disenfranchised. Maturational loss can thus feature in many ways in parent–child relations.

Historical Precedent? The Nuclear Threat

Some insight to the difficult situation around climate anxiety and grief may be gained by taking a deep look at another global-scale worry, the threat of nuclear war. One of the ardent researchers of nuclear psychology has been Robert Jay Lifton, who is still active. Lifton, a famous psychologist and writer since the latter half of the twentieth century (for an overview, see Lifton 2019), integrates insights from many different psychological traditions, including psychoanalytic ones. He considered the impacts of the nuclear war threat to children and families in the 1980s, and these dynamics merit attention also in relation to the climate crisis:

> Undermined now is the fundamental parental responsibility, that of "family security." In the face of the threat of nuclear extinction, parents must now doubt their ability to see their child safely into some form of functional adulthood. And the child must also sense, early on, not only those parental doubts but the general inability of the adult world to guarantee the safety of children. In fact, there is growing evidence of significant impairment to the overall parent-child bond, to the delicate balance between protection and love on the one hand and the inner acceptance of authority on the other. With nuclear subversion of that authority, the always-present ambivalence on both sides can be expected to intensify, perhaps subverting feelings of love.
>
> (Lifton 1982, emphasis added)

Lifton has later applied his thinking also explicitly to climate change (Lifton 2017), focusing on the threats that these kind of crisis can produce to people's conceptions of their future and their meaning in life. Lifton wrote already in the 1980s about the threat of "radical futurelessness," which is much present in the quote above.

Can Lifton's ideas be applied also to the present context of climate change, and if so, to what extent? As phenomena, there are many similarities between the threat of nuclear war and threats posed by the ecological crisis, but of course a major difference is that the ecological crisis and the climate crisis are proceeding constantly. Much damage has already been done and is loaded into the system, which affects people both physically and psychologically.

Similarities in emotional dynamics between Lifton's descriptions and the contemporary climate situation can be easily discerned. In the global research about climate emotions, where I also participated (Hickman et al. 2021), the feelings of unsafety among children became very clear, as well as their feelings of having been betrayed by the older generations of decision-makers. We discussed the possible moral injury that the situation has caused to children. Personally, I see much connections-between our results and Lifton's thoughts. Amidst the climate crisis, children sense, in Lifton's words, "the general inability of the adult world to guarantee the safety of children," and it is possible that "the always-present ambivalence on both sides can be expected to intensify, perhaps subverting feelings of love" (Lifton 1982). This links with the arguments of Weintrobe (2021) about the "culture of uncare" in contemporary societies, and studies about intergenerational tensions in relation to climate issues (e.g. Roy & Ayalon 2022).

This suggests that having even a few trustworthy and climate-aware adults in the children's lifeworld can significantly help intergenerational dynamics. It will not remove all the disappointment and feelings of unsafety, but it manifests that some adults care, working towards maintaining connections between generations. Psychologist Molly Young Brown tells of a child during the Cold War who was not worried about nuclear war when others were, and when the teacher asked why this was so, the child replied that her parents were participating in anti-nuclear activities (Brown 2016). Indeed, many scholars have pointed out that it would be very important to have people of all generations participating together in climate action, both for psychological and political reasons (e.g. Ayalon et al. 2022). In intergenerational encounters amidst various kinds of climate action, different generations can give embodied messages of a culture of care (Weintrobe 2021), or of "love" as Lifton puts it.

Lifton himself discusses the importance of "witnessing professionals": in the case of nuclear threat, these were scientists, and in the case of climate change, these prominently include climate scientists (Lifton 2017, esp. ch. 6). Psychological professionals can become important witnessing professionals, too; both in relation to climate change in general and also in relation to the existence, severity, and complicated character of climate anxiety. However, this requires still much effort in various schools and communities of psychology, including psychodynamic ones (e.g. Orange 2017; Kassouf 2017).

The Challenge of Growing Together

Adversarial Growth and Survivor Missions

For children and youth, growing up has always had the potential to include both felt gains and felt losses; think of sadly leaving the toys of childhood behind you, but at the same time enthusiastically greeting the new opportunities which come with age. Now in the climate crisis, children and youth often have to grow up very early, because the severity of the climate crisis shocks them into a more mature view of suffering in life. This doesn't happen to all of them, of course, and some manage to stay in denial and disavowal, or do not care. For example in Finnish studies about young people and climate issues, roughly 5–6% of youth display anti-climate attitudes (Pihkala, Sangervo, & Jylhä 2022). But many children find themselves to be more mature than many adults in their approach to the reality of the climate crisis. This maturational change should be recognized as including a stronger-than-before loss aspect: growing up includes even more tragic aspects now than before. Since the loss aspects of even common maturational changes have not always been recognized before, this newly needed recognition may not be an easy task, especially given the social and political disputes around climate change.

However, as grief scholars Walter and McCoyd (2016) point out, recognizing and validating this loss aspect can lead to growth, especially with social support. It needs to be asked: what kind of positive aspects may the early maturation of children because of the climate crisis include? In other words, there is a need to be aware of both possible negative and positive aspects, without resorting to binary interpretations. That being said, it is evident that there is very much negative in the destruction around the climate crisis, and any brightsiding must be resisted.

The noted scholar of climate anxiety and children, Caroline Hickman, contributed a highly interesting thought when we discussed these dynamics during the preparation of this chapter (personal communication). She was reminded of the wide discussion around so-called young carers: young people who find themselves caring for other persons who have some kind of disabling condition. Young carers often have to mature very early, which can cause many kinds of losses. However, young carers also often experience many kinds of positive consequences because of their situation: for example, they may feel honor and become in many ways more resilient.

Perhaps children and youth who have to mature early because of the climate crisis experience some similar things? While their "climate maturation" brings many losses, it can also bring many skills, new comrades, a feeling of doing honorable things, and an experience of being able to channel one's caring and empathy into practice.

Broadly, an important research venture would be to study possible "adversarial growth" (Blackie et al. 2023) or post-traumatic growth (Tedeschi et al. 2018) among climate-aware young people. The framework of post-traumatic growth (PTG) has been applied to climate issues by some psychologists, most notably Doppelt (2016), but adversarial growth as a whole would require much more attention in relation to climate anxiety. In these studies, it is noted that major stressors or "seismic events" often cause both growth and negative impacts, which are called "post-traumatic depreciation" in that research (e.g. Taku et al. 2021).

PTG studies usually use five domains and these can be helpful in discerning various impacts of climate anxiety for people, including children: personal strength, relating to others, new possibilities, appreciation of life, and spiritual or existential change (Tedeschi et al. 2018). From literature about climate anxiety, it is easy to find examples of both PTG and post-traumatic depreciation in relation to all these five domains. Personal strengths may emerge or be invigorated, but people may also regress and suffer damage. For many children and young people, becoming climate-aware has brought new social relations and feelings of togetherness, but also social disruptions. Many have developed whole new life paths, some prominently in climate activism (e.g. Halstead et al. 2021), but also paths leading to depression and suicidality are a possibility (e.g. Halstead et al. 2021; Hoggett & Randall 2018; Nairn 2019). Realizing the fragility of existence may cause people to appreciate daily life more, including the natural world (e.g. Zaremba et al. 2022; Marczak et al. 2021), but this is by no means the only possible impact and nihilism is very possible (see discussion in Scranton 2015). This domain is closely linked with childhood, since the ability to wonder and appreciate daily occurrences in the natural world are skills which children can have in optimal circumstances. Youth and adults who feel so strong climate anxiety that they find it difficult to appreciate life face the task of reinvigorating the childlike ability to wonder (for a classic discussion of wonder and nature, see Carson 1965). Changes in worldviews and meaning systems, including spiritual and/or existential aspects, are also evident possibilities in processes of climate anxiety (Jamail 2019; Bell, Dennis, & Brar 2021).

I have spent some time here discussing PTG and broader adversarial growth because that could be a major asset in encounters between children and adults in the context of climate anxiety. It is evident that climate distress can feel terribly bad, but people could explore also simultaneous aspects of growth, and this is naturally closely connected with the themes of maturational loss and developmental tasks. Thus, I am making a two-fold argument in this chapter: pointing out that both the loss and the growth need more attention and nuance. Adults, including therapists, could help children and youth to see that their climate anxiety contains the seeds of maturation. Adults could – and I think *should* – validate young people's climate emotions and explore their dynamics (Mosquera & Jylhä 2022; Pihkala 2022a; Kałwak & Weihgold 2022). The moral and adaptive aspects of "practical eco-anxiety" can be fleshed out, including the ways that eco-anxiety as an emotion can lead to gathering of new information and behavioral changes (Kurth & Pihkala 2022).

People need examples of other people who have survived feelings of climate anxiety, and here both adults and other children and youth may function as role models. Lifton uses the term "prospective survivor" to describe people who have imagined a very traumatic ending and survived that, and who can then engage in a survivor mission of supporting others (Lifton 2017, esp. pp. 153–154). At their best, survivors can mirror both vulnerability and strength, and be able to show their grief and loss. These are no easy tasks, however, and they may complicate traditional dynamics of therapeutic encounters (Lewis, Haase, & Trope 2020; Silva and Coburn 2022; Budziszewska & Jonsson 2022).

For some children and youth, there emerges an ethical and psychological problem of potentially feeling that they have to "carry" the adults. Empathic people, including children, may become such "carriers," while others remain bystanders (Greenspan 2004). To name a practical example: what should an adolescent climate anxiety survivor do about their relatives who are still in climate denial, some of whom are suffering from repressed or suppressed climate grief? Is it ethical or psychologically bearable that the young person engages in an effort to support the older person's emotional journey towards more adaptive coping? These are difficult questions which always demand analysis of contextual factors and dynamics, but awareness of these kind of dilemmas and the ability to speak about them with safe others may at least help.

Psychodynamic and psychosocial thinkers could contribute much-needed depth of analysis in relation to the complexity of people's processes. Climate change is reality, and humans need to encounter that reality in order to adapt, mitigate and behave ethically. If people realize that the process of encountering climate reality includes also possibilities for ethical and psychological growth, this may bring some comfort (and perhaps motivation?) to the pain of maturing.

Many contemporary adults have survived the anxieties about the threat of nuclear war, which many of them felt strongly when they were children or young (e.g. Goldberg et al. 1985; Smith 1988). My experience is that in workshops and lectures about eco-anxiety, adults regularly raise the issue of their experience with the nuclear threat, resulting in conflicts between them and climate-aware youth. The youth may feel that the ongoing character of climate crisis is not given enough recognition when older people tell them of how they survived coping with the nuclear threat. Some older people indeed seem to resort to denial of the gravity of the climate crisis via over-emphasizing successful coping with the nuclear threat. However, I also feel in many of these comments there is genuine desire to draw from past experiences for contemporary coping, and one important aspect of growing together would be to find ways to connect the global coping experiences and challenges of people of various ages (see also Heglar 2020). This is also one possibility for engaging with Lifton's work.

Adults' Developmental Tasks

The depth of the ecological crisis – or the comprehensive crisis which can be called for example a polycrisis (Henig & Knight 2023) – challenges all kinds of developmental tasks, but also the very models of human lifespan development (for an useful overview, see Ivtzan et al. 2015, pp. 31–54). Things which have been considered exemplary may not be so if the environmental impacts are considered. For example, having an esteemed career is ethically compromised if the career is in the fossil industry, which destroys the common climate.

Alternative paradigms have been suggested. Eco-psychologist Bill Plotkin contrasts two models, an ego-centered and an eco-centered pattern of human developmental stages (Plotkin 2008). In the ecologically sensitive model, people in different stages have different tasks, but also different gifts to give for the community. In Plotkin's view, transformative journeys or periods are essential for enabling deeper growth, and he practically works at organizing wilderness-based options for those (Plotkin 2003).

Plotkin's ecological depth psychology may sound radical for many, but at least it needs to be asked how adults could engage constructively with their own developmental tasks and needs. This is crucial also in relation to children, in many ways. Adults can provide role models, and often adults need to grow together with children in general maturity and climate maturity. It is not possible here to engage extensively with the developmental tasks of adults, but certain important issues stand out.

Many common developmental tasks are related to human relations, such as finding partners, learning to live in a relationship, and managing family life (Ivtzan et al. 2015, pp. 31–54). There is a need for sensibility about the plurality of possible life paths and some models of developmental tasks are quite Western and heteronormative. However, ecological issues have increasingly started to affect all kinds of developmental tasks. People searching for partners may use opinions in environmental politics as one criteria, and some dating apps have even created filters for climate opinions (Godin 2022). Partners may find out only later that they have different environmental values, and this may cause conflicts (e.g. Kaplan 2023). If and when children appear in relationships, they are affected by the environment-related psychological dynamics between the adults.

Overall, parents (and grandparents) are a group of people who also need special attention and social support. Climate change and the even broader ecological crisis is rapidly changing the dynamics of "good-enough parenting" (see also Weintrobe 2021, pp. 89–90), and parents may easily feel anxious about how to act – or even overwhelmed and helpless (e.g. Holmes, Natalier, & Leahy 2022; Baker, Clayton, & Bragg 2021). There is growing literature about parenting in the ecological crisis, but this literature is sometimes rather simplistic in its emphasis on pro-environmental action, and not inclusive enough of the deeper psychological dynamics that need attention (for various kinds of literature, see Bechard 2021; Cripps 2023; Sanson, Burke, & Van Hoorn 2018).

Parents can themselves experience a plurality of tangible and intangible losses related to climate change. The loss of earlier, easier models (and possibly times) of parenting is one of them, and many dark emotions may emerge in relation to this, such as envy towards earlier generations or bitterness. Becoming a parent can include maturational loss, and amidst the ecological crisis, reproduction decisions are increasingly difficult for many people (e.g. Schneider-Mayerson & Leong 2020). Regardless of what choices people end up making, there may be complex combinations of feelings of loss and gain (e.g. Wray 2022). The concept of role loss (Mitchell & Anderson 1983) can be helpful here, since it pinpoints felt losses of important roles, such as the role of a parent or a grandparent.

Guilt and feelings of inadequacy, of not being enough, are common among parents, and these may intensify during the profound demands of the ecological crisis. It is easy to see how guilt dynamics may prevent parents or grandparents from constructive engagement with children's climate emotions: it may simply feel too much (Weintrobe 2021; for eco-guilt, see Jensen 2019). Commitment to pro-climate attitudes and action can bring counterforces to feelings of guilt and shame, but ambivalence should be kept in mind, so that people do not end up in problematic forms of behavior, such as desperate efforts to alleviate guilt by constant action (Hoggett & Randall 2018) or self-deceptive emphasis on tokenistic environmental actions (Sapiains, Beeton, & Walker 2015). Shame and depressive moods are also possibilities. Psychodynamic thought has much to offer in relation to understanding these dynamics better (e.g. Haseley 2019; Dodds 2011; Randall 2005).

Parents would very much like to, and need to, feel that they are regarded as good-enough parents by their children and by their communities. Furthermore, developmental psychologists have often emphasized that people in late adulthood, including grandparents, want to leave a positive legacy and a good memory of themselves (for a classic discussion, see Erikson & Erikson 1998; for various theories, see Ivtzan et al. 2015, pp. 31–54). All this can easily become increasingly complicated amidst the ecological crisis. For example, if older people and grandparents are seen by climate-aware youth as perpetrators of fossil fuel lifestyles, they can feel threatened of losing their possibilities for positive legacy and respect among later generations (see also Lifton's framework of symbolic immortality, e.g. Lifton 1979). It would not be surprising if some of them feel existential dread because they fear that the generation of grandchildren will not take care of them when they are old and fragile, also for climate-related reasons.

These kind of psychosocial threats can have many outcomes. One potential result is an increased tendency to use defenses related to denial and disavowal, resulting also in disenfranchised grief. Another option would be to grow caring relations, where shortcomings are understood, but acceptance is also offered (Weintrobe 2021). I often wonder what would happen if climate-aware children and young people were able to offer their parents and grandparents both the message of "humanity needs to change" and "I still love you, even though I'm critical of these lifestyles."

For parents and other people in midlife, the ecological crisis intensifies the need to develop into more mature adulthood, where facing mortality and finitude is one key issue (e.g. Hollis 1993; Vaillant 2003). The ecological crisis itself can remind people of mortality (for an overview, see Smith et al. 2022), and it would seem that the developmental tasks of "ecological maturity" and death-aware ("second") adulthood can be closely interconnected. If adults are able to engage in meaning reconstruction (Neimeyer 2019) and life story retelling (McAdams 1996), this also provides children and youth with role models and incentive. Examples of such meaning reconstruction caused by increasing climate awareness (and anxiety) are found in several books (Jamail 2019; Newby 2021; Wray 2022; Gillespie 2020).

Adults, including parents and grandparents, need places to engage with their own feelings, just between adults. Psychological professionals, including psychodynamic therapists, could contribute to this engagement both via their therapeutic practice and as members of community. Many have already done so, for example via organizations such as Climate Psychology Alliance (www.climatepsychologyalliance.org).

Finally, the challenge and opportunity of encountering children's climate anxiety and grief is a necessary community task: it includes parents, relatives, neighbors, educators, social and health care workers, and so on. In this communal task, various kinds of psychological expertise, including grief theory and psychodynamic theories, are an essential part of the team of people and range of knowledge needed to address the emotional side of our ecological crisis.

References

Attig, Thomas. 2004. "Disenfranchised Grief Revisited: Discounting Hope and Love." *Omega: Journal of Death and Dying* 49(3): 197–215. doi:10.2190/P4TT-J3BF-KFDR-5JB1.

Ayalon, Liat, Senjooti Roy, Omer Aloni, & Norah Keating. 2022. "A Scoping Review of Research on Older People and Intergenerational Relations in the Context of Climate Change." *The Gerontologist*, February, gnac028. doi:10.1093/geront/gnac028.

Babbott, Margaret. 2023. "Pretraumatic Climate Stress in Psychotherapy: An Integrated Case Illustration." *Ecopsychology*, February. doi:10.1089/eco.2022.0076.

Baker, Cambry, Susan Clayton, & Eshana Bragg. 2021. "Educating for Resilience: Parent and Teacher Perceptions of Children's Emotional Needs in Response to Climate Change." *Environmental Education Research* 27 (5): 687–705. doi:10.1080/13504622.2020.1828288.

Barnett, Joshua Trey. 2022. *Mourning in the Anthropocene: Ecological Grief and Earthly Coexistence.* East Lansing: Michigan State University Press.

Bechard, Elizabeth. 2021. *Parenting in a Changing Climate: Tools for Cultivating Resilience, Taking Action, and Practicing Hope in the Face of Climate Change.* 1st edition. Brasstown, NC: Citrine Publishing.

Bell, Finn McLafferty, Mary Kate Dennis, & Glory Brar. 2021. "'Doing Hope': Ecofeminist Spirituality Provides Emotional Sustenance to Confront the Climate Crisis." *Affilia: Journal of Women and Social Work* 36. doi:10.1177/0886109920987242.

Blackie, Laura E.R., Nic M. Weststrate, Kit Turner, Jonathan M. Adler, & Kate C. McLean. 2023. "Broadening Our Understanding of Adversarial Growth: The Contribution of Narrative Methods." *Journal of Research in Personality* 103 (April): 104359. doi:10.1016/j.jrp.2023.104359.

Boss, Pauline. 2020. "Understanding and Treating the Unresolved Grief of Ambiguous Loss: A Research-Based Theory to Guide Therapists and Counselors." In *Non-Death Loss and Grief: Context and Clinical Implications*, edited by Darcy L.Harris, 73–79. New York: Routledge.

Brown, Molly Young. 2016. "Supporting Children Emotionally in Times of Climate Disruption." In *Education in Times of Environmental Crises*, edited by Ken Winograd, 195–209. New York: Routledge.

Brugger, Julie, K. W. Dunbar, Christine Jurt, & Ben Orlove. 2013. "Climates of Anxiety: Comparing Experience of Glacier Retreat across Three Mountain Regions." *Emotion, Space and Society* 6 (1): 4–13. doi:10.1016/j.emospa.2012.05.001.

Budziszewska, Magdalena, & Sofia Elisabet Jonsson. 2021. "From Climate Anxiety to Climate Action: An Existential Perspective on Climate Change Concerns Within Psychotherapy." *The Journal of Humanistic Psychology* 61 (Ahead of Print). doi:10.1177/0022167821993243.

Budziszewska, Magdalena, & Sofia Elisabet Jonsson. 2022. "Talking about Climate Change and Eco-Anxiety in Psychotherapy: A Qualitative Analysis of Patients' Experiences." *Psychotherapy.* doi:10.1037/pst0000449.

Burke, Susie, Ann Sanson, & Judith Van Hoorn. 2018. "The Psychological Effects of Climate Change on Children." *Current Psychiatry Reports* 20 (35). doi:10.1007/s11920-018-0896-9.

Carson, Rachel. 1965. *The Sense of Wonder.* New York: Harper & Row.

Clayton, Susan. 2020. "Climate Anxiety: Psychological Responses to Climate Change." *Journal of Anxiety Disorders* 74: 102263. doi:10.1016/j.janxdis.2020.102263.

Coppola, Isabel, & Panu Pihkala. 2023. "Complex Dynamics of Climate Emotions among Environmentally Active Finnish and American Young People." *Frontiers in Political Science* 4. www.frontiersin.org/articles/10.3389/fpos.2022.1063741.

Cripps, Elizabeth. 2023. *Parenting on Earth: A Philosopher's Guide to Doing Right by Your Kids-and Everyone Else.* Cambridge, MA: The MIT Press.

Cunsolo, Ashlee, & Neville R. Ellis. 2018. "Ecological Grief as a Mental Health Response to Climate Change-Related Loss." *Nature Climate Change* 8 (4): 275–281.

Cunsolo Willox, Ashlee, & Karen Landman, eds. 2017. *Mourning Nature: Hope at the Heart of Ecological Loss & Grief.* Montreal: McGill-Queen's University Press.

Diffey, James, 2022. "'Not about Us without Us' – the Feelings and Hopes of Climate-Concerned Young People around the World." *International Review of Psychiatry* 34 (5): 499–509. doi:10.1080/09540261.2022.2126297.

Dodds, Joseph. 2011. *Psychoanalysis and Ecology at the Edge of Chaos: Complexity Theory, Deleuze/Guattari and Psychoanalysis for a Climate in Crisis.* New York: Routledge.

Doka, Kenneth. 2020. "Disenfranchised Grief and Non-Death Losses." In *Non-Death Loss and Grief: Context and Clinical Implications*, edited by Darcy L.Harris, 25–35. New York: Routledge.

Doppelt, Bob. 2016. *Transformational Resilience: How Building Human Resilience to Climate Disruption Can Safeguard Society and Increase Wellbeing.* Saltaire: Taylor & Francis.

Erikson, Erik H., & Joan M. Erikson. 1998. *The Life Cycle Completed.* Extended version. New York: W.W. Norton.

Galway, Lindsay P., & Ellen Field. 2023. "Climate Emotions and Anxiety among Young People in Canada: A National Survey and Call to Action." *The Journal of Climate Change and Health* 9 (January): 100204. doi:10.1016/j.joclim.2023.100204.

Gillespie, Sally. 2020. *Climate Crisis and Consciousness: Re-imagining Our World and Ourselves.* New York: Routledge.

Gillespie, Sally. 2022. "Maturing Conversations: Developing Climate Engagement through Group Dialogues." *Journal of Analytical Psychology* 67 (5): 1452–1474. doi:10.1111/1468-5922.12867.

Godin, Melissa. 2022. "'I Couldn't Date a Climate Change Denier!' The Couples Who Bond – and Split – over Love for the Planet." *The Guardian*, September 5. www. theguardian.com/lifeandstyle/2022/sep/05/i-couldnt-date-a-climate-change-de nier-the-couples-who-bond-and-split-over-love-for-the-planet.

Goldberg, Susan, Suzanne Lacombe, Dvora Levinson, K. Ross Parker, Christopher Ross, & Frank Sommers. 1985. "Thinking about the Threat of Nuclear War." *American Journal of Orthopsychiatry* 55 (4): 503–512. doi:10.1111/j.1939-0025.1985.tb02701.x.

Goldman, Linda. 2022. *Climate Change and Youth: Turning Grief and Anxiety into Activism*. New York: Routledge.

Greenspan, Miriam. 2004. *Healing Through the Dark Emotions: The Wisdom of Grief, Fear, and Despair*. Boulder: Shambhala.

Halstead, Florence, Lucie R. Parsons, Ally Dunhill, & Katie Parsons. 2021. "A Journey of Emotions from a Young Environmental Activist." *Area* 53 (4): 708–717. doi:10.1111/area.12745.

Harris, Darcy L., ed. 2020a. *Non-Death Loss and Grief: Context and Clinical Implications*. New York: Routledge.

Harris, Darcy L., ed. 2020b. "Tangible and Intangible Losses." In *Non-Death Loss and Grief: Context and Clinical Implications*, edited by Darcy L.Harris, 237–242. New York: Routledge.

Haseley, Dennis. 2019. "Climate Change: Clinical Considerations." *Int J Appl Psychoanal Studies* 16: 109–115. doi:10.1002/aps.1617.

Heglar, Mary Annaïse. 2020. "Climate Change Isn't the First Existential Threat." *Medium*, February 18. https://zora.medium.com/sorry-yall-but-climate-change-a in-t-the-first-existential-threat-b3c999267aa0.

Henig, David, & Daniel M. Knight. 2023. "Polycrisis: Prompts for an Emerging Worldview." *Anthropology Today* 39 (2): 3–6. doi:10.1111/1467-8322.12793.

Hickman, Caroline. 2019. "Children and Climate Change: Exploring Children's Feelings About Climate Change Using Free Association Narrative Interview Methodology." In *Climate Psychology: On Indifference to Disaster*, edited by Paul Hoggett, 41–59. Cham: Palgrave Macmillan.

Hickman, Caroline. 2020. "We Need to (Find a Way to) Talk about … Eco-Anxiety." *Journal of Social Work Practice* 34 (4): 411–424. doi:10.1080/02650533.2020.1844166.

Hickman, Caroline. 2023. "Feeling Okay with Not Feeling Okay: Helping Children and Young People Make Meaning from Their Experience of Climate Emergency." In *Holding the Hope: Reviving Psychological And Spiritual Agency In The Face Of Climate Change*, edited by Linda Aspey, CatherineJackson, & Diane Parker, 183–198. Monmouth: PCCS Books.

Hickman, Caroline, Elizabeth Marks, Panu Pihkala, Susan Clayton, R. Eric Lewandowski, Elouise E. Mayall, Britt Wray, Catriona Mellor, & Lise van Susteren. 2021. "Climate Anxiety in Children and Young People and Their Beliefs about Government Responses to Climate Change: A Global Survey." *The Lancet Planetary Health* 5 (12): e863–873. doi:10.1016/S2542-5196(21)00278-00273.

Hoggett, Paul. 2019. *Climate Psychology: On Indifference to Disaster*. Cham: Palgrave Macmillan. doi:10.1007/978-3-030-11741-2.

Hoggett, Paul, & Rosemary Randall. 2018. "Engaging with Climate Change: Comparing the Cultures of Science and Activism." *Environmental Values* 27 (3): 223–243. doi:10.3197/096327118X15217309300813.

Hollis, James. 1993. *The Middle Passage: From Misery to Meaning in Midlife.* Toronto: Inner City Books.

Holmes, Mary, Kristin Natalier, & Carla Pascoe Leahy. 2022. "Unsettling Maternal Futures in Climate Crisis: Towards Cohabitability." *Families, Relationships and Societies*, October, 1–17. doi:10.1332/204674321X16621119776374.

Horwitz, Allan V., & Jerome C. Wakefield. 2007. *The Loss of Sadness: How Psychiatry Transformed Normal Sorrow into Depressive Disorder.* Oxford: Oxford University Press.

Ivtzan, Itai, Tim Lomas, Kate Hefferon, & Piers Worth. 2015. *Second Wave Positive Psychology: Embracing the Dark Side of Life.* 1st edition. New York: Routledge.

Jamail, Dahr. 2019. *End of Ice: Bearing Witness and Finding Meaning in the Path of Climate Disruption.* New York: The New Press.

Jensen, Tim. 2019. *Ecologies of Guilt in Environmental Rhetorics.* Cham: Palgrave Macmillan.

Jones, Charlotte A., & Aidan Davison. 2021. "Disempowering Emotions: The Role of Educational Experiences in Social Responses to Climate Change." *Geoforum* 118: 190–200. doi:10.1016/j.geoforum.2020.11.006.

Jones, Owain, Kate Rigby, & Linda Williams. 2020. "Everyday Ecocide, Toxic Dwelling, and the Inability to Mourn: A Response to Geographies of Extinction." *Environmental Humanities* 12 (1): 388–405. doi:10.1215/22011919-8142418.

Kałwak, Weronika, & Vanessa Weihgold. 2022. "The Relationality of Ecological Emotions: An Interdisciplinary Critique of Individual Resilience as Psychology's Response to the Climate Crisis." *Frontiers in Psychology* 13 (April). doi:10.3389/fpsyg.2022.823620.

Kaplan, Alison. 2023. "When Climate Change Melts Your Relationship." *The New York Times*, March 31. www.nytimes.com/2023/03/31/style/modern-love-relationship-climate-change.html.

Kassouf, Susan. 2017. "Psychoanalysis and Climate Change: Revisiting Searles's 'The Nonhuman Environment,' Rediscovering Freud's Phylogenetic Fantasy, and Imagining a Future.(Harold Searles's Book 'The Nonhuman Environment in Normal Development and Schizophrenia,' Sigmund Freud)(Critical Essay)." *American Imago* 74 (2): 141–171. doi:10.1353/aim.2017.0008.

Kevorkian, Kriss A. 2020. "Environmental Grief." In *Non-Death Loss and Grief: Context and Clinical Implications*, edited by Darcy L. Harris, 216–226. New York: Routledge. www.routledge.com/Non-Death-Loss-and-Grief-Context-and-Clinical-Implications/Harris/p/book/9781138320826.

Kretz, Lisa. 2017. "Emotional Solidarity: Ecological Emotional Outlaws Mourning Environmental Loss and Empowering Positive Change." In *Mourning Nature: Hope at the Heart of Ecological Loss & Grief*, edited by Ashlee Cunsolo Willox & Karen Landman, 258–291. Montreal: McGill-Queen's University Press.

Kurth, Charlie, & Panu Pihkala. 2022. "Eco-Anxiety: What It Is and Why It Matters." *Frontiers in Psychology* 13. doi:10.3389/fpsyg.2022.981814.

Lertzman, Renée Aron. 2015. *Environmental Melancholia: Psychoanalytic Dimensions of Engagement.* New York: Routledge.

Levine, Megan. 2017. *It's Ok That You're Not Ok: Meeting Grief and Loss in a Culture That Doesn't Understand.* Boulder: Sounds True.

Lewis, Janet, Elizabeth Haase, & Alexander Trope. 2020. "Climate Dialectics in Psychotherapy: Holding Open the Space Between Abyss and Advance." *Psychodynamic Psychiatry* 48 (3): 271–294.

Lifton, Robert Jay. 1979. *The Broken Connection: On Death and the Continuity of Life.* New York: Simon and Schuster.

Lifton, Robert Jay. 1982. "The Psychic Toll of the Nuclear Age." *The New York Times*, 26 September.

Lifton, Robert Jay. 2017. *The Climate Swerve: Reflections on Mind, Hope, and Survival.* New York: The New Press.

Lifton, Robert Jay. 2019. *Losing Reality: On Cults, Cultism, and the Mindset of Political and Religious Zealotry.* New York: The New Press.

Marczak, Michalina, Małgorzata Winkowska, Katia Chaton-Østlie, & Christian A. Klöckner. 2021. "'It's like Getting a Diagnosis of Terminal Cancer.' An Exploratory Study of the Emotional Landscape of Climate Change Concern in Norway." Preprint. doi:10.21203/rs.3.rs-224032/v1.

McAdams, Dan P. 1996. *The Stories We Live by: Personal Myths and the Making of the Self.* New York: The Guilford Press.

Mitchell, Kenneth R., & Herbert Anderson. 1983. *All Our Losses, All Our Griefs: Resources for Pastoral Care.* 1st ed. Philadelphia: Westminster Press.

Mitscherlich, Alexander, & Margarete Mitscherlich. 1984. *The Inability to Mourn: Principles of Collective Behavior.* Translated by Robert Jay Lifton and Beverley R. Placzek. 1st Evergreen ed. New York: Grove Press.

Mosquera, Julia, & Kirsti M. Jylhä. 2022. "How to Feel About Climate Change? An Analysis of the Normativity of Climate Emotions." *International Journal of Philosophical Studies*, November, 1–24. doi:10.1080/09672559.2022.2125150.

Nairn, Karen. 2019. "Learning from Young People Engaged in Climate Activism: The Potential of Collectivizing Despair and Hope." *Young* 27 (5): 435–450. doi:10.1177/1103308818817603.

Neimeyer, Robert A. 2019. "Meaning Reconstruction in Bereavement: Development of a Research Program." *Death Studies* 43 (2): 79–91. doi:10.1080/07481187.2018.1456620.

Newby, Jonica. 2021. *Beyond Climate Grief: A Journey of Love, Snow, Fire, and an Enchanted Beer Can.* Sydney: NewSoundBooks.

Nicholsen, Shierry Weber. 2002. *The Love of Nature and the End of the World: The Unspoken Dimensions of Environmental Concern.* Cambridge: MIT Press.

Ogunbode, Charles A., 2022. "Climate Anxiety, Wellbeing and pro-Environmental Action: Correlates of Negative Emotional Responses to Climate Change in 32 Countries." *Journal of Environmental Psychology* 84 (December): 101887. doi:10.1016/j.jenvp.2022.101887.

Ojala, Maria. 2007. *Hope and Worry: Exploring Young People's Values, Emotions, and Behavior Regarding Global Environmental Problems.* Örebro: Örebro University: Universitetsbiblioteket.

Ojala, Maria. 2016. "Facing Anxiety in Climate Change Education: From Therapeutic Practice to Hopeful Transgressive Learning." *Canadian Journal of Environmental Education* 21: 41–56.

Ojala, Maria. 2022. "How Do Children, Adolescents, and Young Adults Relate to Climate Change? Implications for Developmental Psychology." *European Journal of Developmental Psychology* online ahead of print (August). doi:10.1080/17405629.2022.2108396.

Ojala, Maria, Ashlee Cunsolo, Charles A. Ogunbode, and Jacqueline Middleton. 2021. "Anxiety, Worry, and Grief in a Time of Environmental and Climate Crisis: A Narrative Review." *Annual Review of Environment and Resources* 46 (1). doi:10.1146/annurev-environ-012220-022716.

Orange, Donna. 2017. *Climate Change, Psychoanalysis, and Radical Ethics.* New York: Routledge.

Passmore, Holli-Anne, Paul K. Lutz, & Andrew J. Howell. 2022. "Eco-Anxiety: A Cascade of Fundamental Existential Anxieties." *Journal of Constructivist Psychology*, May, 1–16. doi:10.1080/10720537.2022.2068706.

Pihkala, Panu. 2017. "Environmental Education After Sustainability: Hope in the Midst of Tragedy." *Global Discourse* 7 (1): 109–127.

Pihkala, Panu. 2018. "Eco-anxiety, Tragedy, and Hope: Psychological and Spiritual Dimensions of Climate Change." *Zygon* 53 (2): 545–569.

Pihkala, Panu. 2020a. "Anxiety and the Ecological Crisis: An Analysis of Eco-Anxiety and Climate Anxiety." *Sustainability* 12 (19): 7836. doi:10.3390/su12197836.

Pihkala, Panu. 2020b. "Climate Grief: How We Mourn a Changing Planet." www.bbc. com/future/article/20200402-climate-grief-mourning-loss-due-to-climate-change? ocid=ww.social.link.email.

Pihkala, Panu. 2020c. "Eco-Anxiety and Environmental Education." *Sustainability* 12 (23): 10149. doi:10.3390/su122310149.

Pihkala, Panu. 2020d. "The Cost of Bearing Witness to the Environmental Crisis: Vicarious Traumatization and Dealing with Secondary Traumatic Stress among Environmental Researchers." *Social Epistemology* 34 (1): 86–100. doi:10.1080/ 02691728.2019.1681560.

Pihkala, Panu. 2022a. "Toward a Taxonomy of Climate Emotions." *Frontiers in Climate* 3. doi:10.3389/fclim.2021.738154.

Pihkala, Panu. 2022b. "The Process of Eco-Anxiety and Ecological Grief: A Narrative Review and a New Proposal." *Sustainability* 14 (24): article number 16628. doi:10.3390/su142416628.

Pihkala, Panu. 2023. "Climate Sorrow: Discerning various forms of climate grief and responding to them as a therapist." In *Being a Therapist in a Time of Climate Breakdown*, edited by Judith Anderson, Caroline Hickman & Tree Stanton. London: Routledge.

Pihkala, Panu, Julia Sangervo, & Kirsti M. Jylhä. 2022. "Nuorten Ilmastoahdistus Ja Ympäristötunteet." In *Kestävää Tekoa: Nuorisobarometri 2021*, edited by Tomi Kiilakoski, 95–116. Helsinki: Valtion nuorisoneuvosto & Nuorisotutkimusseura & Opetus- ja kulttuuriministeriö.

Plotkin, Bill. 2003. *Soulcraft: Crossing into the Mysteries of Nature and Psyche.* Novato, CA: New World Library.

Plotkin, Bill. 2008. *Nature and the Human Soul: Cultivating Wholeness and Community in a Fragmented World.* Novato, CA: New World Library.

Randall, Rosemary. 2005. "A New Climate for Psychotherapy?" *Psychotherapy and Politics International* 3 (3): 165–179. doi:10.1002/ppi.7.

Randall, Rosemary. 2009. "Loss and Climate Change: The Cost of Parallel Narratives." *Ecopsychology* 1 (3): 118–129.

Rehling, Joseph T. C. 2022. "Conceptualising Eco-Anxiety Using an Existential Framework." *South African Journal of Psychology* 52 (4): 472–485. doi:10.1177/ 00812463221130898.

Ross, Susan. 2020. "Chronic Sorrow." In *Non-Death Loss and Grief: Context and Clinical Implications*, edited by Darcy L. Harris, 192–204. New York: Routledge. www.routledge.com/Non-Death-Loss-and-Grief-Context-and-Clinical-Implications/Harris/p/book/9781138320826.

Roy, Senjooti, & Liat Ayalon. 2022. "Intergenerational Relations in the Climate Movement: Bridging the Gap toward a Common Goal." *International Journal of Environmental Research and Public Health* 20 (1): 233. doi:10.3390/ijerph20010233.

Sangervo, Julia, Kirsti M. Jylhä, & Panu Pihkala. 2022. "Climate Anxiety: Conceptual Considerations, and Connections with Climate Hope and Action." *Global Environmental Change* 76 (September): 102569. doi:10.1016/j.gloenvcha.2022.102569.

Sanson, Ann V., Susie Burke, & Judith Van Hoorn. 2018. "Climate Change: Implications for Parents and Parenting." *Parenting* 18 (3): 200–217. doi:10.1080/15295192.2018.1465307.

Sapiains, Rodolfo, Robert J. S. Beeton, & Iain A. Walker. 2015. "The Dissociative Experience: Mediating the Tension Between People's Awareness of Environmental Problems and Their Inadequate Behavioral Responses." *Ecopsychology* 7 (1): 38–47.

Schneider-Mayerson, Matthew, & Kit Ling Leong. 2020. "Eco-Reproductive Concerns in the Age of Climate Change." *Climatic Change* 163 (2). doi:10.1007/s10584-020-02923-y.

Schultz, Cynthia L., & Darcy L. Harris. 2011. "Giving Voice to Nonfinite Loss and Grief in Bereavement." In *Grief and Bereavement in Contemporary Society: Bridging Research and Practice*, edited by Robert A.Neimeyer, Darcy L.Harris, Howard R.Winokuer, and Gordon Thornton, 235–245. New York: Routledge.

Scranton, Roy. 2015. *Learning to Die in the Anthropocene: Reflections on the End of a Civilization*. San Francisco: City Lights Publishers.

Silva, Jules F. B., & Jennifer Coburn. 2022. "Therapists' Experience of Climate Change: A Dialectic between Personal and Professional." *Counselling and Psychotherapy Research*. doi:10.1002/capr.12515.

Smith, Lauren K. M., Hanna C. Ross, Stephanie A. Shouldice, & Sarah Elizabeth Wolfe. 2022. "Mortality Management and Climate Action: A Review and Reference for Using Terror Management Theory Methods in Interdisciplinary Environmental Research." *WIREs Climate Change* 13 (4): e776. doi:10.1002/wcc.776.

Smith, Tom W. 1988. "A Report: Nuclear Anxiety." *The Public Opinion Quarterly* 52 (4): 557–575.

Taku, Kanako, 2021. "Posttraumatic Growth (PTG) and Posttraumatic Depreciation (PTD) across Ten Countries: Global Validation of the PTG-PTD Theoretical Model." *Personality and Individual Differences* 169: 110222. doi:10.1016/j.paid.2020.110222.

Tedeschi, Richard G., Jane Shakespeare-Finch, Taku Kanako, & Lawrence G. Calhoun. 2018. *Posttraumatic Growth: Theory, Research, and Applications*. New York: Routledge.

Tschakert, Petra, Neville R. Ellis, C. Anderson, A. Kelly, & J. Obeng. 2019. "One Thousand Ways to Experience Loss: A Systematic Analysis of Climate-Related Intangible Harm from around the World." *Global Environmental Change* 55: 58–72. doi:10.1016/j.gloenvcha.2018.11.006.

Vaillant, George E. 2003. *Aging Well: Guideposts to a Happier Life*. New York: Warner.

Vergunst, Francis, & Helen L. Berry. 2021. "Climate Change and Children's Mental Health: A Developmental Perspective." *Clinical Psychological Science*, September. doi:10.1177/21677026211040787.

Verlie, Blanche, Emily Clark, Tamara Jarrett, & Emma Supriyono. 2020. "Educators' Experiences and Strategies for Responding to Ecological Distress." *Australian Journal of Environmental Education* 37 (2): 132–146. doi:10.1017/aee.2020.34.

Walter, Carolyn Ambler, & Judith L. M. McCoyd. 2016. *Grief and Loss across the Lifespan: A Biopsychosocial Perspective*, 2nd edition. New York: Springer.

Wardell, Susan. 2020. "Naming and Framing Ecological Distress." *Medicine Anthropology Theory* 7 (2): 187–201. doi:10/17157/mat.7.2.769.

Weber, Jack Adam. 2020. *Climate Cure: Heal Yourself to Heal the Planet*. Woodbury: Llewellyn Publications.

Weintrobe, Sally, ed. 2013. *Engaging with Climate Change: Psychoanalytic and Interdisciplinary Perspectives*. London: Routledge.

Weintrobe, Sally, ed. 2021. *Psychological Roots of the Climate Crisis: Neoliberal Exceptionalism and the Culture of Uncare*. New York: Bloomsbury.

Worden, J. William. 2018. *Grief Counseling and Grief Therapy: A Handbook for the Mental Health Practitioner*. 5th ed. New York: Springer.

Wray, Britt. 2022. *Generation Dread: Finding Purpose in an Age of Climate Crisis*. Toronto: Alfred A. Knopf.

Zaremba, D., M. Kulesza, A. M. Herman, M. Marczak, B. Kossowski, M. Budziszewska, J. M. Michałowski, C. A. Klöckner, A. Marchewka, & M. Wierzba. 2022. "A Wise Person Plants a Tree a Day before the End of the World: Coping with the Emotional Experience of Climate Change in Poland." *Current Psychology*, October. doi:10.1007/s12144-022-03807-3.

Dwelling and the Climate Crisis

A Developmental Perspective and Its Implications

Ryan LaMothe

During the last 10 years the discourse around climate change has shifted to climate crisis/emergency, which reflects the growing concern of scientists and others about the present and looming disasters human and other animals face, including a mass extinction event.[1] The shift in discourse accompanies a mounting literature on eco-emotions. Scholar Panu Pihkala has researched and written extensively about various eco-emotions, such as eco-anxiety, anger, grief, fear, etc.[2] More particularly, Pihkala points out that "a recent global survey about climate change among 10,000 children and youth in 10 countries revealed that 56% of them thought that 'humanity is doomed,' while 75% felt the climate future to be frightening; 42% reported having felt at least some hesitation in having children because of the climate crisis."[3] This research clearly points to a wide range of negative emotions and cognitions. Tarra Léger-Goodes et al. have researched the negative effects of eco-anxiety on children, not the least of which is a loss of hope.[4] Inmaculada Boluda-Verdú et al. explored the negative psychological and physical health effects of eco-anxiety and its relation to eco-depression, which are connected to feelings of helplessness and powerlessness.

Eco-emotions signify not simply the reality of the climate emergency and the sixth extinction event but also an existential crisis of dwelling for human beings, as well as billions of other living beings.[5] Anthropologist Eduardo Kohn, like philosopher Charles Sanders Peirce a century ago,[6] rightly contends that "All life is semiotic and all semiosis is alive."[7] All semiosis depends on a viable habitat or biodiverse Earth. A biodiverse Earth, Terry Eagleton writes, "is the first condition of our (and all other living beings) existence."[8] Moreover, all life, all semiosis is aimed at survival and well-being, which are central features of experiences of dwelling. Semiosis, life, and dwelling are infinitely manifold in appearances. The negative side of dwelling is the trauma of being unhoused, which is evident in the death or diminishment of living of a single living being,[9] as well as the extinction of an entire species. Both death and extinction represent the abyss of existential insignificance – cessation of life and semiosis. If all life is semiotic and all semiosis is alive, then all living beings dwell on this one habitat and face together the precarity of existential

DOI: 10.4324/9781003520757-21

insignificance. Indeed, semiosis takes place against its dialectical partner existential insignificance.[10] Eco-emotions, in short, are the signal affects of the semiotic crisis of being unhoused.

Let me add briefly for now that eco-emotions, as a semiotic crisis of being unhoused, can also be understood as signal affects that point to the reality of Western subjects' disconnection or alienation from other living beings and the Earth. Put another way, the sense of homelessness addressed by Western philosophers and theologians is a symptom of the ontological rift between human beings (and othered human beings) and other species, which accompanies a sense of not being at home in this world upon which all dwelling depends. This rift or disconnection vis-à-vis other species has been noted in varied ways by Sigmund Freud,[11] Giorgio Agamben,[12] Jacques Derrida,[13] Bruno Latour,[14] Isabelle Stenger,[15] and others.[16] For instance, Agamben writes:

> It is as if determining the border between human and animal were not just one question among many discussed by philosophers and theologians, scientists and politicians, but rather a fundamental metaphysico-political operation in which alone something like "man" can be decided upon and produced. If animal life and human life could be superimposed perfectly, then neither man nor animal – and, perhaps, not even the divine – would any longer be thinkable.[17]

Agamben writes that a burning house, which is the climate crisis and Western human apparatuses of dwelling, reveals "the fundamental architectural problem [that] becomes visible for the first time."[18] This epistemological flaw – founded on the illusions of human superiority, exceptionalism, and dominion over other species – is the ontological rift evident in Western philosophies, theologies, and sciences. In relation to psychoanalysis, I suggest that Harold Searles decades ago recognized the consequences of this rift when he observed that psychoanalytic developmental theories overlook other species.[19] More recently, psychoanalyst Susan Kassouf returned to this issue, indicating that what Searles identified continues to be true today.[20]

I add that what we have in common with all living beings is the motivation to dwell in the world and that to dwell entails survival and flourishing, which depend on a biodiverse Earth. Dwelling, then, is dependent on semiosis and all semiosis is aimed at survival and flourishing. These semiotic experiences of dwelling are in dialectical relation to the reality of the existential insignificance – insignificance as the cessation of semiosis – and therefore dwelling. "Death" or extinction is a sign of the eclipse of dwelling and all signifying operations. This precarity is shared by all living beings.

If eco-anxiety is a semiotic crisis of dwelling, as well as an indication of a rift between Western human beings and other species, then how might this alter how we understand psychosocial development vis-à-vis dwelling? This chapter offers a perspective of psychosocial development in light of the notions of dwelling and semiosis against the backdrop of the climate crisis and the existential threat of being unhoused. I begin by briefly explicating the notion of dwelling as it relates to semiosis. This sets the stage for outlining a developmental perspective that aims to overcome the architectural rift. I conclude by identifying several implications of this perspective, which are not a panacea for responding to eco-emotions.

Dwelling

All life dwells on and in the Earth. In general, dwelling is a verb that points to relational experiences of being at home in a place and a noun (e.g., home, house, residence, abode, habitat). This affective-relational dimension of dwelling is tied to experiences of belonging, whether that is to a place, other similar beings, and other inhabitants. In the wider ecological context, there are two features of dwelling. The first relates to Aristotle's notions of potentiality and actuality, which we need not explicate in depth. For my purposes, a seed, a sperm uniting with an egg, represent the potential for dwelling and that actualizing this potential is contingent upon a number of factors depending on the particular existent being, which I will delve into below regarding infants.[21] A second feature of dwelling concerns the principles of multinaturalism and perspectivism, which, according to anthropologist Eduardo Viveiros de Castro, means that "the world [is] composed of a multiplicity of points of view. Every existent is a center of intentionality apprehending other existents according to their respective characteristics and powers."[22] He states further that "Multiplicity can be taken as a kind of plurality [and] Amazonian multinaturalism affirms not so much a variety of natures as the naturalness of variation – variation as nature."[23] There are, in short, manifold ways to dwell in the world, though I will narrow the focus to human beings (and human dwelling is manifold in itself).

There is a long history in Western philosophies and theologies regarding what it means to dwell, which I will only touch on for the sake of identifying other key features of Western dwelling and its relation to the ontological rift. Hegel argued that Socrates, while ensconced in the Athenian polis, gave birth to the idea of the homeless spirit that is liberated (independence of an individual's thought) through his method of questioning.[24] Maybe Socrates, after being tried and sentenced for corrupting the youth, exemplified a liberated spirit by choosing to be unhoused (executed) at home. Over a century after Hegel, two prominent philosophers, Martin Heidegger and Emmanuel Levinas, agreed about existential homelessness and our desire to be at home, but disagreed on its causes and implications. Heidegger, David Gauthier argues, "interprets homelessness as a symptom of the abandonment [or forgetfulness] of Being by beings."[25] Yet, Heidegger focused on place, identity, and tradition to mitigate estrangement and to experience being at home. Heidegger believed, Gauthier argues, that the challenge "is to create a philosophy that will facilitate a return to rootedness to help man [*sic*] to become at home in the world."[26] "I know that everything essential and everything great," Heidegger writes, "originated from the fact that man had a home and was rooted in tradition."[27] Critics of Heidegger argue that his obdurate focus on place, identity, and tradition vis-à-vis existential homelessness fit well with the Nazis' racism, antisemitism, and German exceptionalism, which are examples of an ontological rift in the dwelling between Nazis and Jewish persons.

Emmanuel Levinas's philosophy is, in one sense, a critical response to Heidegger. Levinas contends that Heidegger "effectively subordinates ethics to ontology,"[28] and it is ontology that tends to eliminate "the alterity, or otherness, of the comprehended being."[29] More pointedly, Gauthier states that, in Levinas's view, "the Heideggerian project is ethically problematic because it is oblivious to the needs of strangers."[30] Levinas believed that the presence of Others and *our refusal to be obliged to the Other* is the reason why human beings experience homelessness – rather than the forgetfulness of Being.[31] This source and sense of homelessness "provides the impetus for human fraternity."[32] Gauthier notes that for Levinas "the home achieves its full dignity when the Other is welcomed."[33] In his book, *Totality and Infinity*, Levinas states, "To dwell is not the simple fact of the anonymous reality of a being cast into existence, as a stone one casts behind oneself; it is a recollection, a coming to oneself, a retreat home with oneself as in a land of refuge, which answers to a hospitality, an expectancy, a human welcome."[34] Here is the apparent paradox of Levinas' view: the Other evokes a sense of homelessness and this homelessness is mitigated by the infinite obligation to recognize and welcome Others in their singularity. The Other's alterity is their singularity.

Levinas is particularly helpful here because of his relevance for the discussion below. Levinas, in my view, was addressing the issue of alterity and its relation to experiences of homelessness and being at home. Welcoming the alterity of the Other is essential for human belonging and dwelling and this falls within the principles of multinaturalism and perspectivism, more on this later. Yet, there are three aspects of the idea of existential homelessness to tease out. First, the idea presupposes one either had an original home and has been dispossessed (e.g., garden of Eden, the womb, being thrown into this world) and/or one is created with a telos of eventually finding a home (e.g., kingdom of God). Certainly, human beings can experience a sense of homelessness, existentially or as a matter of fact (fleeing one's homeland), but it is not clear to me that homelessness is a universal, existential feature of being human. I have read hundreds of Indigenous stories and I do not come away with the idea that existential homelessness is an anthropological feature of their philosophies. But what these stories share with Socrates, Heidegger, and Levinas is that human beings dwell in the world and dwelling entails a shared struggle to survive and thrive. Second, for Levinas, human Others depend on the recognition of their singularity for experiences of being at home with themselves and others. Failure to recognize and respect the singularity of Others gives rise to an experience of homelessness, which can be seen as traumas of being unhoused.[35] We do not have to presuppose existential homelessness, then, to hold to the idea of the necessity of recognizing the singularities of others so that they can experience being at home.

Third and relatedly, the homelessness Hegel noted in Socrates and the existential homelessness of Heidegger (and, in part, Levinas because of his leaving out the singularities of other-than-human species) I contend are *symptoms* of the ontological rift. As noted above, the ontological rift is produced by Western semiotic apparatuses that construct human beings as superior, the center or apex of life (exceptionalism, anthropocentrism, exemptionalism), possessing power over other species (dominion or sovereignty), while other species and the Earth are inferior, subordinate, possessing mere use value – their alterity is believed to lack singularity. Put another way, Western political philosophies and theologies, which are semiotic fortresses, do not include other species and the Earth, except instrumentally. Other species, believed to lack singularity, do not belong in the polis, except when they are constructed in terms of their use-value. Western apparatuses of the rift, then, produce ways of dwelling in the world that accompany a sense of *not being* at home in and with the world and its inhabitants.

To carry this further, the term "symptom," in Freud's view, involves displacement, wherein the symptom resembles, though is different from, what it symbolizes.[36] The aims of symptom formation are gratification and repression.[37] What is gratified vis-à-vis the ontological rift is a sense of esteem (dependent on inferiority of species) and control (e.g., power over, dominion). What is repressed or displaced, in my view, is the reality of existential insignificance. Heidegger's and Levinas homelessness partially screens the reality of existential insignificance, in part, by reassuring us of the possibility of being at home, whether that was in the past prior to being thrown into existence (apparently from somewhere) or some future experience of being at home.

While I question the reality of existential human homelessness vis-à-vis all human beings, I contend that the homelessness addressed by Western philosophers and theologians are symptoms of the ontological rift between human beings and other species, which accompanies a sense of not being at home in this world upon which all dwelling depends.[38] I add that what we have in common with all living beings is the motivation to dwell in a precarious world and to dwell entails survival and flourishing, which depend on a biodiverse Earth. Stated differently, for all living beings dwelling is dependent on semiosis and all semiosis is aimed at survival and flourishing. These semiotic experiences of dwelling are in dialectical relation to the reality of the existential insignificance – insignificance as the cessation of semiosis – and therefore dwelling. "Death" or extinction is a sign of the eclipse of dwelling and all signifying operations. There is, then, no universal existential homelessness, but there are semiotic experiences of homelessness (e.g., evictions, traumas) or the existential threat of the eclipse of dwelling, namely death or extinction (e.g., climate crisis[39]). To gain a clearer picture of dwelling vis-à-vis human beings I turn to a psychosocial developmental perspective.

Dwelling: Developmental Dimensions

In this section, I briefly depict what dwelling means vis-à-vis psychosocial development. I believe Levinas's view of our dependence on the hospitality of others for our experience of being at home in the world can be further elaborated from a developmental perspective – a perspective that renders the rift inoperative and does not posit universal existential homelessness. Moreover, the key insight that an individual's alterity is their singularity can be applied to other-than-human beings, though this insight is absent whenever the apparatuses of the ontological rift are present. I begin the discussion with parents, since their care is central to infants' and children's experiencing a sense of being at home in the material world and being at home with other human beings and other-than-human beings. I then move to describe the outlines of infants' and children's experience of dwelling and trauma as experiences of homelessness.

The foundation of the reliable care (e.g., attunements, repair) of good-enough parents is the recognition of their infants as persons – unique, valued (significant), inviolable, and responsive subjects. Philosopher John Macmurray argued that personal recognition or knowing is a matter of intention and a matter of fact.[40] For parents caring for their infants is a matter of intention and infants *qua* persons believed to be a matter of fact. In Levinas's terms, parents in recognizing and welcoming the alterity of infants includes acknowledging and accepting their singularity, which is foundational to their experiences of dwelling. This acknowledgement of infants' singularities (for Levinas, the face of the other) undergirds good-enough parents' sense of the categorical demand to care,[41] which also takes place against the precarity of life – the reality of existential insignificance. Good-enough parents, in other words, know, consciously or unconsciously, that their infants are, because they are altricial, vulnerable and dependent, which gives rise to the categorical demand to care.

Donald Winnicott recognized that in order for parents to handle and hold their infants, they must be recognized as persons and cared for and about by extended family, community, and society.[42] Another way of saying this is that parents' experiences of being at home in the world depend, to a significant degree, on relations and social-political institutions that acknowledge their singularities, which, in turn, means providing psychosocial and material support necessary for their survival and well-being – dwelling. Negative examples, wherein dwelling is undermined, are racism, sexism, classism, and other systemic forms of oppression, which all signify the production of an ontological rift between one group of human beings and another. Apparatuses of racism, for instance, produce forms of depersonalization that undermine the material, political, economic, and social resources needed to flourish and, in the case of parents, this dearth makes it more difficult to care for their children. Put another way, parents whose experiences of dwelling in society are undermined face obstacles in providing their children with experiences of being at home in the world.

Before turning to infants' experiences of dwelling, there is one other point to make regarding parents' good enough care. While the larger social, political, and economic apparatuses shape or impact parental care, they do not determine care. Indeed, I contend that parents' earliest caring behaviors render inoperative apparatuses of the ontological rift (e.g., sovereignty/dominion, superiority, etc.). While there may be some parents (Laius comes to mind) who are imperious, I suspect that good-enough parents do not believe they are superior to and have dominion over their infants. To be sure parents have power and are obliged to provide care, but this power of care is *anarchic* in the sense it is not tied to apparatuses of sovereignty or dominion over the child. Parents' anarchic care, in brief, represents being-for-infants,[43] which means creating a space and relation for infants to appear in their singularity and in appearing dwell.

To shift to infants' experiences of being at home in the world, researchers have shown that a baby develops preferences prior to birth.[44] This suggests psychic-semiotic (pre-representational) organizations of experience that exist prior to birth and the womb is the material condition for the actualization of these earliest organizations of embodied experiences of dwelling. I add here that these embodied-semiotic organizations represent a nascent agency. Decades ago, Winnicott surmised that "birth can easily be felt by the infant" and they participate through "personal effort."[45] Here we see the notion that the infant's pre-representational "belief" that they participates in the birth, implying the presence of agency (an aspect of the ego). In short, the organization of embodied, pre-representational experiences of dwelling points to a nascent bodily agency – agency necessary for dwelling. As Matt Waggoner notes in his discussion of Theodor Adorno's philosophy, "The human experience of dwelling begins with embodiment, with the fact that [nascent] consciousness is inseparable from the somatic and sensorial."[46]

There is another way to depict these early pre-representational organizations of experience, which I develop further below. Eduardo Kohn writes that all life is semiotic and I would add that all living semiotic beings possess a sense of being for-themselves and being-in-themselves. With regard to pre-birth infants, these early pre-representational organizations of preferences reflect these sense of being for-itself and in-itself. Instead of positing narcissism as an early state or stage of development, I opt for these two embodied pre-representational senses of dwelling. While admittedly abstract, these senses take on particular hues once infants are born and continue to develop.

I add here that being-for-itself and being-in-itself are part of infants' drives to survive and thrive – to dwell. Put differently, pre-birth semiotic constructions of preferences vis-à-vis the parent, in other words, can be understood from an evolutionary perspective. Infants' semiotic pre-representational organizations of preferences are inextricably tied to the drives to survive and flourish. The womb is the infant's first home, providing the material conditions for infants to actualize not just physical potentialities, but also their early capacities for semiosis – capacities necessary for survival. Dwelling in this first home is embodied, pre-representational/semiotic organizations of being-for-itself and being-in-itself.

Of course it is birth that disrupts these earliest embodied, presymbolic organizations of experience of dwelling. Initially ensconced in the warmth and dampened quiet of the womb, the newborn is thrust into a sensorial complex world – "one great blooming buzzing confusion."[47] In birth, infants are, if you will, unhoused,[48] which means that there is now a gap between need and the meeting of that need. This gap in time gives rise to emerging consciousness and the semiotic impulse to communicate. John Macmurray writes that the infant "is, in fact, 'adapted,' so to speak paradoxically, to being unadapted, 'adapted to complete dependence' … He can only live through other people."[49] In infants' unadapted, dependent-vulnerable state, they possess an impulse or motivation to communicate – "the impulse to communicate is [their] sole adaptation to the world."[50] It is important to point out that pre-birth semiotic organizations of preferences for the mother's voice are inextricably part of the impulse to communicate. In my view, this is an evolutionary necessity in that infants, to survive, must seek out and connect with the source of sustenance – a source that the infant is dependent on for survival, for being able to dwell in a great blooming, buzzing world. This impulse to communicate is met by parents' personalizing care (attunement, mirroring) that recognizes and affirms infants' singularities. These good-enough parent-infant cooperative interactions, I suggest, contribute to emerging semiotic, embodied organizations of self-esteem, self-confidence, and self-respect that are integral for infants developing agencies and experiences of dwelling in this new complex world. In other words, parents' reliable personalizing care facilitates the actualization of infants' experiences of being-for-itself and being-in-itself – senses of relational-embodied self-esteem, self-respect, and self-confidence, which are foundational for emerging agency.

Winnicott's notion of transitional objects/phenomena can help in further depicting experiences of dwelling. To understand the child's transition from relatively undifferentiated, embodied relation to engaging and using objects, Winnicott posits that a transitional object (TO) in early childhood is the first not-me possession, which entails the infant's rudimentary ability to recognize an object as independent of the infant (e.g., an external object).[51] In the act of possessing and using this object, the infant, Winnicott claims, both retains and partially hands over their belief in and experience of omnipotence.[52]

There is, for Winnicott, a paradoxical interplay of the internal and external reality. This primary[53] TO is "not an internal object – it is a possession. Yet, it is not (for the infant) an external object either."[54] Strictly speaking, the TO, for Winnicott, is neither an internal object nor an external object, and yet it is both. Another related feature of TOs is that they must not be changed unless by the infant, because to do so would challenge the infant's sense of and belief in omnipotent control and, consequently, disrupt his experience of continuity or sense of going on being[55] – dwelling as embodied-being-with-Other. Similarly, the parent is not to challenge the child's omnipotent selection and use of the primary transitional object, because this too would challenge the child's *presymbolic* belief in and experience of omnipotence, which Winnicott contends is needed for a child to have confidence in organizing experience.

Since the primary TO is not simply identified with either the external or internal world, a question is raised regarding the child's selection of the object and what the TO represents to the child. The object, which may be presented to the infant, is chosen by the infant and not shared.[56] This choice of a *primary* TO is in "accordance with its consistency, texture, size, volume, shape, and odor."[57] which is psychologically joined to the "technique of mothering"[58] – the caregiver's handling, holding, comforting, and consoling of the infant – anarchic care.[59] That is, the child unconsciously chooses a TO that represents the parent's anarchic care for the child, which permits the child to manage (regulate emotions) during times of separation from the parent.[60]

All of this can be understood in terms of the experience of dwelling. This first object under the child's omnipotent control (belief) provides embodied soothing, rest, and continuity – experiences of being-at-home-with-an-object – an object that is me and not-me. In other words, *being-for-itself and being-in-itself are joined to being-with-an-other*. This "other" is not recognized as a person, but for the infant vis-à-vis omnipotence the object is living. There are four important points here. First, this is an early form of knowing and relating to an object – believed to be animate. Second and relatedly, this early epistemology, if you will, will continue to develop, becoming more sophisticated or complex, but nevertheless present in later forms of knowing.[61] More precisely, this early method of animate knowing is foundational for engaging in and dwelling with others. Third, there is no ontological rift manifested in this form of knowing and relating. Fourth, it is important to explain that this early embodied experience is not to be understood as if one is housed alone in one's body (in utero being-for-itself and in-itself). In discussing Adorno's work on dwelling, Matt Waggoner writes that "No one ever singularly inhabits a body because embodiment is not…a state of being, a self-sufficient thing. Its existence is inseparable from and can only be constituted as such within a matrix of contact and connectivity."[62] So, from the very first, embodied dwelling is relational – embodied dwelling is belonging. Moreover, he adds "The [nascent ego] is housed by extended realities that are material and social in nature."[63]

As infants develop, more complex symbolic capacities emerge as do relations with other animate and inanimate objects. I differentiate between primary and secondary transitional objects because there are significant differences between the proverbial blanket and the child's use of cultural objects as transitional objects. Perhaps one could say that the question and task with regard to the secondary transitional objects is how to dwell with others *qua* persons. A playful illustration of this is the comic strip *Calvin and Hobbes*. Calvin is a little boy who plays with his stuffed tiger, Hobbes – a cultural object. Together they create and inhabit a world together. Hobbes and Calvin embark on all kinds of adventures, they argue and make up, and they comfort each other when hurt or distressed. The parents are present, but in the background. To Calvin, Hobbes is alive, and not just alive but a tiger-person who, Calvin believes, recognizes and treats Calvin as a human person. In Calvin's imagination, there is, then, a sense of mutual self-esteem, self-confidence, and self-respect that is derived from speaking and acting together, which includes shared repairs of conflict. While it is in Calvin's imagination, they together, in speaking and acting together, create a pre-political space of appearances wherein they reside in cooperative, reciprocal relations.[64] As a secondary transitional object, Hobbes serves as a process toward Calvin's learning to dwell-with-others, whether these others are human persons or other-than-human persons. In my view, secondary transitional objects do not represent the challenge between inner and external reality, but rather a life-long relational challenge of dwelling, namely, how to navigate being-in-itself and for-itself with being-for-others. Calvin, for instance, has symbolically organized experiences of being-for-himself and in-himself in relation to Hobbes who is believed to have similar experiences. The challenge is how these two manage to mutually care (being-for-others) such that they can care for each other while dwelling together.

While this is clearly play, it nevertheless, in my view, represents a key epistemological factor with regard to dwelling. With regard to the secondary transitional object there is no ontological rift. Hobbes is a tiger-person and treated as a person who dwells with Calvin. There are no beliefs in superiority and inferiority between them. Dominion does not exist in their relationship or in his animistic knowing, which now entails personal recognition. In adult life we see glimpses of this form of knowing and dwelling. Many people who live with their dogs and cats recognize and treat them as persons. That is, they recognize that these animals have personalities and are unique. They do not confuse cats for human persons, but recognize their cats as cat-persons and, in so doing, share in their dwelling together. Similarly, many indigenous peoples recognize other species as persons. A crow is a crow person. A bear is a bear person. I wish to stress that personal epistemology is not simply playful or imaginary. It represents an evolutionary epistemological method, which is evident in care and dwelling with other species. Put another way, this animistic personalizing epistemology represents a method of learning about the singularities of others that inhabit the world with us and that this epistemology is necessary for survival and flourishing – not just of human beings, but of other species that dwell with human beings in this particular habitat.

Someone may object, pointing out that children's playful imagination is radically different from adults' more sophisticated rational approaches to dwelling. Granted, children continue to develop and learn more complex ways of knowing and engaging the world. However, the objection also reveals the presence of the ontological rift. What happens to Hobbes, according to Winnicottian thinking, is that he is not mourned, having lost significance as Calvin learns to socialize and play with others. Yet, I suggest that in the West, children begin to internalize the apparatuses of the ontological rift, which includes beliefs in the superiority of reason/science, dominion/mastery, and putative inferiority of animistic epistemologies. Put another way, animistic thinking is infantilized and seen as something to overcome. Of course, Freud believed this early form of knowing is always present in adult life/dwelling, but when operative it is pathologized. Yet, consider the popularity among adults of the comic strip *Calvin and Hobbes*, or the *Toy Story* films, or love of pets. One may argue that this is simply a nostalgic form of adult play, which may be true, but it is not the whole story.

Let me take a short detour before returning to psychosocial development and dwelling. In recent decades, anthropologists have sought to decolonize Western anthropologies with the aim of better understanding Indigenous peoples. Studying various Indigenous peoples in Central and South America, Eduardo Kohn,[65] and Eduardo Viveiros de Castro[66] note that these peoples view other species as persons or potential persons and that this personification, if you will, is, for them, necessary for gaining knowledge of the other creature, the habitat, and themselves. This does not imply that Indigenous people believe other species *qua* persons means other animals are identical to human beings/persons. As stated above, there are crow persons, bear persons, lion persons, etc. In this epistemology there is multiplicity of perspectives and multinaturalism.[67] Animistic personification, then, means accepting and respecting the singularities of other species. That is, Indigenous peoples come to know them in their singularities by way of animistic personification. Moreover, this form of knowing is, for Indigenous peoples, necessary for living/dwelling in and with nature – not trying to control and dominate "Nature," but rather learning to adapt to and cooperate with nature, in which they see themselves as intrinsically a part of. Other species, as persons or potential persons, means that they are considered inextricably part of a common world of belonging. The fact that Indigenous peoples have lived/dwelled sustainably for millennia provides evidence of the practicality and effectiveness of their animistic epistemologies that personify other species.

Critics may point out that this form of knowing is not scientific and they would be right if "science" is Baconian–Cartesian – objectifying and seeking dominion over nature and other species. Nevertheless, like all scientists,

Indigenous peoples observe, assess, and analyze in relation to their environment and its varied inhabitants. Where they may differ from Baconian-Cartesian scientists is their aims. In other words, the aims of their personalizing, animistic epistemologies are survival and flourishing, which necessarily includes adaptation to and cooperation with other beings that dwell with them in the habitat upon which their lives depend. Let me stress here that not all Western science is captive to the ontological rift. There are scientists who recognize and respect the singularities of the species they are investigating.[68] They may not construct them as persons, but they treat them as fellow creatures, recognizing and respecting their singularities, eschewing the Baconian–Cartesian scientific approach of objectification and domination.

To return psychosocial-development and Calvin's dwelling. Is it possible to imagine children's development that does not lead to the infantilization and repression of personalizing-animistic thought and, correspondingly, the repression of the principles of multinaturalism and perspectivism? Are children's subjectivities in the West determined by the apparatuses of the rift – apparatuses that have clearly undermined the dwelling of human beings and millions of other species? Can we find ways to extend anarchic care to other species? In other words, can we, as adults, render inoperative the apparatuses of the rift so that we might recognize the singularities of other species, as well as recognize that all species, sharing the precarity of existential insignificance, seek to survive and thrive – dwell – in and on the Earth? We need to recognize that there are numerous examples where adults are not captive to the rift, whether these are scientists or lay persons. Authors of children's literature use personalizing-animistic thought to convey not simply morals or values, but ways of dwelling with other species. Some parents encourage their children to imagine the singular life of another animal or even of trees, as they visit zoos, go on hikes, etc. Of course, this is over and against the massive systemic apparatuses (e.g., Baconian-Cartesian science, capitalism, petrocultures, imperialism) that produce relations of domination and depersonalize not only other species but many human beings. Nevertheless, there are possibilities of human dwelling where personalizing epistemologies include other species and habitats. In part, these possibilities depend on psychosocial development that renders the rift inoperative, which creates spaces for children and adults to dwell with other species on this one Earth we all share.

Conclusion

Above, I claimed that the climate crisis is a semiotic crisis of dwelling. This crisis is largely due to Western ways of dwelling in the world that are linked to apparatuses that produce an ontological rift between human beings and other species. An implication is that the crisis invites, perhaps even demands, that we critically examine our theories of psychosocial development, which are theories of dwelling in the world. As Clayton Crockett argues, "We need to experiment radically with new ways of thinking and living, because the current [Western] paradigm is in a state of exhaustion, depletion, and death."[69] This chapter is a small step toward reimaging psychosocial development that does not replicate the rift and, in so doing, move us closer to forms of dwelling characterized by cooperation, reciprocity, and adaptation with other species.

Notes

1 Klein, *This Changes Everything*. Kolbert, *The Sixth Extinction*. Wilson, *The Future of Life*.
2 Pihkala, "Environmental Education after Sustainability." Pihkala, "Eco-anxiety, Tragedy, and Hope." Pihkala, "Anxiety and the Ecological Crisis." Pihkala, "The Process of Eco-anxiety and Ecological Grief." Pihkala, "Toward a Taxonomy of Climate Emotions."
3 Pihkala, "Eco-anxiety and Pastoral Care," p.1.
4 Léger-Goodes, "Eco-anxiety in Children."
5 Boluda-Verdú et al. "Fear for the Future: Eco-anxiety and Health Implications, a Systematic Review."
6 Hoopes, *Peirce on Signs*.
7 Kohn, *How Forests Think*, p.117.
8 Eagleton, *Why Marx Was Right*, p.228.
9 LaMothe, "The Silence of Othered Species."
10 Semiosis is more comprehensive than language or linguistics, which was a focus of Ferdinand de Saussure. All of this is quite complex, which is not necessary or possible to delve into in this paper. For my purposes semiosis includes all types of signifying operations, whether that be signs (e.g., bee dance), symbols, language, ritual, narrative, etc. These signifying operations, according to Charles Morris, entail four uses, namely, informative, valuative, incitive, and systemic. Morris later adds that signs' designative dimension…embody man's [sic] knowledge of the world and of himself; in their appraisive dimension they reflect the conceived values which serve as man's goals; in their prescriptive dimension they direct the specific course of human action toward envisaged goals." Semiosis, then, comprise knowledge, meanings, and valuations. A key feature of signifying or semiotic operations is valuation, which is implicit in the term "significance." Insignificance can mean an object is without value. Existential insignificance, however, does not indicate an object or living being has no significance. It means, rather, that (1) the cosmos or universe is not a semiotic entity assigning significance and (2) death is the absence of semiosis. Morris, *Signification and Significance*, pp.15, 81.
11 Freud, "Why War?," p.204.
12 Agamben, *The Open: Man, and Animal*.
13 Derrida, *The Animal That Therefore I am*.
14 Latour, *We Were Never Modern*.
15 Stengers, *Making Sense in Common*.
16 Roland Faber, exploring the work of Alfred North Whitehead, locates the philo-sophical source of the rift in Plato's philosophy. He writes that "the platonic divide becomes unmasked as an abstraction from the process of experience and not its condition." Whitehead also said that all Western philosophy is but a footnote to Plato's work, which suggests that this abstraction, this platonic divide is evident in Western philosophies and attending social imaginaries like the human sciences. Faber, *The Mind of Whitehead*, p.18.
17 Agamben, *The Open: Man, and Animal*, p.92.

18 Agamben, *Potentialities: Collected Essays in Philosophy*, p.118.
19 Searles, *The Nonhuman Environment in Normal Development and Schizophrenia.*
20 Kassouf, "Psychoanalysis and Climate Change: Revisiting Searles' the Nonhuman Environment."
21 For philosopher Giorgio Agamben, the Western philosophical tradition has largely "subordinated potentiality to actuality: so, we begin with the actual, speaking humans and their political and artistic productions, and we see potentiality at present as a capacity or skill that is defined by the final action. We see potentiality as secondary or accidental." This is derived, in part, from Aristotle's notion that actuality is prior to potentiality, though for Agamben "the very essence of humanity lies in a potentiality that is expressed when it does not unfold into actuality." Colebrook and Maxwell, *Agamben*, pp.188, 289.
22 Viveiros de Castro, *Cannibal Metaphysics*, p.55.
23 Viveiros de Castro, *Cannibal Metaphysics*, p.74.
24 Gauthier, *Martin Heidegger, Emmanuel Levinas, and the Politics of Dwelling*, pp.3–5.
25 Gauthier, *Martin Heidegger, Emmanuel Levinas, and the Politics of Dwelling*, p.129.
26 Gauthier, *Martin Heidegger, Emmanuel Levinas, and the Politics of Dwelling*, p.9.
27 In Gauthier, *Martin Heidegger, Emmanuel Levinas, and the Politics of Dwelling*, p.92.
28 Gauthier, *Martin Heidegger, Emmanuel Levinas, and the Politics of Dwelling*, p.104.
29 Gauthier, *Martin Heidegger, Emmanuel Levinas, and the Politics of Dwelling*, p.105.
30 Gauthier, *Martin Heidegger, Emmanuel Levinas, and the Politics of Dwelling*, p.97.
31 Gauthier, *Martin Heidegger, Emmanuel Levinas, and the Politics of Dwelling*, p.113.
32 Gauthier, *Martin Heidegger, Emmanuel Levinas, and the Politics of Dwelling*, p.114.
33 Gauthier, *Martin Heidegger, Emmanuel Levinas, and the Politics of Dwelling*, p.131.
34 Levinas, *Totality and Infinity*, p.156.
35 This chapter focusses on dwelling and being unhoused in relation to human beings. Implicit in this discussion is the ontological rift between human experiences of dwelling and the dwelling of other species. Industrial slaughter factories, factory farms, experimentation on other species, etc. are some examples of human beings unhousing other species, which, as Peter Singer and Martha Nussbaum note, fall under the heading of injustice. Singer, *Liberation for Animals*. Nussbaum, *Justice for Animals*.
36 Freud, "Inhibitions, Symptoms, and Anxiety," pp.110–111.
37 Freud, "Inhibitions, Symptoms, and Anxiety," pp.94–95.
38 An example of this sense of homelessness or not-being-at-home in this world was noted by philosopher Hannah Arendt. Decades ago, when the Soviet Union launched Sputnik, Hannah Arendt was struck by a reporter's response. Instead of "pride and awe at the tremendousness of human power and mastery" there "was relief about the first 'step toward escape from men's imprisonment on the earth.'" The Earth as a prison strikes me as not simply an expression of the alienation and homelessness that are spawned by the ontological rift, it is also fantastical by which I mean that to escape the material conditions for human life would mean paradoxically creating these very same conditions in space or on other planets. It is also fantastical in the unstated belief that life on other planets will support human life and even if another planet does why would we not also experience that as a

prison. A more current sentiment is expressed in Elon Musk's and NASA's hope that human beings will become an interplanetary species. Arendt, *The Human Condition*, p.1.

39 The sixth extinction event, brought about by the climate crisis, does not mean human beings, in becoming extinct, will experience existential homelessness. It will mean, however, the Earth as the material basis for all dwelling, will not be habitable for human beings. The Earth will no longer be habitable for human beings and human beings will, like our dinosaur ancestors, sink into the abyss of insignificance.

40 Macmurray also pointed to object knowing, which is necessarily a part of personal knowing. However, to be ethical, object knowing must be subordinate to personal knowing. He uses an example of a psychologist diagnosing a patient. This diagnosis is ethical as long as diagnostic (object and instrumental) knowing is subordinate to personal knowing. Macmurray, *Persons in Relation*. See also Løgstrup, *The Ethical Demand*.

41 See Engster, *The Heart of Justice: Care Ethics and Political Theory*. Hamington, *Embodied Care*.

42 Winnicott, *Playing and Reality*.

43 Space does not allow me to write in depth about parents' anarchic care and its relation to their own development. I will say, however briefly, that parents also possess experiences of being-for-themselves and being-in-themselves. These are often secondary in moments of care – a care manifested as being-for-infants.

44 See DeCasper and Fifer, "Of Human Bonding: Newborns Prefer Their Mothers' Voices." DeCasper and Spence, "Prenatal Maternal Speech Influences Newborns' Perception of Speech Sounds." Kumin, *Pre-Object Relatedness*. Beebe and Lachmann, "Representation and Internalization in Infancy." Beebe and Lachmann, *Infant Research and Adult Treatment*. Trevarthen, "First Things First: Infants Make Good Use of the Sympathetic Rhythm of Imitation, Without Reason or Language."

45 Winnicott, "Through Paediatrics to Psychoanalysis," p.186.

46 Waggoner, *Unhoused: Adorno and the Problem of Dwelling*, p.105.

47 James, *The Principles of Psychology, vol.1*, p.488.

48 Otto Rank argued that "analysis turns out to be a belated accomplishment of the incompleted mastery of the birth trauma." Rank, I believe, indicates that the nascent, pre-birth ego is unable to "master" the experience of this new reality, hence his use of the term "trauma." I am not convinced that this first experience of being unhoused is traumatic. Yes, birth is, as William James noted, is confusing, painful, etc., but not necessarily traumatic. If we follow Rank, then it would seem that every human beings needs therapy, unless they have found other means to master the birth trauma. It is more likely that the preferences for the warmth of the mother's body and swaddling sooth infants' entry into this new world. Rank, *The Trauma of Birth*, p.5.

49 Macmurray, *Persons in Relation*, pp.8, 51.

50 Macmurray, *Persons in Relation*, p.60.

51 Winnicott, *Playing and Reality*, pp.3–4.

52 Winnicott, *Playing and Reality*, pp.9–11.

53 Winnicott did not differentiate between the transitional objects of childhood and those of adulthood, which raises all kinds of questions. The developmental achievements and complex psychological and relational realities that take place between infancy and adulthood are huge. This in itself would demand differentiation between types of objects. In this case, I associate the first transitional object as primary TO – an object that is associated with pre-symbolic modes of organizing

experience. For critiques of Winnicott's theory of development, see Applegate, "The Transitional Object Reconsidered: Some Sociocultural Variations and Their Implications." Brody, "Transitional Objects: Idealization of a Phenomenon." Flew, "Transitional Objects and Phenomena: Interpretations and Comments." Litt, "Theories of Transitional Object Attachment: An Overview."

54 Winnicott, *Playing and Reality*, p.9.

55 Winnicott, *Playing and Reality*, p.4.

56 Paul Pruyser suggests that the transitional object is shared to the extent that the object is often one that is "found" in the cultural realm of family life. Moreover, parents and siblings tolerate and accept the child's use of the object. While I agree with this view, it is clear that Winnicott does not believe that the earliest transitional objects or what I call primary transitional objects are shared or intersubjectively held, at least during infancy. Pruyser, *The Play of the Imagination: Toward a Psychoanalysis of Culture*.

57 Kestenberg and Weinstein. "Transitional Objects and Body Image Formation," p.89.

58 Winnicott did not assign care simply to the mother. A mothering environment could be accomplished by fathers as well.

59 Winnicott, *Playing and Reality*, p.11.

60 One might wonder what happens to primary (and secondary) TOs as the child grows. In the course of human development Winnicott argued that the TO "is not forgotten and it is not mourned. It loses its meaning, and this is because transitional phenomena have become diffused: they have become spread out over the whole intermediate territory between 'inner psychic reality' and 'the external world as perceived by two persons in common,' that is to say, over the whole cultural field." Winnicott, *Playing and Reality*, p.5.

61 Since Freud, the psychoanalytic tradition has tended to associate animate thinking with early childhood and so-called primitive people or savages. It is a type of thinking and relating that is to be overcome, surpassed by reason. An entire article could be taken up pointing out the varied problems with this perspective and its propagation of the ontological rift. But I will say, it overlooks the evolutionary and adaptive function of animism. Moreover, so-called primitive people who "animate" other beings and objects successfully dwell sustainably in their habitats, many for thousands of years.

62 Waggoner, *Unhoused: Adorno and the Problem of Dwelling*, p.107.

63 Waggoner, *Unhoused: Adorno and the Problem of Dwelling*, p.109.

64 Hannah Arendt, in her political philosophy, uses the terms space of appearances and speaking and acting together. In my view, the secondary TO represents a prepolitical space of appearances and this space is free of the ontological rift. Arendt, *The Human Condition*.

65 Kohn, *How Forests Think*.

66 Viveiros de Castro, *Cannibal Metaphysics*.

67 In explicating perspectivism and multinaturalism, anthropologist Eduardo Viveiros de Castro contends that "virtually all peoples of the New World share a conception of the world as composed of a multiplicity of points of view. Every existent is a center of intentionality apprehending other existents according to their respective characteristics and powers." He states further that "Multiplicity can be taken as a kind of plurality [and] Amazonian multinaturalism affirms not so much a variety of natures as the naturalness of variation – variation as nature." Viveiros de Castro, *Cannibal Metaphysics*, pp.55, 74.

68 See Goodall, *In the Shadow of Man*. Montgomery, *The Soul of an Octopus*. Schlanger, *Light Eaters*.

69 Crockett, *Radical Political Theology*, p.165.

References

Agamben, Giorgio. *Potentialities: Collected Essays in Philosophy*, translated by Daniel Heller-Roazen. Stanford: Stanford University Press, 1999.

Agamben, Giorgio. *The Open: Man, and Animal*, translated by Kevin Attell. Stanford: Stanford University Press, 2004.

Applegate, Jeffery. "The Transitional Object Reconsidered: Some Sociocultural Variations and Their Implications," *Child and Adolescent Social Work*, 6 (1) (1989): pp. 38–51.

Arendt, Hannah. *The Human Condition*. Chicago: University of Chicago Press, 1958.

Beebe, Beatrice and Frank Lachmann. *Infant Research and Adult Treatment*. Hillsdale: Analytic Press, 2002.

Beebe, Beatrice and Frank Lachmann. "Representation and Internalization in Infancy: Three Principles of Salience," *Psychoanalytic Psychology*, 11 (1994): pp.127–165.

Boluda-Verdú, Inmaculada, Marina Senent-Valero, MariolaCasas-Escolano, Alicia Matijasevich, and Maria Pastor-Valero. "Fear for the Future: Eco-anxiety and Health Implications, a Systematic Review." *Journal of Environmental Psychology*, 84 (2022): 1–17.

Brody, Sylvia. "Transitional Objects: Idealization of a Phenomenon," *Psychoanalytic Quarterly*, 49 (1980): pp.561–605.

Colebrook, Claire and Jason Maxwell. *Agamben*. New York: Polity Press, 2016.

Crockett, Clayton. *Radical Political Theology*. New York: Columbia University Press, 2012.

DeCasper, Anthony and William Fifer. "Of Human Bonding: Newborns Prefer Their Mothers' Voices," *Science*, 208 (1980): pp.1174–1176.

DeCasper, Anthony and Melanie Spence. "Prenatal Maternal Speech Influences Newborns' Perception of Speech Sounds," *Infant Behavior and Development*, 4 (1986): pp.19–36.

Derrida, Jacques. *The Animal That Therefore I am*. New York: Fordham University Press, 2008.

Eagleton, Terry. *Why Marx Was Right*. New Haven: Yale University Press, 2016.

Engster, Daniel. *The Heart of Justice: Care Ethics and Political Theory*. Oxford: Oxford University Press, 2007.

Faber, Roland. *The Mind of Whitehead*. Eugene: Pickwick, 2023.

Flew, Anthony. "Transitional Objects and Phenomena: Interpretations and Comments." In *Between Reality and Fantasy*, edited by Simon Grolnick and Leonard Barkin, pp.483–502. Northvale, NJ: Aronson, 1978.

Freud, Sigmund. "Inhibitions, Symptoms, and Anxiety," *Standard Edition*, 20 (1926): pp.87–175.

Freud, Sigmund. "Why War?" *Standard Edition*, 22 (1933): pp.195–216.

Gauthier, David. *Martin Heidegger, Emmanuel Levinas, and the Politics of Dwelling*. Lanham: Lexington Books, 2011.

Goodall, Jane. *In the Shadow of Man*. New York: Mariner Books, 2000.

Hamington, Maurice. *Embodied Care*. Urbana: University of Illinois, 2004.

Hoopes, James. *Peirce on Signs*. Durham: University of North Carolina Press, 1991.

James, William. *The Principles of Psychology, vol.1*. New York: Henry Holt, 1918/1956.

Kassouf, Susan. "Psychoanalysis and Climate Change: Revisiting Searles' the Nonhuman Environment." *American Imago*, 74 (2) (2017): pp.141–171.

Kestenberg, Judith and Joan Weinstein. "Transitional Objects and Body Image Formation." In *Between Reality and Fantasy*, edited by Simon Grolnick and Leonard Barkin, pp.75–96. Northvale: Aronson, 1978.

Klein, Naomi. *This Changes Everything: Capitalism vs. the Climate*. New York: Simon and Schuster, 2014.

Kohn, Eduardo. *How Forests Think*. Berkeley: University of California Press, 2013.

Kolbert, Elizabeth. *The Sixth Extinction: An Unnatural History*. New York: Henry Holt and Company, 2014.

Kumin, Ivri. *Pre-Object Relatedness*. London: Guilford Press, 1996.

LaMothe, Ryan. "The Silence of Othered Species: The Anthropocene Age, Trauma, and the Ontological Rift." *Pastoral Psychology* 72 (2023): 385–402.

Latour, Bruno. *We Have Never Been Modern*. Harvard University Press, 1993.

Léger-Goodes, Tarra, Catherine Malboeuf-Hurtubise, Trinity Mastine, Malissa Généreux, Pier Paradis, and Chantal Camden. "Eco-anxiety in Children: A Scoping Review of the Mental Health Impacts of the Awareness of Climate Change." *Frontiers in Psychology*, 13 (2022): 872544.

Levinas, Emmanuel. *Totality and Infinity*. Pittsburgh: Duquesne University Press, 1969.

Litt, Carole. "Theories of Transitional Object Attachment: An Overview," *International Journal of Behavioral Health*, 9 (1986): pp.383–399.

Løgstrup, K. (1997). *The ethical demand*. Notre Dame: Notre Dame University Press.

Macmurray, John. *Persons in Relation*. London: Humanities Press, 1991.

Montgomery, Sy. *The Soul of an Octopus*. New York: Atria Books, 2016.

Morris, Charles. *Signification and Significance*. Cambridge: MIT Press, 1964.

Nussbaum, Martha. *Justice for Animals*. New York: Simon & Schuster, 2022.

Pihkala, Panu. "Environmental Education after Sustainability: Hope in the Midst of Tragedy." *Global Discourse*, 7 (2017): 109–127.

Pihkala, Panu. "Eco-anxiety, Tragedy, and Hope: Psychological and Spiritual Dimensions of Climate Change." *Zygon*, 53 (2018):, 545–569.

Pihkala, Panu. "Anxiety and the Ecological Crisis: An Analysis of Eco-anxiety and Climate Anxiety." *Sustainability*, 12 (2020): 7836.

Pihkala, Panu. "The Process of Eco-anxiety and Ecological Grief: A Narrative Review and a New Proposal." *Sustainability*, 14 (24) (2022): 1–53, doi:10.3390/su142416628.

Pihkala, Panu. "Eco-anxiety and Pastoral Care: Theoretical Considerations and Practical Suggestions." *Religions*, 13 (2022): 1–19.

Pihkala, Panu. "Toward a Taxonomy of Climate Emotions." *Frontiers in Climate*, 3 (2022): doi:10.3389/fclim.2021.738154.

Pruyser, Paul. *The Play of the Imagination: Toward a Psychoanalysis of Culture*. New York: International Universities Press, 1983.

Rank, Otto. *The Trauma of Birth*. New York: Routledge, 1929/2014.

Schlanger, Zoé. *Light Eaters*. New York: Harper, 2024.

Searles, Harold. *The Nonhuman Environment in Normal Development and Schizophrenia*. Washington: International Universities Press, 1960.

Singer, Peter. *Animal Liberation*. New York: HarperCollins, 1975.

Stengers, Isabelle. *Making Sense in Common: A Reading of Whitehead in Times of Collapse*. Minneapolis: University of Minnesota Press, 2023.

Trevarthen, Colwyn. "First Things First: Infants Make Good Use of the Sympathetic Rhythm of Imitation, Without Reason or Language." *Journal of Child Psychotherapy*, 31 (2005): pp. 91–113.

Viveiros de Castro, Eduardo. *Cannibal Metaphysics.* Minneapolis: Minnesota University Press, 2017.

Waggoner, Matt. *Unhoused: Adorno and the Problem of Dwelling.* New York: Columbia Books, 2018.

Wilson, Edward. *The Future of Life.* New York: Abacus, 2005.

Winnicott, Donald. *Playing and Reality.* London: Routledge, 1971.

Winnicott, Donald. "Through Paediatrics to Psychoanalysis." *The International Psychoanalytic Library,* 100 (1975): pp.1–325.

Chapter 19

The Climate Crisis
The Impact of Fragile Identificatory Models on Adolescence

Christine Franckx

Introduction

Adolescence is a critical time in life. Under the pressure of intense drives in puberty, body, and mind undergo tremendous changes. The impact of these changes will have a decisive influence on the maturational process of transforming the identity of the young person. It is a true catastrophic change, in the Bionian sense, of transformation. It involves a deconstructing and subverting of an old system to make way for something new. The environmental catastrophe superimposes onto the adolescent process of finding new solutions to conflicts between Eros and Thanatos, between sexual and aggressive drives (Franckx 2023).

For some teenagers, adolescence can be a traumatic experience that tips instinctual life and psychic defenses out of balance, leading to a regression to more primitive defenses to prevent being overwhelmed and internally disorganized. As a consequence, the adolescent could lose all symbolizing capacity and remain attached to concrete reality, foreclosing the important process of integrating internal and external reality. A crucial condition for a successful outcome is the capacity to mourn one's childhood, which means abandoning the position of infantile omnipotence and starting to face reality. Such psychic movement instills a sense of having some control over the external world and imagining one's place in it. This is why adolescents must challenge parental objects to be able to take a subjective position on issues that matter in their identificatory process. Historically, adolescents have always found an ideology in society to make their own, and they have turned to groups as transitional "families" for containment of the self during this transitional period of life. Their new, creative, and often rebellious ideas have very often resulted in radical changes in society, politics and culture. For this process to be constructive, it is important that the illusion of being able to radically change the world is well understood by adults and tactfully accompanied for a healthy disillusionment is to gradually take place. Winnicott's saying "there is no such thing as a baby," according to Philippe Gutton (2013), can be adjusted to "there is no such thing as an adolescent." He needs others to

DOI: 10.4324/9781003520757-22

acknowledge him, take him seriously, and provide an identificatory framework for growth and development.

This chapter will concentrate on the specifics of a superposition of the catastrophic change of adolescence and the climate catastrophe (Franckx 2024).

Climate Warming, an External Catastrophe

The anxiety of the threat of extinction of all human and non-human life has become real and tangible in everyone's life. The environmental catastrophe impacts undoubtedly all developmental processes, but it intensely affects adolescence as young people are preparing to enter an active position in the world of tomorrow, their world. The future is no longer guaranteeing progress and more welfare for all. The divide between northern and southern parts of the planet deepens, with an unequal distribution of riches and environmental conditions for a safe and prosperous life. The Global Trends report 2024 of UNHCR indicates that 3 out of 4 of the 120 million refugees are climate refugees, with alarmingly raising numbers. A number of the red to dark-red climate emergency zones in Africa and South-America are also conflict situations.

Historically, the environmental crisis started with the geo-political exploitation of the Earth's resources by Western colonialism, justified by a dismissal of the respectful and reciprocal relationships humans could entertain with their natural surroundings as "primitive superstition." Man started to see himself as the ruler of Earth, Nature and all its creatures. These had no other reason of existence except as a resource for humans to satisfy their own needs.

Global warming, during what some call the Anthropocene, is a consequence of this new geological regime and is setting in motion a highly complex process of physical, biological, social and psychological interactions. Historians describe new ways to think more globally about historical regimes at the crossroads of ecology and the world. These imply a radically different historiography that situates man and nature in a reciprocally interactive relationship on a much larger time-scale (Chakrabarty et al. 2017). The environmental catastrophe is transforming the course of human history and may in itself provoke a sixth massive extinction.

The Challenge to Adolescents

Will the parental and societal identificatory internal objects survive the adolescent's attacks and can the former generation, in the external world, confidently model a universal transformational process? The anger of young people reveals the violence they feel subjected to in discovering not only the scientific facts of global warming and the disappearance of a high number of animal species, but even more in realizing the incapacity of their parents' and grandparents' generations to react appropriately. The anxiety of the baby's infantile dependency on the person of his/her mother for survival, which is always reactivated in the separation process of adolescence, receives an extra burden by the insight of the fragility of Mother Earth who is signaling that the climate may not provide a safe home for future life. In all human relations, we can expect some level of interdependency, but Nature is truly ruthless and it will continue its course without us.

For the first time in human history, the fear of breakdown is projected in the future and is not something that has already taken place but not yet internalized, as Winnicott theorized. This impacts the adolescent's existential questioning and identification of a position in the chain of generations. The adolescent facing this situation is confronted with a reality without an external object to challenge, with a temporality that is unreliable and therefore threatening, and with internal parental objects that cannot provide the necessary containment of protection and sense-making. We live in the middle of a highly complex state of affairs that questions intergenerational relationships, the role of the individual in the group, and the social contract. Ethical demands on the next generation are enormous, with no adult authority to lead the way. As a 14-year old clever girl, told her mother: "we are the first generation to really having to deal with the climate crisis and we are also the last one who can do something about it."

Adolescence and Internal Catastrophic Change

The specific phase of adolescence as a developmental stage in human life is rather recently recognized in history. In his famous work *Emile* (1762), Jean-Jacques Rousseau advocated for this phase of life to be an education in line with the child's natural development instead of training children to move to an adult role and function in society. The discovery of the psychic qualities of adolescence as essential to the process of growing up from child to adult is crucial to our vision of the human being in many respects. Adolescence is one of the things that differentiates us from other mammals. Two major changes take place during adolescence, both of which involve intense psychic work to find a new balance in a short period of time: the body matures sexually and childhood objects must be mourned.

The Sexually Maturing Body

The hormonal maturation of the body initiates a chain of biological changes. The drives, libidinal and aggressive, are at the boundary of psychic and somatic dimensions, and the psyche has to work to bind the impulses coming from the body. One could think of the drive as a vehicle for instincts to "breathe" and a provision to the psyche of something to "think," to imagine, and to elaborate. The richness of our inner world is the reflection of the extent to which the somatic material has become mentalized. Cognitive functioning is influenced by the development of the brain with a leap from concrete to formal logical thinking capacity. This impacts the way young people can begin to formulate the complexities of the surrounding world in a systematic way.

Freud (1905) described the role of infantile sexuality in adult object choice during adolescence. Psychosexual evolution takes shape in two times, interrupted by a latency phase of waiting for bodily maturation and the release of the first love objects. This by no means implies that younger children would not experience and be driven by sexual and aggressive impulses or engage in an individuation-separation process. Yet, the concomitant biological and social (r)evolution of the adolescent age brings the individual into a new reality with different needs, wishes and expectancies.

Adolescence is a truly new era! Many parents and educators will surely recognize how the reasonable fairness of their latency age children has faded as snow in the sun, almost overnight. The same psychic mechanisms of infancy are reappearing in this developmental phase, albeit in a different context. The proximity of the early psychic world demonstrates the challenge for youngsters and the intensive mental work to be accomplished. No question this explains the frequent psychopathological reactions, mostly at the beginning or the end of this period. Puberty, with the onset of biological maturation, as well as the closure of adolescence, with the choice of a sexual partner, are moments of helplessness, despair and instability.

Why a return to the infantile? The helplessness of a baby, driven by the instinctive need to stay alive, mobilizes primitive defense mechanisms to reach out for help from a primary object whose main task it is to maintain the baby in an omnipotent illusion of being fully in control. His delusional perception will be delicately contained by a good enough object who can gradually help him take in the external world and disillusion his omnipotent belief in small, digestible steps. The infantile theories about the mysteries of life and death, of the generational differences and of the primal scene are revisited in adolescence because the more mature capacities demand an acknowledgment of external reality. Adolescence is therefore primarily a state of acquiring a new version of the old established schemes.

Some of our earliest internal psychic theories however remain active throughout life. For example, feelings of dependence continue to be painful and anxiety-laden because they confront us once again with the well-known infantile helplessness. With respect to climate change and humanity's hand in this, could we think that we consider our planet ruthlessly as a toilet breast (Meltzer 1967), a receptive wasteland where all unwanted thrash can be disposed of? After all, the infantile feelings of dependence give rise to intense hatred and a wish to punish, control and destroy the inside of Mother Earth. Of course, this is followed by primitive guilt and a reparative movement because of the fear of retaliation. The cycle of paranoid-schizoid/depressive position continues endlessly in an ongoing dynamic.

Mourning Childhood Objects

The changing relationship between parents and adolescent is signified by a strong pull towards full independence. Simultaneously, the regressive qualities of this stage in life come to the foreground. The Italian writer, Erri de Luca writes in his book *Grandeur Nature* about the pain of separation:

> The no-man's-land of adolescence suddenly emerges between parents and children. Adults are the past. Their voices of reproach, even when shouted, are muffled. This desert must disappear on its own, no direction can shorten it. Bereavements reinforce it, first love prolongs it.
>
> (De Luca, 2021)

We recognize an essential loneliness of being cut off from all meaningful connections without the hope of finding a trustworthy way out. Mourning is an individual process that the young person has to travel alone to break free and create his own destiny. At the same time, parents must be able to identify retrospectively with the endeavor they once encountered themselves, and they are thrown back into an era of despair, depressive anxiety and aloneness. The fortunate outcome of this process depends on both parties being able to let go of old ideals and safe beacons, and to rely confidently on the potential of new opportunities.

This means that parents and educators have to allow enough freedom and trust in their teenagers who are coming to terms with this inner conflict of life and death, of the battle between Eros and Thanatos. A desire to stop time and dwell in a garden of Eden of eternal happiness, a bubble of omnipotent illusion, lurks for all parties. Parents are suddenly confronted with their advancing age and with the fact that the next generation is ready to take a first row in society. In adolescents, we often witness more anxiety towards the closing of this phase in life, for example, in the difficulty of finishing university successfully. Fortunately, Eros thrusts all forward, as adult life beckons for the young people who are eager to participate, find a sexual partner, and contribute to society. Ideally, the older generation can calm down the terror of the finality of life in hoping for a new generation to come.

With regard to the ecological crisis and the intensity with which adolescents engage with this problem worldwide, the risks of not finding the right parental and social support and reliable identificatory objects are real. After all, the threat to humanity is present, and evidence of former generations not having taken sufficient precautions to prevent this from happening is everywhere. Indeed, adolescents are anxious about the uncertainty of the future, and they may feel haunted by their primary objects not being able to adequately protect them. Constituting a strong, confident adult self is challenging.

It is important for adolescents to have sufficient freedom to explore many aspects of the world and reconcile inner turmoil with the demands of external reality. This is a delicate and intensive process, in which the opposition to authority in strong political and social ideals can often offer adolescences a positive source of identification.

Global warming and environmental catastrophe are threatening realities that will impact future generations. Adolescents are keenly aware of the fact that former generations have neglected to prevent this evolution, despite clear warnings. In their anger for this state of affairs, adolescents remind adults of their responsibilities, which is the world upside down. The internal turbulence of the adolescent passage and their search for safe and reliable reference points are reproduced in our contemporary society. Today, humanity is in a transitional phase with the loss of established certainties and the extremely rapid evolution of new and unknown realities (Lebon 2022).

The evolution of one generation to the next is also complicated by the accelerated technological revolution and the recent and quickly increasing know-how in younger generations. Adolescents are virtually on their own when "connected" via multiple social media channels, without the possibility of adult guidance. Adults are lacking experience, knowledge and often an interest in the new technological developments with artificial intelligence. One of the signs, or symptoms, of the changing social contracts between the generations is the impact on language structure. We also notice a uniformization in vocabulary and a loss of the grammatical structure that differentiates and hierarchizes language. This shift can be understood as a sign of changing relationships, but easily becomes a symptom of confused psychic states.

Hartmut Rosa (2021), a German sociologist, describes how the image of the adult as an authoritarian educator has been transformed into that of a friendly, collaborative partner, which adds a dark side to the adolescent process. In their struggles to relate to a sense that time is accelerating, many adults nowadays are striving for eternal youth and do not always pick up on their children's need for differentiation and guidance. The psychic fragility that is inherent in puberty and adolescence calls for the supportive presence of adults to contain the anxiety and excitement raised by new contacts with the wider general society. Heightened psychosocial emotional turbulence and the pressing need for new safe beacons of reliable reference points are characteristics of our changing times. Adolescents are often alone in their search for a containing model.

The Experience of Time and Mourning in a Changing World

Mourning is closely connected with a subjective feeling of existing in time and being able to let transformational processes take their course. As a young man in his early twenties struggling to leave home for university in another city so sensitively formulated: "my memory functions like a sieve-I cannot hold on to things, they disappear before I can even think about them." It was impossible for him to keep internally in touch with his parental figures because of his highly ambivalent feelings towards them. Discovering the intensity of his resentment and even hateful emotions towards both his parents was necessary to free the loving aspects of his relationship with them. He needed to mourn an intimate belief of an idealized childhood and accept that there would not be a second turn for him or his parents, who were preoccupied by their retiring plans and some emerging health problems.

Freud wrote the essay "On Transience" in the same year as his metapsychological paper "Mourning and Melancholia" (1916), showing the psychological importance of being able to integrate the pleasure and reality principles:

> No! it is impossible that all this loveliness of Nature and Art, of the world of our sensations and of the world outside, will really fade away into nothing. It would be too senseless and too presumptuous to believe it. Somehow or other this loveliness must be able to persist and to escape all the powers of destruction. But this demand for immortality is a product of our wishes too unmistakable to lay claim to reality: what is painful may none the less be true.
>
> (Freud, 1916)

David Bell (2006) argues that "feeling oneself as existing in time is an important developmental achievement. For some, however, it is felt as a fixed, imminent catastrophe to be evaded by the creation of a timeless world in which, apparently, nothing ever changes, where there is an illusion of time standing still." As the main character of the novel by J. D. Salinger, *The Catcher in the Rye*, 16-year-old Holden Caulfield exclaims: "Certain things, they should stay the way they are. You ought to be able to stick them in one of those big glass cases and just leave them alone" (Salinger 1951). The story movingly describes his passionate search for the feeling of belonging to a world with adult people he can try to identify with, while being pushed forwards and away from his childhood identificatory figures. The book has become an icon of teenage rebellion but also feared by educators for the sexual and societal liberal themes that they fear could spoil young people.

Saving the Planet

Global warming is undoubtedly affecting the adolescent developmental process. The internal psychic task of youngsters facing this global catastrophe is complex and delicate. Will internal objects survive the adolescent's attacks and can the previous generation be relied on to truly engage with a process of responsible transformation of the outside world? Facing this catastrophic state, there always seem to be three possible ways of reacting: denying or disavowing reality and continuing as usual; protesting out of a feeling of injustice and anger; or accepting the real threat which brings anxiety and psychic suffering. These reactions are universal and all present at different moments for everybody: we need most of the time some denial and disavowal in the form of defense of functional splitting to protect a vital sense of going-on-being. At other moments we can be very angry and revolted. All of us experience moments of anxiety. Adolescents show an impressive lucidity in this matter and the reality principle seems to be more accessible to the young generation. They are better informed and ready to take action because of their sensitivity to what is going on in the world of tomorrow. This is their world, and they are eager to participate.

In recent years, we have witnessed many forms of ecological activism on the part of teenagers committed to saving the planet, who urgently feel the stakes of this crisis. They passionately aspire to a reliable and promising future to fuel their ideals and projects, while at the same time they are confronted with the pressure of their individual pubertal crisis in their private natural environment (Bernateau 2021).

It is not a surprise that the ecological crisis mobilizes adolescents worldwide, but this carries to my mind two additional risks for the adolescent process. First, there is a risk of them not finding the appropriate parental and societal containment necessary to transform their illusion of infantile omnipotence into a clear and mature appreciation of reality. Because of a real threat to survival, young people can become imprisoned in a state of continuously trying to repair the damaged parental objects instead of entering the depressive position of mourning. This could lead to a despairing state of helplessness with clinical depression as a result. The iconic adolescent, Greta Thunberg, refused the role of repairing guilty internal objects and addressed the world leaders with the words "How dare you!" thereby pointing out the catastrophe she felt former generations have created as well as their inability or unwillingness to take joint action and limit the damage (Robin 2021). This is an upside down world of adolescents reminding adults they need to contain their insatiable infantile drives of greed and must face the facts of reality. Luc Magnenat (2019) suggests that Greta Thunberg could be understood as Antigone who wants to bring back the order that was violated by the former generation. The climate crisis brings to mind the oedipal myth, according to which the biological parents of Oedipus tried to kill their infant son in the vain hope of being released from the threatening prophecy of one day being murdered by him. Did the former generation sacrifice their future offspring in order to, as Sally Weintrobe (2021) put it, to remain in a bubble of denial and magical omnipotent thinking of rearranging reality with fraudulent arguments? We can see why Greta Thunberg succeeded in creating a worldwide movement of furious school age children and youngsters, by providing them with a perspective of protest. The school strikes are a clear sign of refusal to be imprisoned in a guilt-laden repair of parental objects.

Secondly, the inner, libidinal crisis of catastrophic change in puberty may be overtaken by the external real catastrophic environmental threat, occupying all space in climate protesting adolescents. Any internal adolescent drive scene can be potentially traumatic if the psyche does not succeed in bringing life and death drives working together. The condensation of the two times in the libidinal elaboration of the infantile sexual theories, not allowing a "latency" time, thrusts the child into adulthood and "prevents that the climate threat acts on the adolescent according to the catastrophics of a traumatic neurosis" (Glas 2022). This author underlines the danger for adolescents, focusing on saving the planet of "not being able to construct a psychic organization under the reign of a super-ego organized by oedipal conflict." The real external threat of survival can have an impact on the resolution of the oedipal conflict as it colludes with the internal castration anxiety facing the super-ego.

Clinical vignette

Jennifer, 15 years old, was referred because of a serious suicide attempt which her parents and friends had not been able to anticipate. They were shocked to realize that their perfectly happy girl, always a good student and mother's confidante, broke down suddenly and apparently for "no reason." Jennifer showed me in a first interview how exhausted she had become in trying to meet all the demands of her parents and teachers. She had offered an external shining image while being extremely dark and preoccupied internally. She felt there was no future for her in this world, and she could not think of a way of finding happiness. Lonely and feeling isolated from her peers, she had engaged in compulsively seeking satisfying contacts through social media. This, however, had only brought more disappointment. In the psychotherapy that followed, Jennifer demonstrated how profoundly depressed and hopeless she felt. The psychoanalytic work moved on with great difficulty and internal conflicts emerged slowly. At some point, she told me how painfully disappointed she was in her mother, with whom as a child she had felt very close, as she had discovered the extramarital relationship of her mother. It became clear how Jennifer had felt more herself to be more of a partner to her mother, making up for the emotional coldness between her parents. Now, she was replaced by a real lover. What was most painful to Jennifer was the fact that she had suddenly woken up from a state of "illusion" with the reality thrown in her face. She felt deceived and lied to. At last, some anger and feelings of injustice could be voiced. She then brought a dream in which she was visiting a shopping mall with an unknown young woman, when all of a sudden they were trapped in the building on fire. She could not get out and experienced an unbearable suffocating heat. On top of that, when at last the firemen arrived, they flooded the building without noticing the people inside. She lost hold of the other woman's hand and nearly drowned herself. She woke up in fear.

Although this dream seems to point to some repressed homosexual feelings, awakened in the transference, and to her anxiety of being literally overwhelmed and nearly drowned by the intense libidinal experience, her immediate associations refer to the effects of global warming. I interpreted this as a resistance to engage with her internal turmoil and formulated something about her anxiety of feeling overwhelmed in close relationships. She did not accept this and became angry with me for not taking up the environmental explanation. It turned out that she had become very involved with a group on social media that focused on climate activism. There was no other option, for the moment, then to take her association to the environmental seriously and to accept it as a constructive and real attempt to start taking part in adult life. In the countertransference, I realized that the images of her dream had made me very anxious and that my interpretation on an oedipal libidinal level had been a way out for me. This example made me think how important it is in analysis with adolescents nowa-days to fully understand the superposition of eco-anxiety and anxiety raised by a neurotic conflict as a symptom of our contemporary collective suffering. Interpreting our shared concern, helplessness, and despair about the future before relocating the transferential meaning of the dream was necessary and paved the way to more, very necessary intrapsychic work.

The difficulty in our contemporary society is that internal representations of life and death and of reproduction and attachment are being reproduced in the external world. We hear young people hesitate over whether they even want to have children and whether they will die of old age. While these concerns are legitimately related to climate change, they may also obscure some internal hesitancy over taking an adult, parental position in society. The identificatory models have become fragile. Nevertheless, it is only through mobilizing internal phantasy life and the symbolizing capacity that real life can become possible. Although more complicated in our time, helping adolescents to reach this mature psychic state is a necessity for their future.

Conclusion

In taking the anxiety about climate change seriously we can find a way to connect with the adolescent that conveys a shared collective feeling about the uncertainty of the future, while at the same time maintaining the intrapsychic conflict and transference dimension in tension. The scientific reality of global warming and its potential consequences for the survival of the human species can bring adolescents into contact with the fear of collapse, which, according to Winnicott (1974), refers to a catastrophe that has already occurred but remains unresolved intrapsychically. It can cast a shadow over intimate reality and, in a way, steal the internal conflictuality of the adolescent's ambivalence, which is so necessary for growing up and/or recovering from archaic anxieties.

Psychoanalysis and the psychoanalytic approach to adolescent problems can be helpful to help young people reconnect with good internal figures and strengthen ego-functions in the libidinal chaos of adolescence. Helping to find and create a subjective meaning to feelings, experiences and thoughts is what psychoanalysis is all about. After all, they are the future and will need a healthy psychic structure to deal with the new problems of humanity.

References

Bell, D. 2006. Existence in Time: Development and Catastrophe. *The Psychoanalytic Quarterly*, 75: 783–805.

Bernateau, I. 2021. Menace sur la terre et vulnérabilité adolescente. *Adolescence*, 39, p 31–42.

Charabarty, D, Haber, S, & Guillibert, P. 2017. Réécrire l'histoire depuis l'anthropocène. *Actuel Marx*, 61: 95–105.

De Lucca, E. 2021. *Grandeur nature*. Paris: Gallimard.

Franckx, C. 2023. Malaise dans la Nature. La psychanalyse face à la crise climatique. *Rev.Belge de Psychanalyse*, 82: 111–136.

Franckx, C. 2024. Climate Change and Adolescence: A Dangerous Collusion of Internal and External. In C. Schinaia (Ed.), *Against Catastrophism: Building our Future in a Changing World*. London: Routledge.

Freud, S. 1905. Three Essays on Infantile Sexuality. In J. Strachey (Ed. and Trans.), *The Standard Edition of the Complete Psychological Works of Sigmund Freud*, vol. 7. London: Hogarth Press.

Freud, S. 1916. On Transience. In J. Strachey (Ed. and Trans.), *The Standard Edition of the Complete Psychological Works of Sigmund Freud*, vol. 14. London: Hogarth Press.

Glas, J. 2022. L'investissement massif de la décroissance écologique, un avatar inconscient de la jouissance dans le dépouillement de la castration. In D. Bourdin & D. Tabone-Weil, *Planète en détresse. Fantasmes et réalités*. Paris: PUF.

Gutton, P. 2013. *Le pubertaire*. Paris: PUF.

Lebon, C. 2022. L'écologie familiale ou la famille à l'épreuve de la crise écologique mondial. *Divan Familial*, 49: 69–83.

Magnenat, L. 2019. *La crise environnementale sur le divan*. Paris: In Press.

Robin, M. 2021. How Dare You? La jeunesse en mode survie. *Adolescence*, 3, p 15–30.

Rosa, H. 2021. *Resonance. A Sociology of our Relationship to the World*. Cambridge: Polity Press.

Salinger, J.D. 1951. *The Catcher in the Rye*. New York: Little, Brown and Company.

Weintrobe, S. 2021. *The Psychological Roots of the Climate Crisis*. London: Psychoanalytic Horizons.

Winnicott, DW. 1974. Fear of Breakdown. *Int. Rev. Psychoanalysis*, 1: 103–107.

Index

Note: Locators in bold refer table respectively.

For Product Safety Concerns and Information please contact our EU
representative GPSR@taylorandfrancis.com
Taylor & Francis Verlag GmbH, Kaufingerstraße 24, 80331 München, Germany

9 781032 835822